THE HUMAN MICROBIOME IN EARLY LIFE

THE HUMAN MICROBIOME IN EARLY LIFE

Implications to Health and Disease

Edited by

Omry Koren

Azrieli Faculty of Medicine, Bar-Ilan University, Safed, Israel

Samuli Rautava

Department of Pediatrics, University of Helsinki, Helsinki, Finland
Helsinki University Hospital, Helsinki, Finland

ELSEVIER

ACADEMIC PRESS

An imprint of Elsevier

Academic Press is an imprint of Elsevier
125 London Wall, London EC2Y 5AS, United Kingdom
525 B Street, Suite 1650, San Diego, CA 92101, United States
50 Hampshire Street, 5th Floor, Cambridge, MA 02139, United States
The Boulevard, Langford Lane, Kidlington, Oxford OX5 1GB, United Kingdom

British Library Cataloguing-in-Publication Data
A catalogue record for this book is available from the British Library

Library of Congress Cataloging-in-Publication Data
A catalog record for this book is available from the Library of Congress

ISBN: 978-0-12-818097-6

For Information on all Academic Press publications
visit our website at https://www.elsevier.com/books-and-journals

Publisher: Andre Wolff
Acquisitions Editor: Linda Versteeg-buschman
Editorial Project Manager: Kristi Anderson
Production Project Manager: Maria Bernard
Cover Designer: Mark Rogers

Typeset by MPS Limited, Chennai, India

Contents

List of Contributors

Kjersti M. Aagaard Division of Maternal-Fetal Medicine, Department of Obstetrics and Gynecology, Baylor College of Medicine, Texas Children's Hospital, Houston, TX, United States; Center for Microbiome and Metagenomics Research, Baylor College of Medicine, Texas Children's Hospital, Houston, TX, United States; Department of Molecular and Human Genetics, Baylor College of Medicine, Texas Children's Hospital, Houston, TX, United States; Department of Molecular and Cell Biology, Baylor College of Medicine, Texas Children's Hospital, Houston, TX, United States

Phillip R. Bennett Parturition Research Group, Institute of Reproductive and Developmental Biology, Imperial College London, London, United Kingdom; March of Dimes European Preterm Birth Research Centre, Imperial College London, London, United Kingdom

Avital Cher Faculty of Medicine, The Hebrew University of Jerusalem, Jerusalem, Israel

Jose C. Clemente Department of Genetics and Genomic Sciences, Icahn Institute for Data Science and Genomic Technology, Immunology Institute, Icahn School of Medicine at Mount Sinai, New York, NY, United States

Maria Carmen Collado Institute of Agrochemistry and Food Science, Spanish National Research Council, Valencia, Spain

Gwenny M. Fuhler Department of Gastroenterology and Hepatology, Erasmus MC University Medical Center, Rotterdam, The Netherlands

Lindsay J. Hall Gut Microbes and Health, Quadram Institute Bioscience, Norwich, United Kingdom; Chair of Intestinal Microbiome, School of Life Sciences, Technical University of Munich, Freising, Germany

Erika Isolauri Department of Paediatrics, Institute of Clinical Medicine, University of Turku, Turku, Finland; Department of Paediatric and Adolescent Medicine, Turku University Hospital, Turku, Finland

Michael D. Jochum Division of Maternal-Fetal Medicine, Department of Obstetrics and Gynecology, Baylor College of Medicine, Texas Children's Hospital, Houston, TX, United States

Omry Koren Azrieli Faculty of Medicine, Bar Ilan University, Safed, Israel

Magdalena Kujawska Gut Microbes and Health, Quadram Institute Bioscience, Norwich, United Kingdom

David A. MacIntyre Parturition Research Group, Institute of Reproductive and Developmental Biology, Imperial College London, London, United Kingdom; March of Dimes European Preterm Birth Research Centre, Imperial College London, London, United Kingdom

Collado Maria Carmen Institute of Agrochemistry and Food Technology-National Research Council, Valencia, Spain

Eugenie Martineau Department of Genetics and Genomic Sciences, Icahn Institute for Data Science and Genomic Technology, Immunology Institute, Icahn School of Medicine at Mount Sinai, New York, NY, United States

Lybil Mendoza Department of Pediatrics, College of Medicine, University of Florida, Gainesville, FL, United States

Josef Neu Department of Pediatrics, College of Medicine, University of Florida, Gainesville, FL, United States

Hadar Neuman Zefat Academic College, Safed, Israel

Samuli Rautava Department of Pediatrics, University of Helsinki, Helsinki, Finland; New Children's Hospital, Helsinki University Hospital, Helsinki, Finland

Ehud Rinott Faculty of Health Sciences, Ben-Gurion University of the Negev, Beer-Sheva, Israel

Maxim D. Seferovic Division of Maternal-Fetal Medicine, Department of Obstetrics and Gynecology, Baylor College of Medicine, Texas Children's Hospital, Houston, TX, United States

Salminen Seppo Functional Foods Forum, Faculty of Medicine, University of Turku, Turku, Finland

Hania Szajewska Department of Paediatrics, Medical University of Warsaw, Warsaw, Poland

Max Van Belkum Department of Pediatrics, College of Medicine, University of Florida, Gainesville, FL, United States

Gabriel Vinderola Dairy Products Institute (INLAIN, UNL-CONICET), Faculty of Chemical Engineering, National University of Litoral, Santa Fe, Argentina

Moran Yassour Faculty of Medicine, The Hebrew University of Jerusalem, Jerusalem, Israel; The Rachel and Selim Benin School of Computer Science and Engineering, The Hebrew University of Jerusalem, Jerusalem, Israel

Ilan Youngster Pediatric Infectious Diseases Unit and Center for Microbiome Research, Shamir Medical Center, Be'er Ya'akov, Israel; Sackler School of Medicine, Tel-Aviv University, Tel-Aviv, Israel

Introduction

The inspiration for this book stems from the fruitful intersection of two important theoretical and clinical advances occurring in the health sciences. First, we have learned to appreciate that the exposures during fetal life, birth, and early infancy have a profound impact on not only neonatal and infant health, but may bear long-term consequences in the form of disease risk into adulthood and even old age. It is well-established that pre- and perinatal complications including fetal growth restriction and preterm birth are associated with an increased risk of cardiovascular disease in later life. The associations between cesarean section delivery and early antibiotic exposure and increased risk for obesity and allergic disease are an active area of epidemiological and experimental research. Second, recent methodological and theoretical advances in microbiology have considerably widened our view of the role of microbes in human health. Bacteria are no longer regarded merely as potential pathogens, but also as vitally important commensals interacting with the human host and modulating physiological functions ranging from digestion and immune responsiveness to growth, metabolism, and neurophysiology. At the crossroads of these two major insights, we found ourselves in need of a comprehensive overview of the current scientific knowledge—and the gaps therein—on the human microbiome during pregnancy, birth, and early infancy and its implications to host physiology and health.

Humans, and our earliest mammalian, vertebrate, and eukaryotic ancestors have evolved in a world that is already inhabited by bacteria. From the beginning, microbes have formed an essential component of the natural environment in which more complex organisms live. The interactions between bacteria and multicellular forms of life are far more complex than mere coexistence. Our coevolution with microbes has produced intricate networks of interdependence. Indeed, interaction with indigenous and environmental microbes is essential for host physiology. Recent advances in understanding host–microbe interactions have markedly increased our appreciation of the importance of our microbiome for human health. Perturbations in the intestinal microbiome have been associated with a variety of noncommunicable chronic conditions, ranging from allergic and autoimmune disorders and inflammatory bowel disease to obesity, type 2 diabetes mellitus, and even a range of neurological or psychiatric conditions. Innovative experimental studies have provided mechanistic insights into these associations, and in some instances, they suggested a causal role for the aberrant gut microbiome in disease pathogenesis. In line with this notion, meticulously conducted human studies have provided evidence indicating that changes in the gut microbiome may be detected as early as the first weeks and months of life in individuals who later develop atopic disease or become overweight. We interpret these intriguing data to not only suggest causality but also to underscore the importance of microbial contact and colonization in early life for long-term health and disease.

Continuous interaction with microbes appears to be necessary for healthy immune, metabolic, and intestinal development during early infancy. The establishment of the gut microbiome during the perinatal period and infancy may be prone to perturbations from events such as cesarean section delivery, preterm birth, lack of breastfeeding, or antibiotic use. Several chapters in the current volume are dedicated to reviewing and interpreting the current body of scientific evidence regarding the complex interactions between early microbial contact and its disturbances, host physiology, and finally, health and disease.

Whether the human fetus comes into contact with live microbes during a healthy pregnancy has been a focus of intense research and scientific debate during the past years. As discussed here in detail, the precise nature and extent to which live bacteria may exist in the intrauterine compartment remains an open question. It is well-established, however, that the maternal gut and vaginal microbiome undergo significant changes during pregnancy, and that disturbances in the maternal microbiome in the form of bacterial vaginosis or chorioamnionitis may have detrimental effects on the fetus and cause fetal infection or preterm delivery. Continued studies are investigating how microbial contact during pregnancy—be it a direct effect or an indirect one via the pregnant mother—may affect pregnancy outcomes and neonatal health. Given the recently discovered associations between fetal exposures and long-term health alluded earlier, it is perhaps even more intriguing to hypothesize that microbial exposures might be among the fetal and early-life determinants of future health and disease.

We sincerely hope that this book offers the reader a broad and critical view of what is currently known about the early microbiome and its implications to human health. We also expect that this book will stimulate new ideas and hypotheses by highlighting the wide gaps in not only the available data, but also in our theoretical understanding. In order to move forward, we must know where we stand.

Samuli Rautava and Omry Koren

SECTION I

Pregnancy and fetal life

CHAPTER

1

The microbiome in a healthy pregnancy

Hadar Neuman[1] *and Omry Koren*[2]

[1]Zefat Academic College, Safed, Israel, [2]Azrieli Faculty of Medicine, Bar Ilan University, Safed, Israel

Introduction

The human microbiome, comprising hundreds of microbial species residing within the human body, has been shown to have essential roles in promoting and maintaining health. These include effects on immunity, metabolism, and even involvement in the endocrine system and behavior.

Drastic changes to microbial composition and diversity are typically correlated with disease conditions and are termed dysbiosis. Such changes have been observed in autoimmune diseases, diabetes, obesity, and metabolic syndrome [1–4]. However, during healthy pregnancies, distinct changes are observed in the microbiome, some of which resemble the changes observed in disease states. This phenomenon is not coincidental, but rather essential for the progression of a healthy pregnancy. Thus pregnancy is unique as a positive and physiological example of dysbiosis. In this chapter, we will describe the overall microbial changes during pregnancy and relate these changes to their functional significance. The spectrum of functional effects ranges between influencing the maternal immune system, promoting weight gain, and altering metabolism. Pregnancy complications are correlated with additional distinct microbial alterations, emphasizing the potential roles of the microbiome in both healthy pregnancies, and in the case of complications (discussed in the following chapters). Finally, the maternal microbiome has clear effects on the offspring, as manifested by the initial infant microbiome, and the consequences of antibiotic exposure during gestation.

The Human Microbiome in Early Life
DOI: https://doi.org/10.1016/B978-0-12-818097-6.00001-8

Understanding the functional changes in the pregnancy microbiome is important for obtaining a full picture of the physiological processes that occur during pregnancy, for evaluation of the state of the pregnancy, and for deciphering potential risk factors for the pregnant mother and developing fetus. It appears evident that the complex immune, hormonal, metabolic, and microbial alterations that occur during pregnancy are correlated with one another, forming a network of physiological processes.

The importance and roles of a healthy microbiome

In the healthy state, our microbiome includes highly diverse populations of bacteria residing in a variety of body sites and niches. These bacterial populations have different characteristics—some are aerobic, whereas others are anaerobic; they can thrive at a variety of pH values and within a range of nutritional milieus [5]. The bacterial populations are also greatly affected by immune and hormonal factors. This explains why the gut microbiota significantly differs from the vaginal or the oral one, why the microbial populations are not identical between individuals, and emphasizes the importance of rich populations of microbes adapting to the changing surroundings during dynamic homeostatic conditions. On a functional level, it is assumed that microbes residing in different body sites also serve different roles in maintaining overall health. While the gut microbiome may play the most essential roles regarding human metabolism, as well as affecting immunity and hormones, the microbiome of the skin, vagina, and oral cavity contributes directly to the immune system in its defense against outside pathogens [6].

The many metabolic roles of the microbiota include food digestion, fermentation of carbohydrates into short-chain fatty acids (SCFAs), digestion of proteins into amino acids, production of vitamins, and metabolism of drugs and xenobiotics [7]. The microbiota has also been implicated in lipid metabolism, conversion of bile acids, and breakdown of polyphenols.

The immunological roles of microbiota are numerous. They help form a barrier for repressing pathogens and modulate the host immune response by affecting both innate and adaptive immune system components [8]. The presence of the microbiota maintains a balance in the abundance of immune cell populations, such as Treg cells versus T helper 17 (TH17) cells [9]. Moreover, levels of pro-inflammatory cytokines have been shown to depend on the microbial composition.

Additional effects of the microbiome on their hosts include altering hormone levels and affecting brain function and behavior [10–12]. A close interplay between the microbiome and the host endocrine system has been described [13]. While host hormones affect the growth of certain bacteria and their function, bacteria can produce and modulate hormones influencing behavior, appetite, metabolism, gender, and immunity.

Similarly, the microbiome is both affected by the brain and can affect the brain development as well as impact appetite, emotions, stress response, cognition, and behavior, such as anxiety and aggression [14]. To this end, the gut microbiome has been shown to be altered in various psychiatric disorders including major depression, schizophrenia, and autism [13,15,16]. One potential mechanism for this effect is the capacity of gut microbiota to produce several human brain neurotransmitters [14]. Other components that might link the microbiome and behavior include hormones and immune parameters.

While alterations in microbiome are associated with developmental changes across our lifetime (e.g., from infancy to childhood, and later, old age), with dietary changes, and with other environmental factors, most divergence from steady-state microbial populations (dysbiosis) has been associated with disease states. Dysbiosis usually includes a decline in microbial diversity and may have long-term consequences leading to disease or enhancing a disease state. Examples of diseases associated with dysbiosis include obesity, inflammatory bowel disease, diabetes, and metabolic syndrome [3]. In fact, it has been suggested that almost all diseases are characterized by a perturbation of the healthy microbiome into an unbalanced diseased state, so that our overall fitness and health are closely intertwined with our microbiome composition [17].

Some of the microbial changes that occur during pregnancy may seem identical to those seen in disease states. However, in their precise developmental context, they do not lead to morbidity, but rather promote necessary functions, including a more flexible immune state and altered metabolism, and may therefore be considered beneficial.

The gut microbiome is altered during pregnancy

Pregnancy leads to dramatic changes in the gut microbiome. These alterations include a distinct reduction in bacterial diversity and richness within individuals (decreased α diversity), less similarity in microbial composition between different women (increased β diversity), and an increase in the bacterial load [18,19]. In addition, alterations in species abundance are observed, with some bacterial species being overrepresented and others being underrepresented as compared to the nonpregnant state [19].

Interestingly, most of the pregnancy-related microbial changes are seen at late stages of pregnancy, in the third trimester. This is also the time when other physiological parameters such as weight gain, insulin insensitivity, inflammation, and changes in hormonal levels reach a peak [20,21]. A recent study found that progesterone has a direct effect on the microbial composition, specifically increasing the richness of several species including *Bifidobacterium* in the gut microbial composition of women and mice during late pregnancy [22].

Some of the pregnancy-related microbial alterations seem to correlate with the metabolic state observed in late pregnancy. For example, there is a significant reduction in SCFA producers such as the butyrate-producing bacterium *Faecalibacterium*, similar to the depletion seen in metabolic syndrome patients [23]. On the other hand, an increased abundance of the Actinobacteria and Proteobacteria phyla is observed [19]. Additional changes include increased abundance of *Neisseria, Blautia,* and *Collinsella* as pregnancy progresses and decreased abundance of *Clostridium, Dehalobacterium,* and an unclassified Bacteroidales [22].

Changes to the vaginal microbiome during pregnancy

The healthy vaginal microbiome plays important roles in protection from pathogen intrusion and infections, decreasing vaginosis and sexually transmitted diseases [24]. The healthy vaginal microbiota is predominantly populated by *Lactobacillus crispatus, Lactobacillus jensenii, Lactobacillus johnsonii,* and *Lactobacillus iners* [25]. Functionally, some of the most abundant vaginal species including *Lactobacillus* species produce lactic acid, which lowers the pH of their surroundings, thereby forming a barrier against pathogenic bacteria. In fact, lactic acid has been shown to have broad-spectrum antimicrobial activity [26]. Specifically, *Lactobacillus rhamnosus* and *Lactobacillus reuteri* have been shown to inhibit *Candida albicans* and decrease expression of the latter's NF-κB-related inflammatory genes [27]. During pregnancy, the protective functions of the vaginal microbiome, including lowering of the vaginal pH, are enhanced due to an increase in abundance of *Lactobacillus* species [25,28]. This is accompanied by a significant decrease in overall diversity of the vaginal microbiota [29]. One of the distinct pregnancy-related changes is the development of a dominant *Lactobacillus* species over the range of *Lactobacillus* species generally present [30,31]. Mechanistically, this predominance may occur due to bactericidal effects of some *Lactobacillus* species against competing species [31]. As often occurs in microbiome research, not all studies observed identical microbial changes [32]. This may be due to relatively small cohorts, different detection and analysis methods, and different human subject populations including a variety of ethnic groups studied. Accordingly, it was shown in pregnant African–American women that a significant decrease in vaginal microbial richness and diversity, between the first and second trimesters, is associated with preterm birth [33,34]. A thorough study analyzing the vaginal microbiome of 10 pregnant women sampled every 3 weeks during pregnancy revealed increased diversity of *L. iners*-dominated communities along gestational age [35]. This study also suggested that analysis at the bacterial strain-level may yield more informative data when trying to understand microbial alterations during pregnancy [35]. Additional features found to be associated with the vaginal microbiome during pregnancy are higher bacterial load and lower prevalence of Mollicutes, specifically *Mycoplasma* and *Ureaplasma* [36].

In contrast to the changes observed in late pregnancy in the gut microbial composition, the changes in the vaginal microbiome are most pronounced in early pregnancy [25]. It would be interesting to assess whether this is due to a more direct hormonal effect on the microbiota, or other reasons. Functionally, perhaps the microbial protection from vaginosis is most crucial at early stages of pregnancy, when the developing embryo is most vulnerable. Further indications of hormonal influence on the vaginal microbiome may be seen in studies of the effects of hormonal contraceptives. These studies showed that combined (estrogen and progesterone) oral contraceptives are significantly associated with increased vaginal abundance of lactobacilli and reduced abundance of species associated with bacterial vaginosis [37].

During the postpartum period, the pregnancy profiles are reversed; the vaginal microbiome once again becomes more diverse; *Lactobacillus* species decrease in abundance; and bacteria such as *Peptoniphilus, Prevotella,* and *Anaerococcus* become more abundant as compared to the pregnant state [30].

The oral microbiome of pregnant women

Perhaps more surprising than the changes caused to the gut and vaginal microbial compositions, are the changes observed in the oral microbiome during pregnancy. First, there appears to be an increase in oral microbial growth during pregnancy. This was observed by viable counts of seven of the most common bacterial species in pregnant versus non-pregnant Japanese women [38]. Moreover, in a study of the supragingival microbiome, increased diversity and alterations in microbial composition were found among the samples of pregnant versus non-pregnant women [39]. Specifically, an increased abundance of *Neisseria, Porphyromonas,* and *Treponema,* along with decreased abundance of *Streptococcus* and *Veillonella,* was observed in the pregnant group.

Furthermore, the growth of pathogenic bacterial species including *Porphyromonas gingivalis* and *Aggregatibacter actinomycetemcomitans*, as well as *Candida* appears to be elevated during various stages of pregnancy, as supported by multiple studies and the increased rates of gingivitis during pregnancy [40–42]. The oral microbiome shifts back during the postpartum period to once again resemble the nonpregnant state with fewer pathogenic bacteria [42]. In a study comparing the oral microbial populations of pregnant women with gingivitis versus those with healthy gingiva during their third trimester, significant differences in bacterial taxa were found, although no differences in overall diversity were observed [43]. Nonetheless, as there have been relatively few studies to date on the oral microbiome during pregnancy, most of which have included relatively small cohorts, more research is needed in this area. This will allow further analysis of the potential role of dysbiosis during pregnancy in dental pathologies.

It is yet unclear how pregnancy leads to changes in the oral microbiome composition, although several hypotheses have suggested mechanisms based on the overall suppression of immunity during pregnancy [44]. Other potential mechanisms implicate the hormonal changes during pregnancy, specifically the rise in estrogens, which have been shown in previous studies to promote *Candida* growth [45]. Most intriguing is whether the oral microbial changes have functional significance, or whether they are merely a biomarker for the overall physiological changes.

Microbiome changes correlate with necessary functions during pregnancy

The gut microbial changes observed in late pregnancy correlate with increased weight gain, insulin resistance, and a greater inflammatory response seen in women in the third trimester. In fact, it was shown that this is not merely a correlation, but rather that the gut microbiome during the third trimester can actively enhance the observed physiological effects. When the microbiome from pregnant women in the third trimester was transferred to germ-free (GF) mice, these mice gained more weight, displayed increased insulin resistance, and increased levels of inflammatory serum markers, when compared to mice receiving first trimester microbiome [19]. It is important to realize that the weight gain, reduced insulin sensitivity, and perhaps even low-grade inflammation are all necessary for a healthy pregnancy, in order to enable proper fetal development without harming the mother or the fetus. Reinforcing this notion, maternal gut microbiome in the third trimester of pregnancy has also been associated with proper fetal development, based on correlations with newborn head circumference and weight [46].

The connection between microbial alterations and weight gain and metabolic changes during pregnancy

Perhaps the most dramatic physiological change that occurs throughout pregnancy is weight gain. The weight gain itself is due to the weight of the growing fetus, as well as growth of maternal tissues and elevated fat storage. At early stages of pregnancy, weight gain may only be subtle, whereas the most significant and rapid weight gain occurs during the third trimester.

In many microbiome studies, weight was shown to correlate with microbial populations. One of the most prominent examples of this is a study finding vast differences in microbial compositions in monozygotic twin pairs in which one twin was obese, and the other was of normal weight [47]. The microbial populations of the obese twins were further found to promote weight gain, including both increased body mass and fat mass when introduced into GF mice [48].

The microbiome can induce weight gain, insulin resistance, inflammation, and adiposity by multiple mechanisms. These include stimulation of the immune system or effects on host metabolism including enhanced absorption of glucose and fatty acids, increased fasting-induced adipocyte factor secretion, induction of catabolic pathways, and alterations in host energy harvest, energy expenditure, and fat storage [18,19]. Specific pathways by which microbiota influence weight gain include increased fermentation of polysaccharides, increased SCFA production leading to regulation of metabolic peptide secretion, interaction with TLR5 to regulate appetite, and others [49]. In addition, interaction of the microbiome with endocrine cells can alter signaling of glucagon-like peptides 1 and 2, influencing satiety and insulin secretion [50]. Furthermore, weight gain and the obese state have also been shown to alter gut permeability and increase endotoxemia and bacteremia, thereby promoting systemic inflammation [49].

While weight gain is necessary in all pregnant women, differences in initial weight and body mass index (BMI) of women before pregnancy can further affect the microbial populations during pregnancy, and they even have effects on the microbial compositions of the offspring. Differences in gut microbial composition are seen in overweight versus normal weight women. In overweight as opposed to normal weight pregnant women, there is a significantly higher fecal abundance of *Bacteroides* and *Staphylococcus* [18]. Another study in obese and normal weight mothers revealed a positive correlation between pregestational BMI and inflammatory biomarkers during the third trimester [51]. This study also found higher Firmicutes levels and a higher ratio of Firmicutes/Bacteroidetes in overweight and obese subjects, whereas inflammatory markers were correlated with lower microbial diversity.

Gestational weight gain has also been associated with maternal gut microbial composition and reduced diversity [52,53]. It was further shown that the prepregnancy maternal BMI correlates with neonatal gut microbial composition in vaginally delivered offspring [54].

The effects of BMI on the pregnancy microbiome are likely interconnected with expression of metabolic hormones. Insulin, gastric inhibitory polypeptide, and adipokine levels were all found to be correlated with alterations in bacterial abundance in overweight and obese women [55]. Additional metabolic changes associated with pregnancy include elevated fasting blood glucose levels, insulin resistance, and glucose intolerance [56–59]. This is of particular interest, as it has been shown that during late pregnancy the expression of carbohydrate-related microbial functions is regulated to conform with the high availability of glucose [60]. Finally, during pregnancy, changes occur in metabolic hormone levels including appetite-related hormones such as leptin, ghrelin, and glucagon-like peptide 1 (GLP1) [55–59], and several findings have correlated specific bacterial species and levels of leptin and ghrelin [61].

The changes described earlier are only part of the changes in the gastrointestinal (GI) tract that are required to meet the increased nutritional demands of the developing fetus during pregnancy. Additional adaptations include enhancement of the villus surface area in the GI tract, as well as increased numbers of sugar and signaling receptors [62]. These may have direct implications on gut microbiome.

Potential effects of diet on the healthy microbiome during pregnancy

Among the environmental effects on the microbiome composition, diet appears to have the most pronounced effect. Many studies in humans or in animal models have shown that dietary changes can further alter the microbiome [63]. Compositional differences are observed between meat eating and vegetarian subjects [64]. While some short-term diets may cause reversible microbial changes, long-term dietary changes may cause stable changes in microbiota composition. High-fat diets (HFDs) have been shown in numerous human and rodent studies to alter the microbial composition, as compared to low-fat diets or normal chow [65]. Most commonly, HFD causes the relative abundance of Firmicutes to increase at the expense of Bacteroidetes [66]. These microbial alterations appear to be independent of BMI, and therefore these are not directly related to obesity. In a similar manner, HFD has strong effects on the microbiome during pregnancy. A human study comparing high-fat versus standard maternal diets during pregnancy revealed that these diets influenced the neonatal microbiome in a manner that persisted even at 6 weeks of age [67]. Even when female mice were fed on HFD only before conception, microbiome compositional effects were observed in later pregnancy [68]. More specifically, the pregnancy microbial alterations following HFD predict gene expression differences in lipid metabolism, glycolysis, and gluconeogenic metabolic pathways. Furthermore, maternal obesity is known to result in metabolic and neurodevelopmental abnormalities in the offspring [69]. When transplanting microbiota from mice fed with HFD or control low-fat diet to female mice, the male offspring of the dams receiving the HFD microbiota showed significant disruptions in exploratory, cognitive, and compulsive behavior. Specifically, decreased representation of specific members of the Firmicutes phylum was linked to behavioral abnormalities in male animals. In addition, the female offspring of HFD dams displayed increased body weight and adiposity [69].

Intake of vitamins and other nutritional supplements during pregnancy may also affect the microbiome. In a cohort of 60 pregnant women, higher dietary intake of fat-soluble vitamins, especially vitamin D, was associated with reduced gut microbial alpha diversity and increased relative abundance of Proteobacteria [70]. Moreover, the diversity of the gut microbiota and the intake of vitamins, n-3 PUFAs, and fiber in overweight pregnant women was associated with decreased serum zonulin concentration, a marker of intestinal permeability. These results suggest that dietary balance during pregnancy may affect microbial compositions in a beneficial manner, leading to improved metabolic health of both the mother and fetus [71]. Vitamin D supplementation has been inversely associated with bacterial vaginosis during pregnancy and with alterations in certain vaginal bacteria [72].

Microbial alterations and hormonal changes during pregnancy

Numerous pieces of evidence have correlated host hormones with the gut microbiota. This correlation likely reflects reciprocal effects of the gut microbiota composition and hormonal levels, with each of these affecting the other. Microbes have been shown to produce and regulate hormones and to be affected by existing hormonal levels [13].

As pregnancy is characterized by a dramatic rise in progesterone and estrogen levels correlated with gestational age, it is expected that these hormonal surges would have an effect on the microbial compositions [58]. Several pieces of evidence link microbiota composition to maternal hormone levels. First, it has been reported that manipulation of estrogen receptor (ER-β) affects the intestinal microbiome [73]. Second, several microbes including *Prevotella intermedius* have been shown to utilize estradiol and progesterone for their growth [74]. Moreover, in disease states manifested by hormonal imbalance, such as polycystic ovary syndrome, distinct alterations are seen in the gut microbial populations [75]. Finally, a recent study using both in vitro techniques and hormonal manipulations in mice demonstrated that progesterone leads to gut microbiota alterations, including a dramatic rise in *Bifidobacterium* abundance [22].

The vaginal microbiota has been shown to be strongly affected by hormonal status. Most prominently, ovariectomy in rats was associated with lower total bacterial load primarily due to the absence of *Lactobacillus*, absence of anaerobic bacteria, and predominance of *Clostridium perfringens*, *Bacteroides*, *Staphylococcus aureus*, and *Staphylococcus epidermidis* [76]. These effects were substantially reversed by supplementation of estradiol. It was further shown that estrogen (17β-estradiol) can suppress the inflammatory effects of *C. albicans*, thereby protecting the host from potential infection [27]. The vaginal microbiome in both humans and baboons was also shown to change in response to reproductive state, ovarian cycle phase, and sexual behavior [77].

The precise mechanisms and pathways linking the microbiome and hormone levels during pregnancy remain to be fully deciphered; however, large studies including multiomic approaches may lead to better understanding of the interactions between the multiple pathways that occur during pregnancy. In this regard, an initial multiomic study of 17 pregnant women revealed an association between the steroid hormone pregnanolone sulfate and regulation of myeloid dendritic cells and Tregs [78]. These results may provide initial hints regarding the precise interactions necessary for coordination of the endocrine system, immune system, and microbiome, all of which change throughout pregnancy.

Finally, associations have also been identified between the oral microbiome composition during pregnancy and salivary estradiol and progesterone hormone levels. Species of significant abundance found positively correlated to hormonal levels included *Prevotella* spp. and *Treponema* spp. [39].

The maternal microbiota during pregnancy affects the offspring

The main microbial inoculation of the infant occurs during birth, and thereafter, following contact with the environment. As the first exposure during birth is to maternal tissues, members of the maternal microbiome are the first microorganisms to contact the neonate. Therefore the initial gut microbiome of infants born vaginally resembles the maternal vaginal microbiome, while the microbial communities of infants delivered by C-section resemble those of the maternal skin [79].

However, various pieces of evidence link the maternal microbiome during pregnancy with effects on the offspring, suggesting that even before the fetus acquires its own microbial communities, it may be affected by the maternal bacteria. The proposed effects on the offspring include promoting growth, immune system development, influence on BMI, and influencing propensity to develop asthma and allergic disease. The maternal microbiota has been shown to drive early postnatal innate immunity, including both enteric and peripheral immune development, affecting cell numbers and gene expression [80,81]. Among the main effects of the maternal microbiome on innate immunity of the offspring are regulation of IL17 and IFNγ producing cells, as well as iNKT cells. Depletion of microbial exposure during gestation causes peripheral immune dysfunction, decreasing levels of Th17, Treg, neutrophils, and macrophage cells, while increasing iNKT cellular levels [81]. Effects on cytokine levels include IL17, IL1β, and TNFα expression, further influencing normal neurodevelopment. Gestational colonization experiments in mice in which pregnant females are colonized with specific bacteria (*Escherichia coli*) allow detection of immune effects on the pups. Such experiments revealed that gestational colonization increases expression of genes encoding epithelial antibacterial peptides and metabolism of microbial molecules, and reduced inflammatory responses in the pups [80].

Another aspect of maternal effects passed on to the fetus is neuropsychological factors. The interconnections between emotion, neurocognition, the microbiome, and the immune system are well known [82]. Studies probing these pathways, as part of the gut–brain axis, have demonstrated that depression and anxiety have effects on the gut microbiota composition. An intriguing study showed that maternal anxiety, depression, and stress during pregnancy are associated with reduced diversity of the meconium microbiome and lower abundance of the Enterococcaceae family [83]. The aforementioned effects on the offspring's growth, immune system, and neural development, all highlight the importance of a balanced maternal microbiome during pregnancy.

Effects of maternal antibiotics and probiotics consumed during pregnancy on the offspring

Antibiotics are agents that eliminate or inhibit microbial growth. As such, while they are normally administered to treat pathogenic bacteria leading to infection, they also have effects on the healthy microbiota components. Numerous studies have shown that infant exposure to antibiotics may result in long-term effects including metabolic changes, effects on immune system development, increased risk of metabolic syndrome and allergic disease, and even effects on personality, including anxiety and aggressive behaviors [84,85]. In addition, antibiotics administered during pregnancy were shown to affect the gut microbiome composition and reduce microbial diversity in pregnant rats, increasing the abundance of Proteobacteria and *Enterobacter* and decreasing the abundance of Firmicutes and *Lactobacillus* [86]. However, the effects of maternal antibiotics on the offspring are more intriguing and proving them is more challenging.

Several studies suggest that antibiotic exposure during gestation indeed affects the offspring [87]. In studies of pregnant women receiving intrapartum antibiotic prophylaxis prior to delivery, clear alterations are seen in the gut microbiome composition of their infants with increased relative abundance of *Enterococcus* and *Clostridium* and decreased abundance of *Bacteroides* and *Parabacteroides* [88]. A study in mice receiving low-dose penicillin during gestation and lactation also revealed significant alterations in the offspring's gut microbiome [89]. Furthermore, antibiotics administered to pregnant mice were shown to alter behavior in the offspring, which exhibited reduced locomotor activities and exploratory behaviors [90]. Moreover, antibiotics during pregnancy in mice were also shown to affect immune responses, increase weight gain, and increase risk for allergic disease and asthma in the offspring [88,91,92]. Mechanistically, the microbiome may alter the developing fetal immune system by transplacental passage of microbial metabolites and IgGs and by exposure to lipopolysaccharides (LPSs) and SCFAs, rendering the offspring less prone to develop allergic disease and asthma [93,94]. These effects suggest additional functional roles for the maternal microbiota during gestation.

As expected, maternal exposure to prebiotics or probiotics during pregnancy also led to metabolic effects, somewhat reversed to the changes related to antibiotic exposure, emphasizing the connection between microbial composition and host metabolic factors. In a study testing the effect of oligofructose prebiotic treatment during pregnancy and lactation in diet-induced obese rats, compositional effects were noted on the cecal microbiota of the offspring, as well as in prevention of adiposity [95]. Moreover, when pregnant mice received a lactulose-rich diet, their gut microbiome included significantly higher abundances of *Bifidobacterium* and *Bacteroides*, together with higher SCFA levels, higher progesterone levels, and lower serum glucose and total cholesterol concentrations [96].

In a human study in which pregnant mothers received probiotics compared to placebo controls, bacterial alterations were detected in the infant meconium [97]. In another study of maternal probiotics during pregnancy, the infant microbiota was altered, as expression of Toll-like receptor genes in the placenta and meconium [98]. A meta-analysis including over 1000 pregnant women found probiotics to improve glucose and lipid metabolism [99]. However, probiotic administration to overweight pregnant women did not reduce levels of serum zonulin or LPS associated with pregnancy [100]. While treatment of pregnant mothers with probiotics has also been suggested to decrease rates of autoimmune disease in the offspring, little data has been obtained so far, and this effect was not observed in a study measuring prevalence of celiac disease [101,102]. Importantly, ingestion of probiotics throughout pregnancy does not appear to affect the vaginal microbiome composition [103]. However, probiotics significantly increased the abundance of the gut mycobiota, consisting of the fungal populations, in pregnant mothers [104].

Conclusions

In this chapter, we discussed the changes observed in the microbiome composition at multiple sites (gut, vagina, oral cavity) during healthy pregnancy and in pregnancy complications, and the potential functions of these bacteria. The microbial populations change dramatically, in correlation with additional physiological changes that occur during pregnancy, including changes in body composition, weight gain, hormonal levels, inflammation, and metabolism. Some of the microbial alterations have been directly shown to promote weight gain, inflammatory cytokine expression, and insulin insensitivity, thereby substantiating the roles of bacteria in these processes. These findings emphasize the interconnections between the various physiological processes that occur during pregnancy (microbial, metabolic, and immune) and suggest a united overall regulatory network. Maintaining healthy microbial populations appears to be essential for a healthy pregnancy outcome and has long-term effects on the offspring. As the microbiota is greatly affected by diet, BMI, and metabolic states, distinct alterations are seen when a HFD versus control diet is consumed during pregnancy. Understanding the functional microbiota changes during pregnancy raises concerns regarding the use of antibiotics during pregnancy and supports the importance of probiotics and balanced diets during this important period. Furthermore, deciphering roles of the microbiota in pregnancy complications may explain their etiologies and suggest potential therapies.

References

[1] Boulange CL, Neves AL, Chilloux J, Nicholson JK, Dumas ME. Impact of the gut microbiota on inflammation, obesity, and metabolic disease. Genome Med 2016;8:42.

[2] Chervonsky AV. Microbiota and autoimmunity. Cold Spring Harb Perspect Biol 2013;5:a007294.

[3] Spor A, Koren O, Ley R. Unravelling the effects of the environment and host genotype on the gut microbiome. Nat Rev Microbiol 2011;9:279–90.
[4] Turnbaugh PJ, et al. An obesity-associated gut microbiome with increased capacity for energy harvest. Nature 2006;444:1027–31.
[5] Ursell LK, et al. The interpersonal and intrapersonal diversity of human-associated microbiota in key body sites. J Allergy Clin Immunol 2012;129:1204–8.
[6] Belkaid Y, Naik S. Compartmentalized and systemic control of tissue immunity by commensals. Nat Immunol 2013;14:646–53.
[7] Jandhyala SM, et al. Role of the normal gut microbiota. World J Gastroenterol 2015;21:8787–803.
[8] Kamada N, Seo SU, Chen GY, Nunez G. Role of the gut microbiota in immunity and inflammatory disease. Nat Rev Immunol 2013;13:321–35.
[9] Ostman S, Rask C, Wold AE, Hultkrantz S, Telemo E. Impaired regulatory T cell function in germ-free mice. Eur J Immunol 2006;36:2336–46.
[10] Diaz Heijtz R, et al. Normal gut microbiota modulates brain development and behavior. Proc Natl Acad Sci U S A 2011;108:3047–52.
[11] Gareau MG, et al. Bacterial infection causes stress-induced memory dysfunction in mice. Gut 2011;60:307–17.
[12] Neufeld KM, Kang N, Bienenstock J, Foster JA. Reduced anxiety-like behavior and central neurochemical change in germ-free mice. Neurogastroenterol Motil 2011;23:255–64 e119.
[13] Neuman H, Debelius JW, Knight R, Koren O. Microbial endocrinology: the interplay between the microbiota and the endocrine system. FEMS Microbiol Rev 2015;39:509–21.
[14] Dinan TG, Cryan JF. The microbiome-gut-brain axis in health and disease. Gastroenterol Clin North Am 2017;46:77–89.
[15] Sherwin E, Sandhu KV, Dinan TG, Cryan JF. May the force be with you: the light and dark sides of the microbiota-gut-brain axis in neuropsychiatry. CNS Drugs 2016;30:1019–41.
[16] McVey Neufeld KA, Luczynski P, Dinan TG, Cryan JF. Reframing the teenage wasteland: adolescent microbiota-gut-brain axis. Can J Psychiatry 2016;61:214–21.
[17] Pitlik SD, Koren O. How holobionts get sick-toward a unifying scheme of disease. Microbiome 2017;5:64.
[18] Collado MC, Isolauri E, Laitinen K, Salminen S. Distinct composition of gut microbiota during pregnancy in overweight and normal-weight women. Am J Clin Nutr 2008;88:894–9.
[19] Koren O, et al. Host remodeling of the gut microbiome and metabolic changes during pregnancy. Cell 2012;150:470–80.
[20] Schock H, et al. Hormone concentrations throughout uncomplicated pregnancies: a longitudinal study. BMC Pregnancy Childbirth 2016;16:146.
[21] Soma-Pillay P, Nelson-Piercy C, Tolppanen H, Mebazaa A. Physiological changes in pregnancy. Cardiovasc J Afr 2016;27:89–94.
[22] Nuriel-Ohayon M, et al. Progesterone increases *Bifidobacterium* relative abundance during late pregnancy. Cell Rep 2019;27:730–6 e733.
[23] Haro C, et al. The gut microbial community in metabolic syndrome patients is modified by diet. J Nutr Biochem 2015;27:27–31.
[24] Ravel J, Brotman RM. Translating the vaginal microbiome: gaps and challenges. Genome Med 2016;8:35.
[25] Aagaard K, et al. A metagenomic approach to characterization of the vaginal microbiome signature in pregnancy. PLoS One 2012;7:e36466.
[26] O'Hanlon DE, Moench TR, Cone RA. Vaginal pH and microbicidal lactic acid when lactobacilli dominate the microbiota. PLoS One 2013;8:e80074.

[27] Wagner RD, Johnson SJ. Probiotic lactobacillus and estrogen effects on vaginal epithelial gene expression responses to *Candida albicans*. J Biomed Sci 2012;19:58.
[28] Prince AL, Antony KM, Ma J, Aagaard KM. The microbiome and development: a mother's perspective. Semin Reprod Med 2014;32:14–22.
[29] The Integrative HMP (iHMP) Research Network Consortium. The Integrative Human Microbiome Project. Nature 2019;569:641–8.
[30] MacIntyre DA, et al. The vaginal microbiome during pregnancy and the postpartum period in a European population. Sci Rep 2015;5:8988.
[31] Romero R, et al. The composition and stability of the vaginal microbiota of normal pregnant women is different from that of non-pregnant women. Microbiome 2014;2:4.
[32] DiGiulio DB, et al. Temporal and spatial variation of the human microbiota during pregnancy. Proc Natl Acad Sci U S A 2015;112:11060–5.
[33] Serrano MG, et al. Racioethnic diversity in the dynamics of the vaginal microbiome during pregnancy. Nat Med 2019.
[34] Stout MJ, et al. Early pregnancy vaginal microbiome trends and preterm birth. Am J Obstet Gynecol 2017;217:356 e1–356 e18.
[35] Goltsman DSA, et al. Metagenomic analysis with strain-level resolution reveals fine-scale variation in the human pregnancy microbiome. Genome Res 2018;28:1467–80.
[36] Freitas AC, et al. The vaginal microbiome of pregnant women is less rich and diverse, with lower prevalence of Mollicutes, compared to non-pregnant women. Sci Rep 2017;7:9212.
[37] Brooks JP, et al. Effects of combined oral contraceptives, depot medroxyprogesterone acetate and the levonorgestrel-releasing intrauterine system on the vaginal microbiome. Contraception 2017;95:405–13.
[38] Fujiwara N, et al. Significant increase of oral bacteria in the early pregnancy period in Japanese women. J Investig Clin Dent 2015;8(1).
[39] Lin W, et al. Ecological shifts of supragingival microbiota in association with pregnancy. Front Cell Infect Microbiol 2018;8:24.
[40] Fujiwara N, et al. Significant increase of oral bacteria in the early pregnancy period in Japanese women. J Investig Clin Dent 2017;8.
[41] Borgo PV, Rodrigues VA, Feitosa AC, Xavier KC, Avila-Campos MJ. Association between periodontal condition and subgingival microbiota in women during pregnancy: a longitudinal study. J Appl Oral Sci 2014;22:528–33.
[42] Balan P, et al. Keystone species in pregnancy gingivitis: a snapshot of oral microbiome during pregnancy and postpartum period. Front Microbiol 2018;9:2360.
[43] Yang I, Knight AK, Dunlop AL, Corwin EJ. Characterizing the subgingival microbiome of pregnant African American women. J Obstet Gynecol Neonatal Nurs 2019;48:140–52.
[44] Mor G, Cardenas I. The immune system in pregnancy: a unique complexity. Am J Reprod Immunol 2010;63:425–33.
[45] Kumar PS. Sex and the subgingival microbiome: do female sex steroids affect periodontal bacteria? Periodontol 2000 2013;61:103–24.
[46] Sato Y, et al. Maternal gut microbiota is associated with newborn anthropometrics in a sex-specific manner. J Dev Orig Health Dis 2019;10:659–66.
[47] Turnbaugh PJ, et al. A core gut microbiome in obese and lean twins. Nature 2009;457:480–4.
[48] Ridaura VK, et al. Gut microbiota from twins discordant for obesity modulate metabolism in mice. Science 2013;341:1241214.
[49] Shen J, Obin MS, Zhao L. The gut microbiota, obesity and insulin resistance. Mol Aspects Med 2013;34:39–58.
[50] Ley RE. Obesity and the human microbiome. Curr Opin Gastroenterol 2010;26:5–11.

[51] Zacarias MF, et al. Pregestational overweight and obesity are associated with differences in gut microbiota composition and systemic inflammation in the third trimester. PLoS One 2018;13:e0200305.
[52] Aatsinki AK, et al. Gut microbiota composition in mid-pregnancy is associated with gestational weight gain but not prepregnancy body mass index. J Womens Health (Larchmt) 2018;27:1293–301.
[53] Stanislawski MA, et al. Pre-pregnancy weight, gestational weight gain, and the gut microbiota of mothers and their infants. Microbiome 2017;5:113.
[54] Mueller NT, et al. Birth mode-dependent association between pre-pregnancy maternal weight status and the neonatal intestinal microbiome. Sci Rep 2016;6:23133.
[55] Gomez-Arango LF, et al. Connections between the gut microbiome and metabolic hormones in early pregnancy in overweight and obese women. Diabetes 2016;65:2214–23.
[56] Kumar P, Magon N. Hormones in pregnancy. Niger Med J 2012;53:179–83.
[57] Emanuela F, et al. Inflammation as a link between obesity and metabolic syndrome. J Nutr Metab 2012;2012:476380.
[58] Newbern D, Freemark M. Placental hormones and the control of maternal metabolism and fetal growth. Curr Opin Endocrinol Diabetes Obes 2011;18:409–16.
[59] Fuglsang J. Ghrelin in pregnancy and lactation. Vitam Horm 2008;77:259–84.
[60] Gosalbes MJ, et al. Metabolic adaptation in the human gut microbiota during pregnancy and the first year of life. EBioMedicine 2019;39:497–509.
[61] Queipo-Ortuno MI, et al. Gut microbiota composition in male rat models under different nutritional status and physical activity and its association with serum leptin and ghrelin levels. PLoS One 2013;8:e65465.
[62] Astbury S, Mostyn A, Symonds ME, Bell RC. Nutrient availability, the microbiome, and intestinal transport during pregnancy. Appl Physiol Nutr Metab 2015;40:1100–6.
[63] de Morais MB, et al. Microbiota, environment, and diet. J Pediatr Gastroenterol Nutr 2017;65:e24.
[64] Tomova A, et al. The effects of vegetarian and vegan diets on gut microbiota. Front Nutr 2019;6:47.
[65] Zhang M, Yang XJ. Effects of a high fat diet on intestinal microbiota and gastrointestinal diseases. World J Gastroenterol 2016;22:8905–9.
[66] Hildebrandt MA, et al. High-fat diet determines the composition of the murine gut microbiome independently of obesity. Gastroenterology 2009;137:1716–24 e1711-e1712.
[67] Chu DM, et al. The early infant gut microbiome varies in association with a maternal high-fat diet. Genome Med 2016;8:77.
[68] Gohir W, et al. Pregnancy-related changes in the maternal gut microbiota are dependent upon the mother's periconceptional diet. Gut Microbes 2015;6(5):310–20.
[69] Bruce-Keller AJ, et al. Maternal obese-type gut microbiota differentially impact cognition, anxiety and compulsive behavior in male and female offspring in mice. PLoS One 2017;12:e0175577.
[70] Mandal S, et al. Fat and vitamin intakes during pregnancy have stronger relations with a pro-inflammatory maternal microbiota than does carbohydrate intake. Microbiome 2016;4:55.
[71] Mokkala K, et al. Gut microbiota richness and composition and dietary intake of overweight pregnant women are related to serum zonulin concentration, a marker for intestinal permeability. J Nutr 2016;146:1694–700.
[72] Jefferson KK, et al. Relationship between vitamin D status and the vaginal microbiome during pregnancy. J Perinatol 2019;39:824–36.
[73] Menon R, et al. Diet complexity and estrogen receptor beta status affect the composition of the murine intestinal microbiota. Appl Environ Microbiol 2013;79:5763–73.

[74] Kornman KS, Loesche WJ. Effects of estradiol and progesterone on *Bacteroides melaninogenicus* and *Bacteroides gingivalis*. Infect Immun 1982;35:256–63.
[75] Liu R, et al. Dysbiosis of gut microbiota associated with clinical parameters in polycystic ovary syndrome. Front Microbiol 2017;8:324.
[76] Bezirtzoglou E, et al. Hormone therapy alters the composition of the vaginal microflora in ovariectomized rats. Microb Ecol 2008;55:751–9.
[77] Miller EA, Livermore JA, Alberts SC, Tung J, Archie EA. Ovarian cycling and reproductive state shape the vaginal microbiota in wild baboons. Microbiome 2017;5:8.
[78] Ghaemi MS, et al. Multiomics modeling of the immunome, transcriptome, microbiome, proteome and metabolome adaptations during human pregnancy. Bioinformatics 2019;35:95–103.
[79] Shin H, et al. The first microbial environment of infants born by C-section: the operating room microbes. Microbiome 2015;3:59.
[80] Gomez de Aguero M, et al. The maternal microbiota drives early postnatal innate immune development. Science 2016;351:1296–302.
[81] Pronovost GN, Hsiao EY. Perinatal interactions between the microbiome, immunity, and neurodevelopment. Immunity 2019;50:18–36.
[82] Osadchiy V, Martin CR, Mayer EA. The gut-brain axis and the microbiome: mechanisms and clinical implications. Clin Gastroenterol Hepatol 2019;17:322–32.
[83] Hu J, et al. Microbiota of newborn meconium is associated with maternal anxiety experienced during pregnancy. Dev Psychobiol 2019;61(17):640–9.
[84] Neuman H, Forsythe P, Uzan A, Avni O, Koren O. Antibiotics in early life: dysbiosis and the damage done. FEMS Microbiol Rev 2018;42:489–99.
[85] Leclercq S, et al. Low-dose penicillin in early life induces long-term changes in murine gut microbiota, brain cytokines and behavior. Nat Commun 2017;8:15062.
[86] Khan I, et al. Metagenomic analysis of antibiotic-induced changes in gut microbiota in a pregnant rat model. Front Pharmacol 2016;7:104.
[87] Kuperman AA, Koren O. Antibiotic use during pregnancy: how bad is it? BMC Med 2016;14:91.
[88] Azad MB, et al. Impact of maternal intrapartum antibiotics, method of birth and breastfeeding on gut microbiota during the first year of life: a prospective cohort study. BJOG 2016;123:983–93.
[89] Cox LM, et al. Altering the intestinal microbiota during a critical developmental window has lasting metabolic consequences. Cell 2014;158:705–21.
[90] Tochitani S, et al. Administration of non-absorbable antibiotics to pregnant mice to perturb the maternal gut microbiota is associated with alterations in offspring behavior. PLoS One 2016;11:e0138293.
[91] Gonzalez-Perez G, et al. Maternal antibiotic treatment impacts development of the neonatal intestinal microbiome and antiviral immunity. J Immunol 2016;196:3768–79.
[92] Abrahamsson TR, Wu RY, Jenmalm MC. Gut microbiota and allergy: the importance of the pregnancy period. Pediatr Res 2015;77:214–19.
[93] Gray LE, O'Hely M, Ranganathan S, Sly PD, Vuillermin P. The maternal diet, gut bacteria, and bacterial metabolites during pregnancy influence offspring asthma. Front Immunol 2017;8:365.
[94] Vuillermin PJ, et al. The maternal microbiome during pregnancy and allergic disease in the offspring. Semin Immunopathol 2017;39:669–75.
[95] Paul HA, Bomhof MR, Vogel HJ, Reimer RA. Diet-induced changes in maternal gut microbiota and metabolomic profiles influence programming of offspring obesity risk in rats. Sci Rep 2016;6:20683.
[96] Zhang Z, et al. Effects of a lactulose-rich diet on fecal microbiome and metabolome in pregnant mice. J Agric Food Chem 2019;67(27):7674–83.

[97] Rautava S, Kainonen E, Salminen S, Isolauri E. Maternal probiotic supplementation during pregnancy and breast-feeding reduces the risk of eczema in the infant. J Allergy Clin Immunol 2012;130:1355–60.
[98] Rautava S, Collado MC, Salminen S, Isolauri E. Probiotics modulate host-microbe interaction in the placenta and fetal gut: a randomized, double-blind, placebo-controlled trial. Neonatology 2012;102:178–84.
[99] Han MM, et al. Probiotics improve glucose and lipid metabolism in pregnant women: a meta-analysis. Ann Transl Med 2019;7:99.
[100] Mokkala K, et al. The impact of probiotics and n-3 long-chain polyunsaturated fatty acids on intestinal permeability in pregnancy: a randomised clinical trial. Benef Microbes 2018;9:199–208.
[101] Swartwout B, Luo XM. Implications of probiotics on the maternal-neonatal interface: gut microbiota, immunomodulation, and autoimmunity. Front Immunol 2018;9:2840.
[102] Savilahti EM, Ilonen J, Kukkonen AK, Savilahti E, Kuitunen M. Celiac disease by the age of 13 years is not associated with probiotics administration in infancy. J Pediatr Gastroenterol Nutr 2018;66:937–40.
[103] Husain S, et al. Effects of oral probiotic supplements on vaginal microbiota during pregnancy: a randomised, double-blind, placebo-controlled trial with microbiome analysis. BJOG 2019;127(2):275–84.
[104] Schei K, et al. Early gut mycobiota and mother-offspring transfer. Microbiome 2017;5:107.

CHAPTER

2

The microbiome and pregnancy complications

Gwenny M. Fuhler

Department of Gastroenterology and Hepatology, Erasmus MC University Medical Center, Rotterdam, The Netherlands

Introduction

Pregnancy is accompanied by many anatomical and physiological changes that ensure the successful implantation of a fertilized egg, growth of a major histocompatibility complex (MHC)-mismatched fetus, and timely parturition. An increased plasma volume, cardiac output, and oxygen demand are seen. Changes in renal and alimentary tract functionality occur. Reduced platelet count, an increased iron requirement, and alterations in immune cell presence and functionality are some of the hematological consequences of pregnancy. Pregnant women undergo extensive endocrine changes, including an increased level of thyroid hormones T4 and T3, enhanced levels of adrenal gland-derived steroids, prolactin and oxytoxin production at the pituitary gland, increased insulin secretion by the pancreas, and an increase in triglyceride synthesis by the liver. All these processes are carefully regulated, with the pregnancy hormones human chorionic gonadotropin (hCG), progesterone, and estrogen playing key roles in many of these processes. Thus biologically speaking, pregnancy is an incredibly complicated process with integration of many processes required for a successful outcome. It is therefore not surprising that complications in pregnancy arise in as many as 47% of pregnancies [1]. In 2008, fetal loss occurred in over 16% of pregnancies [2]. Importantly, despite the fact that we may expect health care to improve over the decades, the percentage of pregnancies ending in

The Human Microbiome in Early Life
DOI: https://doi.org/10.1016/B978-0-12-818097-6.00002-X

fetal loss did not decrease but rather increased from 15% in 1990 to 17% in 2008. Obviously, we need a better understanding of the factors contributing to pregnancy outcomes if we aim to reduce the number of unsuccessful pregnancies. In this chapter, we will touch upon the hormonal/immunological aspects occurring during normal pregnancy physiology. We will then discuss the most common complications seen during pregnancy, which may affect either mother or child, and investigate which common factors are thought to contribute to the occurrence of these complications. We will then turn our attention specifically to the potential role of the microbiome therein.

The physiology of normal pregnancy

As mentioned earlier, many physiological changes are required for a successful pregnancy. The first step in any successful pregnancy is implantation of the fertilized egg. This requires a receptive uterus, which is mediated by a high level of pro-inflammatory T helper (Th)-1 cells and cytokines at the uterine lining [3]. Upon implantation, placental trophoblasts start to produce the hormone human chorionic gonadotrophin (hCG), systemic levels of which reach peak levels around week 10 of gestation, and quickly drop back to baseline values thereafter [4]. hCG plays an important role in placental growth and vascularization but is also responsible for causing pregnancy nausea through its receptors in the hippocampus, hypothalamus, and brain stem. Furthermore, hCG directly modulates immunity by reducing macrophage activity at the placenta, thereby preventing the destruction of foreign fetoplacental tissue by the maternal immune system [5]. Another action of hCG is stimulation of the production of progesterone by cells of the corpus luteum, a temporary endocrine structure in the ovary that remains after ovulation. Progesterone biosynthesis is also stimulated by estrogen, which is also initially produced in the corpus luteum. Production of both progesterone and estrogen shifts to the growing placenta after week 9 of gestation. Thus their peripheral levels start rising progressively during the first trimester and reach their peak in the third trimester of gestation (Fig. 2.1) [6]. Progesterone is essential for preparing the endometrium for embryo implantation, but continues to contribute to pregnancy physiology throughout pregnancy [7]. Both progesterone and estrogen play an important role in the regulation of placental angiogenesis, uterine artery vasodilatation, maternal blood pressure decrease, and placental cell proliferation and differentiation [6,8].

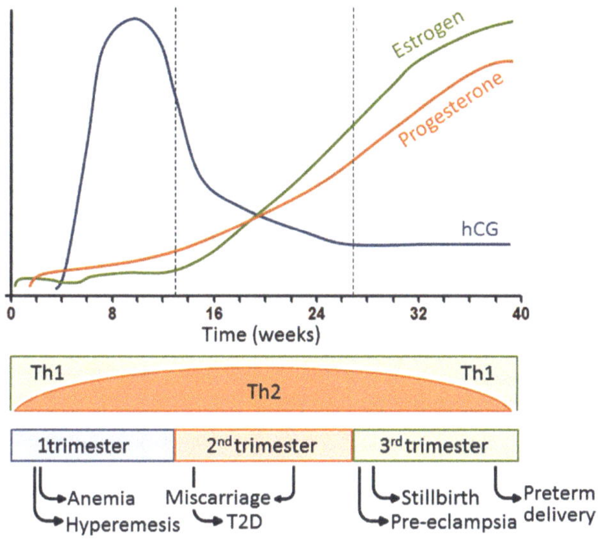

FIGURE 2.1 **Hormonal and immunological changes throughout gestation and the pregnancy complications that may occur.** Human chorionic gonadotropin levels rise during the first trimester and drops at later stages. In contrast, progesterone and estrogen levels keep rising throughout pregnancy. The first stages of pregnancy are associated with a pro-inflammatory Th1 profile at the fetomaternal interface, but to protect the growing fetus from maternal immunological assault, a shift toward a Th2 immunological profile occurs during pregnancy. Toward parturition, inflammatory processes are again required, and a Th1 profile is again observed. Throughout gestation, various pregnancy complications may arise.

The immunological role of sex hormones during pregnancy

A mentioned earlier, implantation of a fertilized egg requires a pro-inflammatory environment, and the early implantation site is characterized by the presence of pro-inflammatory cytokines such as IL6, IL8, and TNFα [9]. These cytokines may be produced by Th1 cells [10], but also by innate lymphoid cells that have been observed in first-term placentae [11]. However, the majority of early infiltrating immune cells ($>70\%$) are natural killer (NK) cells and macrophages. These cells contribute to the pro-inflammatory milieu that favors trophoblast invasion in the surrounding tissue as well as angiogenesis. At later stages of pregnancy, NK cell numbers decline and macrophages shift from an inflammatory phenotype (M1) to a wound-healing phenotype (M2) [12]. M2 macrophages are an important source of the antiinflammatory cytokine IL10 [13]. In addition, these cells produce indolamine 2,3-dioxigenase-1 (IDO1), an enzyme that contributes to the generation of regulatory

T cells (Tregs) and induction of tolerance [14]. Indeed, tolerance against an MHC-mismatched fetus is essential for a successful pregnancy, as activation of the maternal immune defense by paternal antigens crossing the placental barrier may otherwise be detrimental. Part of this tolerance appears to come from a shift from inflammatory Th1 profiles to a more permissive Th2 profile [15,16]. To a large extent, changes in immunoprofiles during pregnancy are regulated by sex hormones. hCG drives early expansion of Tregs and contributes to a reduction of proinflammatory cytokines [17]. Similarly, suppression of the immune system to prevent rejection of the fetus by the maternal immune system appears to be one of the most important roles for progesterone [18,19], while antiinflammatory properties have also been ascribed to estrogen [20,21]. Toward the end of pregnancy, immune patterns at the fetal–maternal interface need to shift again, as parturition appears to require inflammatory signals. In nonhuman mammals, term labor is accompanied by a drop in progesterone levels prior to parturition, although this is less evident in humans [18,22]. Nevertheless, softening of the cervix due to inflammatory mediators has long been established as a requirement for parturition [23]. Thus human pregnancy is accompanied by temporal changes in the immune system, with the current consensus suggesting a shift from Th1 patterns during early pregnancies to a Th2 phenotype for the main duration of the pregnancy, and again Th2-associated pro-inflammatory changes at term.

Complications that may arise during pregnancy

Considering the complicated physiological changes that have to act in concert, successful pregnancy can be considered an astonishing biological feat. Indeed, many complications may arise during conception, pregnancy, and perinatal period. Both maternal and fetal/placental complications can be distinguished.

Fetal risks

The first obstacle to a healthy pregnancy would be a failure of the fertilized egg to implant. This can have many mechanistic reasons, but as without implantation there is technically no pregnancy; we will not consider this as a pregnancy complication here.

The American College of Obstetricians and Gynecologists defines *early pregnancy loss* as "a nonviable, intrauterine pregnancy with either an empty gestational sac or a gestational sac containing an embryo or fetus without fetal heart activity within the first 12 6/7 weeks of gestation" [24]. Although the lines are a bit blurred, *miscarriage* in general (i.e., the spontaneous abortion of a fetus) is defined as the loss of a pregnancy during the first 20–24 weeks of gestation [25]. Many miscarriages occur before a woman is even aware that she is pregnant, with up to 30% of spontaneous abortions occurring within the first days after implantation [26,27] and up to 15% of clinically recognized pregnancies miscarried. Thus miscarriage may occur in up to 50% of pregnancies, making it one of the most commonly recognized pregnancy complications [25]. Half of these spontaneous miscarriages can be attributed to genetic abnormalities of the fetus that are incompatible with life [28]. For the remainder of spontaneous miscarriages, the causes are not always clear. Exposure to environmental factors such as air pollutants (carbon monoxide, particulate matter, and cooking smoke) has been associated to increased pregnancy loss, although the exact mechanisms remain unexplained [29]. Other factors may include inherited uterine abnormalities, maternal thrombophilias, maternal autoantibody production, type 1 diabetes, and alcohol intake [30]. In addition, altered inflammatory cytokine signaling has been detected in decidua from miscarrying women [31], and microbial factors may be involved, in particular during early pregnancy [32].

Preterm delivery, the birth of a living child before gestational week 37, is the principle cause of neonatal death [33,34] with infants born at <week 32 (1%–2% of all births) suffering the most serious illness and mortality [35]. Preterm delivery can be divided into spontaneous preterm labor, preterm labor induced by rupture of membranes (PROM), and assisted deliveries. Preterm birth occurs in approximately 5%–22% of pregnancies, although rates and in particular neonatal death rates vary considerably between high-income and middle-to-low income countries [36]. The etiology of preterm birth is unclear, although known factors that can trigger premature labor include maternal conditions such as uterine trauma, uterine rupture, and (pre) eclampsia (see "Maternal risks" section), fetal conditions such as abnormal fetal heart rate or intrauterine growth restriction, and placental conditions such as abruption (the inappropriate separation of the placenta from the uterus wall) [37]. Placental/fetal infection represents one of the most common causes of preterm birth and will be further discussed later.

Globally, *stillbirth*, defined as a baby born without signs of life after week 28 of gestation, occurs in around 1.8% of pregnancies [38]. Of all stillbirths, fetal death occurs around the time of labor in over 50% [39]. The causes of stillbirth are similar to those of preterm birth, and as such, stillbirth is associated with preterm labor or low birthweight, with infections causing stillbirth in 10%–25% of high-income countries and up to 50% in low-income countries [40].

Maternal risks

In addition to the complications, which threaten fetal life, pregnancy is also associated with maternal health complications. While human reproduction is remarkably inefficient, with a large proportion resulting in early pregnancy failure, maternal deaths are fortunately less common. A recent analysis of maternal deaths in south Asia and sub-Saharan Africa reported an overall maternal mortality rate of 0.34% in life births [41]. Obstetric hemorrhage, nonobstetric complications, hypertensive disorders, and pregnancy-related infections accounted for over 80% of all these deaths. While most pregnancy complications are nonlife threatening, they can nevertheless be debilitating and pose a global health and economic burden that cannot be ignored [1].

As discussed earlier, a rise in hCG levels during early pregnancy is associated with nausea and vomiting in pregnant women. In ~1% of cases, this results in an extreme form, called *Hyperemesis Gravidarum (HG)*, which is a common cause of antenatal hospitalization with an estimated cost of 5.2 million USD per 171,000 deliveries per year (based on Israel data) [42]. The etiology of HG remains obscure, but one of the risk factors associated with this condition is gastric colonization by *Helicobacter pylori* [43,44]. *H. pylori* is widely known for its role in the development of gastric ulcers and atrophic gastritis [45], and infection can cause dyspeptic complaints. It has been suggested that pregnancy leads to an increased susceptibility to infection with this helical bacterium, although others have suggested that the tolerogenic immunological state induced by pregnancy may reactivate a latent *H. pylori* infection [46].

Another common (maternal) cause for hospitalization during pregnancy is hypertension, which can complicate up to 10% of pregnancies. Several *hypertensive disorders* can be recognized, including *gestational hypertension* (GH), *preeclampsia* (PE), and *eclampsia*, in order of increasing severity. Maternal death from hypertensive disorders is estimated to be around 0.05% [47]. GH is defined as a systolic blood pressure of ≥140 mmHg and/or a diastolic blood pressure of ≥90 mmHg that develops after the 20th week of pregnancy in women without prior high blood pressure. PE is superimposed on GH, and it is associated with the significant elevation of urine protein levels [48]. Eclampsia is the occurrence of seizures in patients with hypertension (i.e., PE), and poses a serious risk for mother and child, with almost 2% of women suffering from eclampsia not surviving pregnancy [49]. In addition to the direct risk during pregnancy, hypertensive disorders during pregnancy can increase the risk of developing cardiovascular disease later in life [50]. The exact etiology of hypertensive disorders, and in particular PE, is not known, although it has been shown that abnormal placentation may underlie the pathology. An impaired placental cytotrophoblast differentiation may contribute to a defective remodeling of placental spiral arteries, resulting in increased uterine arterial resistance and chronic placental ischemia. This in turn results in oxidative stress, which is passed to the maternal system through free radicals, oxidized lipids, cytokines, and vascular endothelial growth factor [51]. The thereby triggered inflammatory responses and endothelial changes in the maternal system contribute to presentation of disease through impairment of the hepatic and cerebral endothelium [52].

Another common complication of pregnancy is *anemia*. Pregnancy increases plasma volumes, resulting in a relative reduction in erythrocyte concentration and thereby hemoglobin levels. Furthermore, a growing iron demand of the fetus results in maternal iron depletion, and it can lead to *iron deficiency anemia*, the most common cause of anemia in pregnancy [53]. Similarly, fetal growth and erythrocyte demand in mothers require DNA and amino acid synthesis, which depend on folates. A lack of folic acid uptake thus results in megaloblastic anemia (the name derives from the fact that erythrocyte precursors are increased in size in this condition). Iron deficiency anemia and folate deficiency anemia are particularly common in countries with poor prenatal care (i.e., poor diet and no prenatal iron and folate supplements) and can cause maternal mortality in up to 7% of cases [54]. Low maternal Hb (<110 g/L) is also associated with an increased risk of adverse child outcomes such as low birthweight, stillbirth, and even neonatal mortality [55]. Interestingly, *H. pylori* infection has been associated with reduced iron stores [56], and pregnant women who are infected with *H. pylori* are at an increased risk of developing iron deficiency anemia [46]. Response to iron supplementation is better in iron deficiency anemia patients who are not infected than in those who are infected with *H. pylori* [57]. This might suggest that *H. pylori* actively competes with its host for iron stores, in particular as this bacterium constitutively expresses iron-uptake regulators [58]. However, other mechanisms may include reduced iron uptake by infected individuals as a result of *H. pylori*-induced morphological changes of the gastric mucosa as *H. pylori* modulates gastric pH, and gastric acidity plays an important role in the reduction of iron to absorbable forms [56].

A last pregnancy complication that deserves attention is *gestational diabetes mellitus* (GDM). The global prevalence of GDM is around 10% [59], and risk factors include obesity, westernized diet, and maternal age. While hyperglycemia developed during pregnancy generally resolves postpartum, it can have severe long-term consequences in terms of drastically increased risk of type 2 diabetes (T2D) and cardiovascular disease in the mother, as well as obesity and T2D in the child [60]. Normal pregnancy is accompanied by dynamic modulation of insulin sensitivity, to accommodate the different energy demands over time. An early increase in insulin sensitivity (allowing establishment of fat stores for later energy demands) is followed by a reduced insulin sensitivity and an increased endogenous glucose production in the third trimester [61]. Glucose homeostasis under these latter conditions is maintained by an increased insulin production by pancreatic β-cells. In patients with GDM, a decreased insulin sensitivity prior to conception appears to be already present, with insulin-stimulated uptake of glucose during pregnancy even further reduced as compared to control pregnancies. Furthermore, a reduced β-cell number and dysfunction preclude the compensation for pregnancy-induced insulin sensitivity seen in normal pregnancy [62].

Infections

As described earlier, many pathophysiological processes contribute to the diverse complications observed during pregnancy. However, endogenous factors are not the only contributing elements to pregnancy complications—external circumstances also contribute to birth outcomes. In particular, environmental factors may determine pregnancy outcomes. Infections with pathogens are a major cause of pregnancy complication and include bacterial, viral, protozoal, and fungal agents. Here, we will focus on the bacterial agents in particular.

Fetal and placental infection can occur via two main routes (Fig. 2.2). The first is *hematogenous bacterial dissemination*: the spread of bloodborne microbes from the maternal circulation to the placenta. These bloodborne pathogens (or their products) reach the placenta via the maternal blood circulation, where they may infect the placenta. The second mechanism of infection is through organisms that ascend from the vagina or cervix to the uterus and cause infection of the fetal membranes [63]. This second route of infection is known as *ascending bacterial infection*. Bacterial invasion can cause chorioamnionitis, an inflammation of the fetal membranes, and contamination of the amniotic fluid, which can cause fetal death when inhaled by the fetus [64]. Furthermore, detrimental effects of placental colonization with pathogens are thought to arise as a result of inflammatory responses at the placental interface, resulting in local release of pro-inflammatory cytokines, damage to the fetal membrane, and a disruption in placental circulation. Bacterial receptors expressed at the maternal–fetal interface may mediate these kinds of threats [65]. Toll-like receptors (TLR) are expressed in the decidua, and activation of these receptors through bacterial products can cause apoptosis of decidual stromal cells and an activation of immunity through the release of cytokines

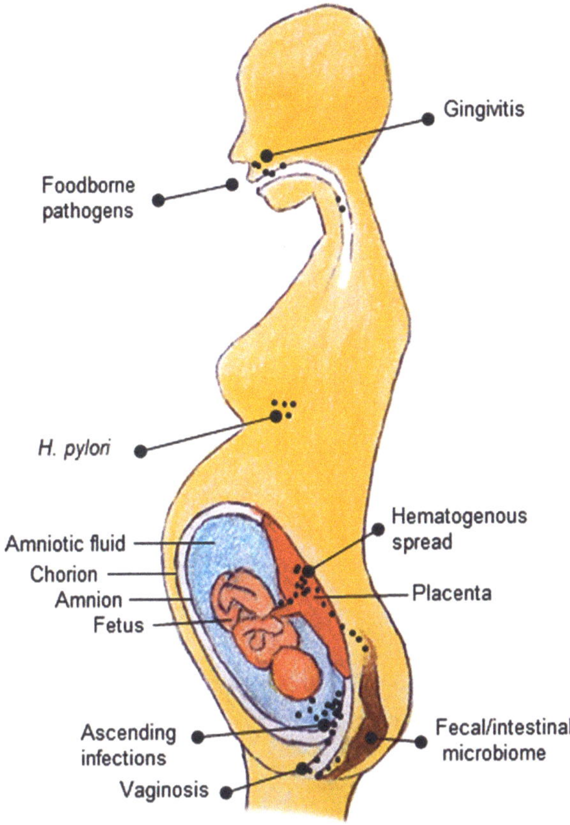

FIGURE 2.2 **Routes of bacterial infections.** Foodborne pathogens as well as oral or gastrointestinal microbes may reach the placenta through hematogenous dissemination. Risk for this route is particularly enhanced when inflammatory processes weakening the epithelial barrier occur, as may be the case in gingivitis or inflammatory bowel diseases. In addition, some bacteria, in particular spirochetes, are known to be able to invade and proliferate in epithelial cells, thus not requiring a breached barrier. Vaginal and rectal microbes may also pose a threat for pregnancy complications through ascending infections, whereby the placental membranes are reached, infected, and chorioamnionitis may ensue.

[21,66]. TLR4 ligation by the bacteria-derived lipopolysaccharide causes production of a range of pro-inflammatory cytokines and prostaglandin E2 in human amniotic epithelial cells [67]. Prostaglandin mediates induction of labor, while an increase in cytokine-attracted neutrophils and production of enzymes weakening cellular contacts, such as metalloproteases, can subsequently lead to cervical ripening and chorioamnion weakening and rupture, preceding preterm delivery [35]. An increased expression of TLR4 was observed in the syncytiotrophoblasts and vascular endothelium of

placentae from patients with early onset preeclampsia, suggesting that bacteria-induced inflammatory responses may contribute to this complication of pregnancy [68]. Indeed, increased levels of IL6, IL8, and TNFα were measured in amniotic fluid from patients with chorioamnionitis [69].

Foodborne pathogens

In recent years, increasing knowledge regarding dietary components and contaminants has resulted in a list of products, the consumption of which is discouraged during early pregnancy. For instance, some fish species may accumulate high amounts of mercury, which in a methylated form (MeHg) can be toxic to maternal nerve and immune cells and may cause developmental problems in children. The intestinal microbiome is thought to modulate the methylation and demethylation of mercury. As the gut microbiome changes during pregnancy (see Chapter 1: The Microbiome in a Healthy Pregnancy), it might be expected that mercury processing alters during pregnancy. However, little evidence was found for relationship between fecal MeHg and bacterial composition [70].

Fish is also a potential source of bacterial contaminants such as *Listeria monocytogenes* and *Salmonella*. Pregnant women are over 17 times more likely to be infected by *Listeria* as compared to healthy nonpregnant individuals [71]. This is most likely due to the fact that *Listeria* is able to invade and spread via epithelial and immune cells and its eradication is dependent on cellular immunity. Thus individuals with a compromised immune system, as is the case for patients with immunodeficiencies, the elderly, and pregnant women during their fetal-tolerant stage are at an increased risk of failing to clear this bacterium [72]. While symptoms of *Listeria* infection are generally mild in pregnant women, the bacterium is able to cross the placental barrier due to its cell-to-cell method of spreading and can cause severe harm to the fetus. *Listeria* infections can cause premature delivery, miscarriage, and stillbirth in up to 40% of infected cases [73]. In addition to *Listeria*, *Salmonella*, *Brucella*, and *Campylobacter* have all been shown to cause adverse pregnancy outcomes. Not all of these have the same route of dissemination within the body, and not all appear equally destructive. Infection with *Salmonella* is quite common, causes gastroenteritis or enteric fever, but is generally self-limiting. While pregnant women may experience more severe infections due to their Th2 immune status, their rate of infection is not different from the general population. When vertical transmission of *Salmonella* to the placenta/fetus takes place, fetal loss may occur. However, despite its fearsome reputation (being the reason for pregnant women being advised against eating uncooked eggs for instance), vertical transmission of *Salmonella enteritis* appears to be rare, with only a few cases reported [74]. *Campylobacter* spp. is another self-limiting agent causing gastroenteritis, which is known to be able to cross the placenta and cause fetal damage. Again, however, vertical transmission was only reported in a few cases [75]. In contrast, pregnancy complications of Brucellosis in pregnant women (spontaneous abortion, preterm birth or fetal death) have been reported to be as high as 10%–45% [76,77].

Hematogenous infection: bacteria at the placental interface

Considerable controversy surrounds the question as to whether or not the healthy placenta contains an active bacterial component. On morphological examination of the desidual basal plate, what appeared to be intracellular bacteria were seen in 27% cases [78], suggesting that at least the maternal side of the placenta may contain bacteria. While hematogenous dissemination of bloodborne pathogens to the placenta could account for this, with the existence of a physiological barrier between mother and child in the form of a continuous layer of villous synctiotrophoblasts, the fetal side of the interface (i.e., the fetal placenta, amnion, chorion, and chorionic fluid) should be sterile under normal conditions. However, even in normal pregnancy, breaks in the trophoblastic surface may exist [79]. Subsequently, sequencing efforts have hinted at the presence of a unique microbiome in healthy placentae [80]. The residing microbiome has also been suggested to affect pregnancy outcome, as in normal term pregnancies; microbial signatures were found to correlate with birthweight: low birthweight was associated with a higher presence of Fusobacteria and a reduced abundance of Cyanobacteria and *Lactobaccillus* [81]. However, the fetal or maternal origin of the placental tissue is not always clear in each publication, making it difficult to assess whether healthy fetal placental tissue is truly colonized by bacteria, and other studies were unable to detect a consistent bacterial signature in placental samples from either term or preterm deliveries [82,83].

It is perhaps more likely that the presence of bacteria in the placenta occurs only under pathological conditions. Bacteria that propagate and distribute via cell-to-cell routes (such as *Listeria*) can potentially cross the placental barrier. Furthermore, inflammation induced by bacteria at the maternal placenta may result in a weakened barrier and allow bacterial translocation to the fetus. Bacterial presence, including *Bacillus cereus, Listeria, Salmonella, Escherichia, Klebsiella pneumoniae, Anoxybacillus, Variovorax, Prevotella, Porphyromonas,* and *Dialister,* was seen in placental tissue from women with preeclampsia but not controls [84]. Spirochetes such as *Borrelia* are known to be able to cross the placenta and infect the fetus [85]. Another pathogen that is able to cross the placenta and infect the fetus is the spirochete *Treponema pallidum*, the causative agent of syphilis. In expecting mothers with active syphilis, this bacterium can cause fetal death in ~40% of cases, and another 30%–40% of babies will be born with congenital syphilis [40].

Chorioamnionitis: ascending infections

While sterility of the fetal site of the placenta has been questioned, theoretically no such controversy should exist for chorionic fluid, as this is exclusively derived from the fetus and theoretically fully separated from the maternal circulation. However, while the field leaned toward the hypothesis of a sterile womb for a long time, with the advent of new sequencing techniques, the detection of microbiomes in previously thought sterile environments has taken flight [86,87]. Studies investigating the microbial presence in amniotic fluid in healthy pregnancies demonstrated a prevalence of Proteobacteria with a high abundance of *Enterobacter* and *Escherichia/Shigella* [88]. However, others were unable to repeat these results, and concerns regarding contamination issues have been raised [87,89]. Nevertheless, mouse studies have suggested that DNA of orally ingested, genetically labeled bacterial strains can be detected in amniotic fluid [90]. Furthermore, probiotic supplementation of expecting human mothers was associated with changes in placental TLR mRNA expression, whereas the presence of *Lactobaccillus* DNA in human chorionic fluid was associated with significant changes in TLR expression in the intestine of the infant [91]. Thus whether or not living bacteria reach the placenta, their effects appear to be noted there.

As described earlier for a placental microbiome, bacterial presence in placental membranes and amniotic fluid has most frequently been described during complications of pregnancy. Intrauterine infections account for ~30% of preterm births [92], and bacterial presence in placental membranes is more often seen in preterm deliveries as compared to term deliveries [93], with a distinct microbiome marked by an increased presence of the genera *Fusobacterium, Streptococcus, Mycoplasma, Aerococcus, Garderella,* and *Ureaplasma* and the family Enterobacteriaceae [94]. Most of the bacteria observed during chorioamnionitis are derived from ascending bacterial infections, with *Ureaplasma urealyticum* and *Mycoplasma hominis* presenting as the most common causes of infection [40]. These bacteria have furthermore been associated with preterm birth, spontaneous abortion, and low birthweight (for *Ureaplasma* spp.) [95]. Other ascending bacteria observed in intrauterine infections are, among others, *Escherichia coli*, Group B *Streptococcus*, *Klebsiella*, *Enterococcus*, *Neisseria gonorrhoeae*, *Chlamidia trachomatis*, and *Bacteroides* spp. [35,96,97].

Biosynthesis of prostaglandin E2, which triggers uterine contraction and induces labor, is mediated by the release of fatty acids from the plasma membrane by phospholipases (PLAs). In particular, type II PLA2 is involved in the production of prostaglandins in the uterus. In patients with preterm labor, levels of PLA2 are increased in the amnion and choriodecidua, but not the placenta, when compared to term labor tissues [98]. Interestingly, many of the bacteria involved in preterm delivery (*Bacteroides fragilis*, *Peptostreptococcus*, *Fusobacterium necrophorum*, *Streptococcus viridans*, *Streptococcus fecalis*, *Streptococcus* A and B, *E. coli*, *Klebsiella*, *Staphylococcus epidermidis*, *Pneumococcus*, *Lactobacillus*, and *M. hominis*) have PLA2 activities that exceeds those of the amnion or chorion [99], presenting a possible mechanism of preterm delivery induction by these microbial agents. Furthermore, as has been demonstrated for Group B *Streptococcus*, not only does bacterial invasion recruit innate immune cells, which produce proteases weakening chorioamniotic membrane stability and contributing to PROM [100], these bacteria themselves also produce these proteases [101].

Bacterial vaginosis

As described earlier, several vaginal/cervical bacteria have been associated with pregnancy infections and associated outcomes. But to what extent do these bacteria represent the normal vaginal microbiome? The healthy vaginal mucosa is predominantly colonized by *Lactobacillus* spp., which produce lactic acid and thereby create a low pH that is nonpermissive for pathogenic growth [102]. Five distinct microbial profiles can be recognized that are each dominated by a different lactobacillus species: group I (26.6% of women) is associated with *Lactobaccillus crispatus*, group II (6.3%) with *Lactobaccillus gasseri*, group II (34.1%) with *Lactobaccillus iners*, and group V (5.3%) with *Lactobaccillus jensenii*. Group IV (27%) is rather heterogeneous and shows a higher bacterial diversity. This group is not dominated by *Lactobaccillus*, but shows higher levels of strictly anaerobic bacteria, including *Prevotella*, *Dialister*, *Atopobium*, *Gardnerella*, *Megasphaera*, *Peptoniphilus*, *Sneathia*, *Eggerthella*, *Aerococcus*, *Finegoldia*, and *Mobiluncus* [103]. Most of the aforementioned infectious agents, including *Bacteroides* spp., Group B *Streptococcus*, *Staphylococcus*, *E. coli*, and *Klebsiella* have also been found in the vaginal microbiome [104], and thus appear to be a part of the normal vaginal microbiome at low abundances.

Bacterial vaginosis (BV), clinically characterized by milky discharge, vaginal fluid pH > 4.5, odor changes upon addition of 10% potassium hydroxide to vaginal secretions, and bacteria-covered epithelial cells seen by microscopy of vaginal smears, is a relatively common phenomenon, occurring in 20%–50% of women [105,106]. BV is associated with an increased risk of adverse pregnancy outcomes, including preterm birth and miscarriage [107]. However, up to 85% of women with BV are asymptomatic, and thus this risk factor is not commonly recognized [108].

BV is accompanied by a microbial dysbiosis associated with an increased bacterial diversity and a reduced presence of *Lactobacillus* spp., in particular *L. crispatus*. While some bacteria are found only in patients with BV, a shift in colony structure rather than a single bacterial feature appears to underlie this disease [109]. Species consistently found in patients with BV are *Prevotella, Eggerthella, Dialister, Megasphaera, Sneathia, Leptotrichia, Peptostreptococcus, Veillonella, Bacteroides, Mobiluncus, Porphyromonas, Mycoplasma, Ureaplasma, Streptococcus, Staphylococcus, Gemella,* and *Escherichia/Shigella* [110], and thus it appears that BV is more associated with a group IV type microbiome. Racial differences in BV rates exist, with non-Hispanic black women more often diagnosed with BV than Caucasians [108]. This may in part be related to the fact that black women more often have communities that are not dominated by *Lactobaccillus* spp., which may make these communities more susceptible to perturbations [111,112].

Women who suffer from vaginal infection/vaginitis are almost three times more likely to be colonized by Group B *Streptococcus* [113]. Similarly, *U. urealyticum* and *M. hominis*, known to be associated with pregnancy complications, were detected in vaginal samples of only 5.9% and 8.9% of healthy women, but increased to 13.5% and 26.8%, respectively, in patients with vaginal dysbiosis [114]. In a large Belgian cohort, up to 40% of patients delivering prematurely were infected with *Ureaplasma* spp. [115]. An abnormal vaginal microbiome (independent of *Ureaplasma* spp.) was seen in 22.7% of patients, but this did not correlate to preterm delivery. Thus it appears that vaginosis is associated with increased risk of colonization with bacteria, which are associated with adverse pregnancy outcomes, but that microbial dysbiosis in itself is not enough to endanger a healthy pregnancy. Complicating matters is the fact that pregnancy alters the vaginal microbiome. Studies have indicated that pregnancy hormones may modulate the vaginal microbiome, and vaginal progesterone taken during pregnancy for the prevention of preterm birth was associated with reduced Group B *Streptococcus* rates [116]. Furthermore, pregnancy induces a lower prevalence of *Ureaplasma* and *Mycoplasma* [117]. In general, a lower richness and diversity, a higher abundance of *Lactobacillus* spp., a higher stability, and reduced BV are seen in pregnant women [118,119], making it tempting to speculate that microbial changes induced by pregnancy hormones are set to minimalize the risk of adverse pregnancy outcomes. Vaginal microbiome changes occurring during pregnancy are not similar between black and Caucasian women, which may perhaps partly account for differences in preterm birthrate, although other factors are likely to contribute [120,121].

Interestingly, the vaginal microbiome also appears to be associated with pregnancy complications that are not directly linked to placental/fetal infection. Patients with GDM showed significant alterations in their vaginal microbiome as compared to patients without GDM, with an increased presence of the genera *Bacteroides, Veillonella, Klebsiella, Escherichia–Shigella, Enterococcus* and *Enterobacter*, and a decreased presence of the genera *Varibaculum, Prevotella, Porphyromonas,* and *Ezakiella* [122]. However, no relationship between BV and PE was observed [123].

Gastrointestinal microbiome

As described in the previous chapter, most of the bacteria that colonize our body reside in the gut. While on the whole, we live in a mutually beneficial symbiosis with our bacteria, it is becoming clear that alterations in the gut microbial composition can significantly contribute to a range of health issues, A role of the gastrointestinal (GI) microbiome in pregnancy complications is also emerging. Mechanistically, GI tract bacteria can enter the blood stream when they cross the epithelial barrier in damaged areas and can thus travel to the placenta through hematogenous dissemination as described earlier. There, they have to be able to attach to, invade, and cross the endothelial and epithelial barriers, something that, unlike vaginal *Fusobacterium gornidiaformans* species, intestinal *Fusobacterium nucleatum* for instance, is able to do [124,125]. In addition, bacterial products can affect the physiological processes involved in pregnancy. In addition to short-chain fatty acids (SCFAs), which serve as an energy source to the host and provide a nutrient source for intestinal epithelial cells, bacteria produce metabolic compounds that feed into glycolysis, Krebs cycle, oxidative phosphorylation, and amino acid and fatty acid metabolism, thus manipulating many human metabolic systems [126]. Thus a large portion of the metabolites in the human blood may be derived from our interaction with the gut microbiome [127], and modulations of the microbiome may thereby substantially alter our metabolomic profile. This is illustrated for instance by the fact that obesity-associated microbiomes exist, which are characterized by an enhanced energy uptake from dietary components and a low bacterial diversity [128], and persist even after dieting [129]. Furthermore, the human gut microbiome is associated with metabolic syndromes such as T2D. While exact compositional differences between T2D and healthy controls differ per study, the general consensus suggests a reduced presence of butyrate-producing *Roseburia* and *F. prausnitzii* and an increased presence of nonbutyrate-producing *Clostridiales* [130]. As pregnancy is associated with metabolic changes and weight gain, perhaps it is not surprising that the GI microbiome may also play a role in, for instance, pregnancy-associated GDM and potentially other pregnancy complications.

The oral microbiome in pregnancy complications

When one thinks about the role of the GI tract microbiome in pregnancy outcomes, the oral microbiome is perhaps not the niche that first comes to mind. Nevertheless, as also described in the previous chapter, there is some evidence that the oral microbiome is of importance in pregnancy. While the oral microbiome appears to be relatively stable during pregnancy, there is an increase in the presence of pathogenic bacteria (e.g., *Prevotella* species, *Porphyromonas, Dialister invisus, Kingella oralis, Veillonella parvula, Selenomonas sputigena, Streptococcus anginosus, Rothia dentocariosa*) during pregnancy, which returns to normal postpartum [131]. This adverse microbial signature may contribute to the enhanced frequency of gingivitis seen during pregnancy, which can occur in 30%–100% of pregnant women, depending on the cohort studied [132]. A reduced epithelial barrier function during inflammation may allow oral pathogens to translocate to the blood. This would be in line with studies suggesting that the microbiome seen in the placentas resembles that from oral rather than vaginal or urethral origin [80]. Indeed, gingivitis, and oral health in general, is associated with adverse pregnancy outcomes, predominantly those associated with placental infections, such as PROM and preterm birth, and low birthweight [133,134]. Translocation of periodontal bacteria to the placenta may also play a role in hypertensive disorders such as eclampsia [135,136]. However, the oral microbiome seems to play a smaller role in host metabolic regulation, and no association between the oral microbiome and GDM, for instance, was found in one study [122]. Nevertheless, it is of interest to note that the vaginal microbiome of patients with GDM shows more homology with the microbiome at other bodysites (i.e., oral and intestinal microbiome) as compared to women without GDM, and in this study GDM was associated with an increased abundance of Proteobacteria and a reduced presence of Firmicutes in the oral microbiome [137].

The lower gastrointestinal tract microbiome and pregnancy complications

The most densely colonized niche of the human body is the colon, which can easily be investigated by analysis of the microbial composition in collected feces. Hence, this is one of the best studied human microbial compartments, and many associations between human health and disease states and gut microbiomes have now been identified. While the involvement of the lower GI microbiome in adverse pregnancy outcomes remains relatively underexplored, several interesting observations have been made to date.

Fecal microbiome and preterm delivery

As mentioned earlier, pregnancy hormones modulate the microbiome [138]. In particular, *Bifidobacterium* and *Streptococcus* appear to be upregulated toward late pregnancy [139,140]. Interestingly, the 4-day postpartum fecal microbiome from mothers delivering prematurely is characterized by a reduced presence of *Bifidobacterium* and *Streptococcus*, in addition to a depletion of Clostridiales and a loss of diversity [141]. Another study confirmed the reduced presence of Clostridia species in mothers from preterm infants [142]. Indeed, infants' microbiomes resemble that of their mothers [143,144], and infants born prematurely (<35 weeks) also show a gut microbiome that is characterized by a reduced diversity and lower levels of *Bifidobacterium* and *Bacteroides* compared to infants delivered at full term [145]. *Bifidobacterium* is generally considered to be a beneficial bacterium (and a component of many probiotic formulations) suggesting that modulation in maternal microbiome may be set to transfer a healthy microbiome to the infant [146,147]. Thus it is also conceivable that these data do not point toward a causative effect of altered maternal microbiome in preterm delivery, but rather are a reflection of the fact that the maternal microbiome at the time of preterm birth is not yet fully "matured."

Fecal microbiome and inflammation

Preterm birth has also been associated with maternal inflammatory diseases. Periodontal disease is one example, but immune diseases such as rheumatoid arthritis, systemic lupus erythematosus, and inflammatory bowel disease (IBD) are a few others [148–150]. In particular, IBD is of interest in this regard. IBD, comprised of Crohn's disease and ulcerative colitis, is characterized by a chronic intermittent inflammation at the epithelial lining of the GI tract. Genetic risk factors affecting immune function contribute to these diseases, but it appears that one of the triggers for developing this disease is a dysbiosis of the intestinal microbiome [151–153]. IBD microbiomes are generally characterized by a depletion of the phyla Firmicutes and Bacteroidetes and a reduced α-diversity. Other findings include an increase in pathogenic bacteria (Enterobacteriaceae, adherent invasive *E. coli*, Fusobacteria) and a reduction in SCFA producers such as *Faecalibacterium* [154–156]. Mothers with IBD have an increased chance of preterm delivery, in particular when flaring during pregnancy [157,158], suggesting that bacteria-induced inflammation at the intestinal lining may permit bacterial translocation or triggers placental responses to circulating inflammatory mediators and bacterial products. It has been speculated that physiological changes occurring during pregnancy (i.e., immune tolerance, microbial changes) may ameliorate disease activity in IBD [159,160]. Interestingly, upon conception, levels of circulating pro-inflammatory cytokines decrease in IBD patients, even when in remission, indeed suggesting that pregnancy induces immune-dampening responses [139]. To date, only two studies investigated the microbial changes that occur in IBD patients during pregnancy. Both showed that microbial α-diversity in IBD patients is lower as compared to healthy controls, but only during early pregnancy, as in later pregnancy microbial diversity in healthy pregnancies also decreases [139,161]. Dynamic modeling of the microbial changes taking place during pregnancy in IBD suggests that the microbiome is in particular driven to modulate *Sutterella*. Interestingly, this might point to a modulation of the inflammation–microbiome axis, as C-reactive protein levels, a marker of inflammation, have been shown to be associated with *Sutterella* [162], although exactly what this means in the context of IBD is as yet unclear. Nevertheless, despite dampening effects, dysbiotic features in IBD persist during pregnancy and may contribute to increased risk of adverse pregnancy outcomes.

Fecal microbiome and gestational hypertension

A clear role for the microbiome in hypertension is emerging, with several studies showing that patients with elevated blood pressure have different microbiomes from those with normal blood pressures. However, the bacterial changes differ per study, with an overgrowth of *Prevotella* and *Klebsiella* seen in one study, and a decrease in *Klebsiella* seen in another [163,164]. Several animal experiments indicate that the clinical phenotype of hypertensive disorders can be transferred from one animal to another by fecal transplantation, depends on the microbial signature, and depends on dysbiosis-induced inflammatory processes [165]. Relatively little is known regarding the intestinal microbiome involvement in gestational hypertensive disorders. One study in Chinese women showed that while there were no overall differences in α or β diversity in stools from women suffering from PE and those with a healthy pregnancy, PE was associated with an increased presence of Cyanobacteria. On species level, the pathogenic bacteria *Bulleidia moorei* and *Clostridium perfringens* were increased, whereas the SCFA producer *Coprococcus* was decreased [166]. Hypertension risk is associated with maternal weight.

One study investigated hypertension in obese women during early pregnancy ($T = 16$WK), and it demonstrated that while obesity itself is associated with alterations in the intestinal microbiome, there was an independent negative correlation between the presence of SCFA-producing Odoribacteraceae and Clostridiaceae families and systolic and diastolic blood pressure, and a negative correlation between the Christensenellaceae family and *Blautia* genus and systolic blood pressure [167]. Higher BMI and high blood pressure were also associated with a reduced capacity to produce butyrate as determined by the presence of the butyrate-producing genes *Buk* (butyrate kinase) and *But* (butyryl-CoA:acetate-CoA transferase). Furthermore, another study showed a negative association between diastolic blood pressure and the SCFA-producing bacteria Rikenellaceae and *Oscillospira* [162]. Thus a potential link between the absence of SCFA-producing microbial composition and the risk of hypertensive disorders might exist.

Fecal microbiome and gestational diabetes mellitus

It is now well established that both obesity and T2D are associated with intestinal microbial alterations. Thus it is perhaps not surprising that in terms of the involvement of intestinal microbiome, the best studied pregnancy complication is GDM. Nevertheless, even here, relatively few studies have been published (also reviewed in [168]), and conflicting results have been obtained. One study showed that patients with GDM have a reduced fecal microbial diversity in late pregnancy, which was similar to that in healthy patients [139]. A reduced presence of the genus *Bacteroides* was seen in fecal samples from GDM patients by Cortez et al., who also reported a nonsignificant increase in *Ruminococcus, Eubacterium,* and *Prevotella* in GDM patients compared to controls [122]. Interestingly, another study confirmed that increases in the family Ruminococcaceae were already present in fecal samples of patients before onset of clinical disease suggesting that this feature could potentially serve as a biomarker, with an area under the receiver operating characteristic (ROC) curve of 0.73 [169]. However, Kuang et al. were able to predict GDM development with an area under the curve (AUC) of 0.91 and a model that was predominated by *Parabacteroides,* one of the most discriminatory factors between GDM and non-GDM developing mothers (others included a higher abundance of *Megamonas* and *Phascolarctobacterium,* and a decreased presence of *Ruminiclostridium, Roseburia, Eggerthella, Fusobacterium, Haemophilus, Mitsuokella,* and *Aggregatibacter*) [170]. Another study demonstrated that while there were no differences in α-diversity or relative abundance at phylum level, fecal samples from women developing GDM were characterized by reduced levels of the antiinflammatory *Faecalibacterium* [137]. While not directly comparing GDM to healthy pregnancies, patients with GDM who were put on a diet showed clear reductions in *Bacteroides* [162]. Furthermore, during pregnancy, *Faecalibacterium* levels increased in both patients who adhered to the diet and those who did not, but levels of this bacterium correlated inversely with fasting glucose levels, confirming the notion that antiinflammatory microbiota may improve insulin sensitivity [171]. All in all, different studies have now suggested that fecal microbial alterations exist in patients developing GDM, which may relate to changes seen in patients with T2D. However, more studies are clearly needed to better define the actual changes that occur during GDM.

Concluding remarks

We have seen that pregnancy is associated with substantial risk, for both mother and child. Pregnancy complications are common, and may in part be mediated by bacterial factors that either preexist pregnancy, or are modulated during pregnancy through physiological changes induced by pregnancy, such as hormone fluctuations. Associations between environmental, oral, intestinal, and vaginal microbiomes (and alterations thereof) and diverse pregnancy outcomes have now been shown. For some of these bacterial agents, a direct role in pregnancy outcomes is clear. Point in case are the overt infections causing chorioamnionitis. For other microbial changes, it is less clear whether they are a cause of adverse pregnancy outcomes, a consequence, or even just a confounding factor. Clearly, more research is needed to answer these questions. The Holy Grail would be to predict adverse pregnancy outcomes, or even potentially prevent them through the use of probiotics or other microbiome modulators, although the evidence to date that maternal probiotic intake would reduce complications is scarce [172,173]. Everyone obviously wants to know how to predict poor pregnancy outcome. Several study protocols have now been published that will follow prospective mothers [174,175]. In a study in Japan, 302 pregnant women were enrolled. Continuous lifestyle monitoring, as well as collection of a range of biological samples, including blood, urine, saliva, and dental plaque, was performed at three timepoints during pregnancy [174]. A second cohort study aims to include 400 Asian pregnant females and will monitor stool, vaginal, oral, and placental microbial composition per trimester, as well as peripheral blood transcriptomes sampled fortnightly, to assess risk factors for preterm birth [175]. A third study will investigate the effect of probiotic intake on maternal and fetal health, including GDM, vaginosis, and Group B *Streptococcus* colonization in 400 women [176]. Thus these are exciting times, and we may expect a rapid growth of our understanding of the role of the microbiome in pregnancy and its complications in the upcoming years.

References

[1] Law A, McCoy M, Lynen R, Curkendall SM, Gatwood J, Juneau PL, et al. The prevalence of complications and healthcare costs during pregnancy. J Med Econ 2015;18 (7):533–41.

[2] Ventura SJ, Curtin SC, Abma JC, Henshaw SK. Estimated pregnancy rates and rates of pregnancy outcomes for the United States, 1990-2008. Natl Vital Stat Rep 2012;60 (7):1–21. Available from: <http://www.ncbi.nlm.nih.gov/pubmed/22970648>.

[3] Mor G, Aldo P, Alvero AB. The unique immunological and microbial aspects of pregnancy. Nat Rev Immunol 2017;17(8):469–82. Available from: <http://www.nature.com/doifinder/10.1038/nri.2017.64>.

[4] Korevaar TIM, Steegers EAP, de Rijke YB, Schalekamp-Timmermans S, Visser WE, Hofman A, et al. Reference ranges and determinants of total hCG levels during pregnancy: the Generation R Study. Eur J Epidemiol 2015;30(9):1057–66. Available from: <http://www.ncbi.nlm.nih.gov/pubmed/25963653>.

[5] Cole LA. Biological functions of hCG and hCG-related molecules. Reprod Biol Endocrinol 2010;8(1):102. Available from: <http://www.ncbi.nlm.nih.gov/pubmed/20735820>.

[6] Berkane N, Liere P, Oudinet J-P, Hertig A, Lefèvre G, Pluchino N, et al. From pregnancy to preeclampsia: a key role for estrogens. Endocr Rev 2017;38(2):123–44. Available from: <https://academic.oup.com/edrv/article/38/2/123/3061617>.

[7] Byrns MC. Regulation of progesterone signaling during pregnancy: implications for the use of progestins for the prevention of preterm birth. J Steroid Biochem Mol Biol 2014;139:173–81. Available from: <https://www.sciencedirect.com/science/article/pii/S0960076013000228?via%3Dihub>.

[8] Albrecht ED, Pepe GJ. Estrogen regulation of placental angiogenesis and fetal ovarian development during primate pregnancy. Int J Dev Biol 2010;54(2–3):397–408. Available from: <http://www.ncbi.nlm.nih.gov/pubmed/19876841>.

[9] Mor G, Cardenas I, Abrahams V, Guller S. Inflammation and pregnancy: the role of the immune system at the implantation site. Ann N Y Acad Sci 2011;1221(1):80–7. Available from: <http://www.ncbi.nlm.nih.gov/pubmed/21401634>.

[10] Nancy P, Erlebacher A. T cell behavior at the maternal-fetal interface. Int J Dev Biol [Internet] 2014 [cited 2018 Mar 5];58(2–4):189–98. Available from: <http://www.ncbi.nlm.nih.gov/pubmed/25023685>.

[11] Vacca P, Montaldo E, Croxatto D, Loiacono F, Canegallo F, Venturini PL, et al. Identification of diverse innate lymphoid cells in human decidua. Mucosal Immunol [Internet] 2015 [cited 2018 Mar 5];8(2):254–64. Available from: <http://www.nature.com/articles/mi201463>.

[12] Faas MM, de Vos P. Uterine NK cells and macrophages in pregnancy. Placenta [Internet] 2017;56:44–52. Available from: <http://www.ncbi.nlm.nih.gov/pubmed/28284455>.

[13] Svensson-Arvelund J, Ernerudh J. The role of macrophages in promoting and maintaining homeostasis at the fetal-maternal interface. Am J Reprod Immunol [Internet] 2015 [cited 2018 Mar 5];74(2):100–9. Available from: <http://www.ncbi.nlm.nih.gov/pubmed/25582625>.

[14] La Rocca C, Carbone F, Longobardi S, Matarese G. The immunology of pregnancy: regulatory T cells control maternal immune tolerance toward the fetus. Immunol Lett [Internet] 2014 [cited 2018 Mar 5];162(1):41–8. Available from: <http://www.ncbi.nlm.nih.gov/pubmed/24996040>.

[15] Wegmann TG, Lin H, Guilbert L, Mosmann TR. Bidirectional cytokine interactions in the maternal-fetal relationship: is successful pregnancy a TH2 phenomenon? Immunol Today [Internet] 1993 [cited 2018 Mar 5];14(7):353–6. Available from: <http://www.ncbi.nlm.nih.gov/pubmed/8363725>.

[16] Sykes L, MacIntyre DA, Yap XJ, Teoh TG, Bennett PR. The Th1:th2 dichotomy of pregnancy and preterm labour. Mediat Inflamm [Internet] 2012 [cited 2018 Mar 6];2012:967629. Available from: <http://www.ncbi.nlm.nih.gov/pubmed/22719180>.

[17] Koldehoff M, Katzorke T, Wisbrun NC, Propping D, Wohlers S, Bielfeld P, et al. Modulating impact of human chorionic gonadotropin hormone on the maturation and function of hematopoietic cells. J Leukoc Biol [Internet] 2011 [cited 2018 Mar 6];90(5):1017–26. Available from: <http://www.ncbi.nlm.nih.gov/pubmed/21878542>.

[18] Shah NM, Lai PF, Imami N, Johnson MR. Progesterone-related immune modulation of pregnancy and labor. Front Endocrinol (Lausanne) [Internet] 2019 [cited 2019 May 3];10:198. Available from: <http://www.ncbi.nlm.nih.gov/pubmed/30984115>.

[19] Hardy DB, Janowski BA, Corey DR, Mendelson CR. Progesterone receptor plays a major antiinflammatory role in human myometrial cells by antagonism of nuclear factor-κB activation of cyclooxygenase 2 expression. Mol Endocrinol [Internet] 2006 [cited 2019 May 6];20(11):2724–33. Available from: <https://academic.oup.com/mend/article-lookup/doi/10.1210/me.2006-0112>.

[20] van der Giessen J, van der Woude C, Peppelenbosch M, Fuhler G. A direct effect of sex hormones on epithelial barrier function in inflammatory bowel disease models. Cells [Internet] 2019 [cited 2019 May 6];8(3):261. Available from: <http://www.ncbi.nlm.nih.gov/pubmed/30893871>.

[21] Lou Y, Hu M, Wang Q, Yuan M, Wang N, Le F, et al. Estradiol suppresses TLR4-triggered apoptosis of decidual stromal cells and drives an anti-inflammatory T H 2 shift by activating SGK1. Int J Biol Sci [Internet] 2017 [cited 2019 May 6];13(4):434–48. Available from: <http://www.ncbi.nlm.nih.gov/pubmed/28529452>.

[22] Konopka CK, Morais EN, Naidon D, Pereira AM, Rubin MA, Oliveira JF, et al. Maternal serum progesterone, estradiol and estriol levels in successful dinoprostone-induced labor. Braz J Med Biol Res [Internet] 2013 [cited 2019 May 6];46(1):91–7. Available from: <http://www.scielo.br/scielo.php?script = sci_arttext&pid = S0100-879X2013000100091&lng = en&tlng = en>.

[23] Kelly RW. Inflammatory mediators and parturition. Rev Reprod [Internet] 1996 [cited 2019 May 6];1(2):89–96. Available from: <http://www.ncbi.nlm.nih.gov/pubmed/9414445>.

[24] Mor G, Cardenas I. The immune system in pregnancy: a unique complexity. Am J Reprod Immunol 2010;63(6):425–33.

[25] Rai R, Regan L. Recurrent miscarriage. Lancet [Internet] 2006 [cited 2019 May 9];368(9535):601–11. Available from: <https://www.sciencedirect.com/science/article/pii/S0140673606692040?via%3Dihub>.

[26] Wang X, Chen C, Wang L, Chen D, Guang W, French J. Conception, early pregnancy loss, and time to clinical pregnancy: a population-based prospective study. Fertil Steril [Internet]. 2003 [cited 2019 May 9];79:577–84. Available from: <https://www.fertstert.org/article/S0015-0282(02)04694-0/pdf>.

[27] Wilcox AJ, Weinberg CR, O'Connor JF, Baird DD, Schlatterer JP, Canfield RE, et al. Incidence of early loss of pregnancy. N Engl J Med [Internet] 1988 [cited 2019 May 9];319(4):189–94. Available from: <http://www.nejm.org/doi/abs/10.1056/NEJM198807283190401>.

[28] van den Berg MMJ, van Maarle MC, van Wely M, Goddijn M. Genetics of early miscarriage. Biochim Biophys Acta Mol Basis Dis [Internet] 2012 [cited 2019 May 9];1822(12):1951–9. Available from: <https://www.sciencedirect.com/science/article/pii/S0925443912001494#bb0005>.

[29] Grippo A, Zhang J, Chu L, Guo Y, Qiao L, Zhang J, et al. Air pollution exposure during pregnancy and spontaneous abortion and stillbirth. Rev Environ Health [Internet] 2018 [cited 2019 May 9];33(3):247–64. Available from: <http://www.degruyter.com/view/j/reveh.2018.33.issue-3/reveh-2017-0033/reveh-2017-0033.xml>.

[30] Pinar MH, Gibbins K, He M, Kostadinov S, Silver R. Early pregnancy losses: review of nomenclature, histopathology, and possible etiologies. Fetal Pediatr Pathol [Internet] 2018 [cited 2019 May 9];37(3):191–209. Available from: <https://www.tandfonline.com/doi/full/10.1080/15513815.2018.1455775>.

[31] Pitman H, Innes BA, Robson SC, Bulmer JN, Lash GE. Altered expression of interleukin-6, interleukin-8 and their receptors in decidua of women with sporadic miscarriage. Hum Reprod [Internet] 2013 [cited 2019 May 9];28(8):2075–86. Available from: <http://www.ncbi.nlm.nih.gov/pubmed/23739222>.

[32] van Oostrum N, De Sutter P, Meys J, Verstraelen H. Risks associated with bacterial vaginosis in infertility patients: a systematic review and meta-analysis. Hum Reprod [Internet] 2013 [cited 2019 May 9];28(7):1809–15. Available from: <http://www.ncbi.nlm.nih.gov/pubmed/23543384>.

[33] Lawn JE, Cousens S, Zupan J. Lancet neonatal survival steering team. 4 million neonatal deaths: when? where? why? Lancet [Internet] 2005 [cited 2019 May 10];365(9462):891–900. Available from: <http://www.ncbi.nlm.nih.gov/pubmed/15752534>.

[34] Quinn J-A, Munoz FM, Gonik B, Frau L, Cutland C, Mallett-Moore T, et al. Preterm birth: case definition & guidelines for data collection, analysis, and presentation of immunisation safety data. Vaccine [Internet] 2016 [cited 2019 May 10];34(49):6047–56. Available from: <http://www.ncbi.nlm.nih.gov/pubmed/27743648>.

[35] Goldenberg RL, Hauth JC, Andrews WW. Intrauterine infection and preterm delivery. Epstein FH, editor. N Engl J Med [Internet] 2000 [cited 2019 May 7];342(20):1500–07. Available from: <http://www.nejm.org/doi/10.1056/NEJM200005183422007>.

[36] Simmons LE, Rubens CE, Gravett MG. Preventing preterm birth and neonatal mortality: exploring the epidemiology, causes, and interventions. Semin Perinatol [Internet] 2010 [cited 2019 May 10];34(6):408–15. Available from: <https://www.sciencedirect.com/science/article/pii/S0146000510001059?via%3Dihub>.

[37] Villar J, Papageorghiou AT, Knight HE, Gravett MG, Iams J, Waller SA, et al. The preterm birth syndrome: a prototype phenotypic classification. Am J Obstet Gynecol [Internet] 2012 [cited 2019 May 10];206(2):119–23. Available from: <https://www.sciencedirect.com/science/article/pii/S0002937811021715?via%3Dihub>.

[38] Blencowe H, Cousens S, Jassir FB, Say L, Chou D, Mathers C, et al. National, regional, and worldwide estimates of stillbirth rates in 2015, with trends from 2000: a systematic analysis. Lancet Glob Health [Internet] 2016 [cited 2019 May 10];4(2):e98–e108. Available from: <https://linkinghub.elsevier.com/retrieve/pii/S2214109X15002752>.

[39] Kc A, Wrammert J, Ewald U, Clark RB, Gautam J, Baral G, et al. Incidence of intrapartum stillbirth and associated risk factors in tertiary care setting of Nepal: a case-control study. Reprod Health [Internet] 2016 [cited 2019 May 10];13(1):103. Available from: <http://reproductive-health-journal.biomedcentral.com/articles/10.1186/s12978-016-0226-9>.

[40] Goldenberg RL, McClure EM, Saleem S, Reddy UM. Infection-related stillbirths. Lancet [Internet] 2010 [cited 2019 May 15];375(9724):1482. Available from: <https://www.ncbi.nlm.nih.gov/pmc/articles/PMC3893931/>.

[41] Alliance for Maternal and Newborn Health Improvement (AMANHI) mortality study group TA for M and NHI (AMANHI) mortality study. Population-based rates, timing, and causes of maternal deaths, stillbirths, and neonatal deaths in south Asia and sub-Saharan Africa: a multi-country prospective cohort study. Lancet Glob Health [Internet] 2018 [cited 2019 May 9];6(12):e1297–308. Available from: <http://www.ncbi.nlm.nih.gov/pubmed/30361107>.

[42] Trovik J, Vikanes Å. Hyperemesis gravidarum is associated with substantial economic burden in addition to severe physical and psychological suffering. Isr J Health Policy Res [Internet] 2016 [cited 2019 May 13];5:43. Available from: <http://www.ncbi.nlm.nih.gov/pubmed/27766142>.

[43] Lutomski J, McCarthy FP, Greene RA. Hyperemesis gravidarum: current perspectives. Int J Womens Health [Internet] 2014 [cited 2019 May 13];6:719. Available from: <http://www.ncbi.nlm.nih.gov/pubmed/25125986>.
[44] Sandven I, Abdelnoor M, Nesheim B, Melby KK. *Helicobacter pylori* infection and hyperemesis gravidarum: a systematic review and meta-analysis of case-control studies. Acta Obstet Gynecol Scand [Internet] 2009 [cited 2019 May 13];88(11):1190–200. Available from: <http://doi.wiley.com/10.3109/00016340903284927>.
[45] Mommersteeg MC, Yu J, Peppelenbosch MP, Fuhler GM. Genetic host factors in *Helicobacter pylori*-induced carcinogenesis: emerging new paradigms. Biochim Biophys Acta Rev Cancer 2018;1869(1).
[46] Cardaropoli S, Rolfo A, Todros T. *Helicobacter pylori* and pregnancy-related disorders. World J Gastroenterol [Internet] 2014 [cited 2019 May 13];20(3):654–64. Available from: <http://www.ncbi.nlm.nih.gov/pubmed/24574739>.
[47] Abalos E, Cuesta C, Grosso AL, Chou D, Say L. Global and regional estimates of preeclampsia and eclampsia: a systematic review. Eur J Obstet Gynecol Reprod Biol [Internet] 2013 [cited 2019 May 7];170(1):1–7. Available from: <https://www.sciencedirect.com/science/article/pii/S0301211513001966?via%3Dihub>.
[48] Antza C, Cifkova R, Kotsis V. Hypertensive complications of pregnancy: a clinical overview. Metabolism [Internet] 2018 [cited 2019 May 14];86:102–11. Available from: <https://www.sciencedirect.com/science/article/pii/S0026049517303232?via%3Dihub>.
[49] Munro PT. Management of eclampsia in the accident and emergency department. J Accid Emerg Med [Internet] 2000 [cited 2019 May 14];17(1):7–11. Available from: <http://www.ncbi.nlm.nih.gov/pubmed/10658982>.
[50] Brown MC, Best KE, Pearce MS, Waugh J, Robson SC, Bell R. Cardiovascular disease risk in women with pre-eclampsia: systematic review and meta-analysis. Eur J Epidemiol [Internet] 2013 [cited 2019 May 14];28(1):1–19. Available from: <http://www.ncbi.nlm.nih.gov/pubmed/23397514>.
[51] Uzan J, Carbonnel M, Piconne O, Asmar R, Ayoubi J-M. Pre-eclampsia: pathophysiology, diagnosis, and management. Vasc Health Risk Manag [Internet] 2011 [cited 2019 May 14];7:467–74. Available from: <http://www.ncbi.nlm.nih.gov/pubmed/21822394>.
[52] Mayrink J, Costa ML, Cecatti JG. Preeclampsia in 2018: revisiting concepts, physiopathology, and prediction. Sci World J [Internet] 2018 [cited 2019 May 14];2018:6268276. Available from: <http://www.ncbi.nlm.nih.gov/pubmed/30622442>.
[53] Sifakis S, Pharmakides G. Anemia in pregnancy. Ann N Y Acad Sci [Internet] 2006 [cited 2019 May 14];900(1):125–36. Available from: <http://doi.wiley.com/10.1111/j.1749-6632.2000.tb06223.x>.
[54] Brabin BJ, Hakimi M, Pelletier D. An analysis of anemia and pregnancy-related maternal mortality. J Nutr [Internet] 2001 [cited 2019 May 14];131(2):604S–15S. Available from: <http://www.ncbi.nlm.nih.gov/pubmed/11160593>.
[55] Young MF, Oaks BM, Tandon S, Martorell R, Dewey KG, Wendt AS. Maternal hemoglobin concentrations across pregnancy and maternal and child health: a systematic review and meta-analysis. Ann N Y Acad Sci [Internet] 2019 [cited 2019 May 15]; nyas.14093. Available from: <https://onlinelibrary.wiley.com/doi/abs/10.1111/nyas.14093>.
[56] Muhsen K, Cohen D. *Helicobacter pylori* infection and iron stores: a systematic review and meta-analysis. Helicobacter [Internet] 2008 [cited 2019 May 14];13(5):323–40. Available from: <http://www.ncbi.nlm.nih.gov/pubmed/19250507>.
[57] Nashaat EH, Mansour GM. *Helicobacter pylori* and anemia with pregnancy. Arch Gynecol Obstet [Internet] 2014 [cited 2019 May 14];289(6):1197–202. Available from: <http://link.springer.com/10.1007/s00404-013-3138-8>.
[58] van Vliet AHM, Stoof J, Vlasblom R, Wainwright SA, Hughes NJ, Kelly DJ, et al. The role of the ferric uptake regulator (Fur) in regulation of *Helicobacter pylori* iron uptake.

Helicobacter [Internet] 2002 [cited 2019 May 14];7(4):237–44. Available from: <http://www.ncbi.nlm.nih.gov/pubmed/12165031>.

[59] Behboudi-Gandevani S, Amiri M, Bidhendi Yarandi R, Ramezani Tehrani F. The impact of diagnostic criteria for gestational diabetes on its prevalence: a systematic review and meta-analysis. Diabetol Metab Syndr [Internet] 2019 [cited 2019 May 15];11:11. Available from: <http://www.ncbi.nlm.nih.gov/pubmed/30733833>.

[60] Johns EC, Denison FC, Norman JE, Reynolds RM. Gestational diabetes mellitus: mechanisms, treatment, and complications. Trends Endocrinol Metab [Internet] 2018 [cited 2019 May 15];29(11):743–54. Available from: <https://www.sciencedirect.com/science/article/pii/S1043276018301656?via%3Dihub>.

[61] Catalano PM. Trying to understand gestational diabetes. Diabet Med [Internet] 2014 [cited 2019 May 15];31(3):273–81. Available from: <http://www.ncbi.nlm.nih.gov/pubmed/24341419>.

[62] Plows JF, Stanley JL, Baker PN, Reynolds CM, Vickers MH. The pathophysiology of gestational diabetes mellitus. Int J Mol Sci [Internet] 2018 [cited 2019 May 15];19(11). Available from: <http://www.ncbi.nlm.nih.gov/pubmed/30373146>.

[63] McClure EM, Dudley DJ, Reddy U, Goldenberg RL. Infectious causes of stillbirth: a clinical perspective. Clin Obstet Gynecol [Internet] 2010 [cited 2019 May 15];53(3):635. Available from: <https://www.ncbi.nlm.nih.gov/pmc/articles/PMC3893929/>.

[64] Tita ATN, Andrews WW. Diagnosis and management of clinical chorioamnionitis. Clin Perinatol [Internet] 2010 [cited 2019 May 16];37(2):339–54. Available from: <http://www.ncbi.nlm.nih.gov/pubmed/20569811>.

[65] Koga K, Mor G. Review article: expression and function of toll-like receptors at the maternal–fetal interface. Reprod Sci [Internet] 2008 [cited 2019 May 3];15(3):231–42. Available from: <http://journals.sagepub.com/doi/10.1177/1933719108316391>.

[66] Abrahams VM, Bole-Aldo P, Kim YM, Straszewski-Chavez SL, Chaiworapongsa T, Romero R, et al. Divergent trophoblast responses to bacterial products mediated by TLRs. J Immunol [Internet] 2004 [cited 2019 May 7];173(7):4286–96. Available from: <http://www.ncbi.nlm.nih.gov/pubmed/15383557>.

[67] Motedayyen H, Fathi F, Fasihi-Ramandi M, Sabzghabaee A, Taheri R. Toll-like receptor 4 activation on human amniotic epithelial cells is a risk factor for pregnancy loss. J Res Med Sci [Internet] 2019 [cited 2019 May 14];24(1):1. Available from: <http://www.jmsjournal.net/text.asp?2019/24/1/1/251198>.

[68] Nizyaeva NV, Kulikova GV, Nagovitsyna MN, Shchegolev AI. Peculiarities of the expression of TLR4 and inhibitor of TLR-cascade Tollip in the placenta in early and late-onset preeclampsia. Bull Exp Biol Med [Internet] 2019 [cited 2019 May 14];166(4):507–11. Available from: <http://link.springer.com/10.1007/s10517-019-04383-6>.

[69] Houben ML, Nikkels PGJ, van Bleek GM, Visser GHA, Rovers MM, Kessel H, et al. The association between intrauterine inflammation and spontaneous vaginal delivery at term: a cross-sectional study. Wright L, editor. PLoS One [Internet] 2009 Aug 10 [cited 2019 May 17];4(8):e6572. Available from: <https://dx.plos.org/10.1371/journal.pone.0006572>.

[70] Rothenberg SE, Keiser S, Ajami NJ, Wong MC, Gesell J, Petrosino JF, et al. The role of gut microbiota in fetal methylmercury exposure: insights from a pilot study. Toxicol Lett [Internet] 2016 [cited 2019 May 6];242:60. Available from: <http://www.ncbi.nlm.nih.gov/pubmed/26626101>.

[71] Southwick FS, Purich DL. Intracellular pathogenesis of listeriosis. Epstein FH, editor. N Engl J Med [Internet] 1996 Mar 21 [cited 2019 May 6];334(12):770–6. Available from: <http://www.nejm.org/doi/10.1056/NEJM199603213341206>.

[72] Pohl AM, Pouillot R, Bazaco MC, Wolpert BJ, Healy JM, Bruce BB, et al. Differences among incidence rates of invasive listeriosis in the U.S. FoodNet Population by age, sex, race/ethnicity, and pregnancy status, 2008–2016. Foodborne Pathog Dis

[Internet] 2019 [cited 2019 May 7];16(4):290–7. Available from: <http://www.ncbi.nlm.nih.gov/pubmed/30735066>.

[73] Vázquez-Boland JA, Krypotou E, Scortti M. Listeria placental infection. MBio [Internet] 2017 [cited 2019 May 7];8(3). Available from: <http://www.ncbi.nlm.nih.gov/pubmed/28655824>.

[74] Delcourt C, Yombi JC, Vo B, Yildiz H. *Salmonella enteritidis* during pregnancy, a rare cause of septic abortion: case report and review of the literature. J Obstet Gynaecol (Lahore) [Internet] 2019 [cited 2019 May 7];39(4):554–5. Available from: <https://www.tandfonline.com/doi/full/10.1080/01443615.2018.1529153>.

[75] Kuperman-Shani A, Vaknin Z, Mendlovic S, Zaidenstein R, Melcer Y, Maymon R. *Campylobacter coli* infection causing second trimester intrauterine growth restriction (IUGR): a case report and review of the literature. Prenat Diagn [Internet] 2015 [cited 2019 May 7];35(12):1258–61. Available from: <http://doi.wiley.com/10.1002/pd.4670>.

[76] Khan MY, Mah MW, Memish ZA. Brucellosis in pregnant women. Clin Infect Dis [Internet] 2001 [cited 2019 May 7];32(8):1172–7. Available from: <http://www.ncbi.nlm.nih.gov/pubmed/11283806>.

[77] Inan A, Erdem H, Elaldi N, Gulsun S, Karahocagil MK, Pekok AU, et al. Brucellosis in pregnancy: results of multicenter ID-IRI study. Eur J Clin Microbiol Infect Dis [Internet] 2019 [cited 2019 May 7];1–8. Available from: <http://link.springer.com/10.1007/s10096-019-03540-z>.

[78] Stout MJ, Conlon B, Landeau M, Lee I, Bower C, Zhao Q, et al. Identification of intracellular bacteria in the basal plate of the human placenta in term and preterm gestations. Am J Obstet Gynecol [Internet] 2013 [cited 2019 May 7];208(3):226.e1–7. Available from: <http://www.ncbi.nlm.nih.gov/pubmed/23333552>.

[79] Burton GJ, O'Shea S, Rostron T, Mullen JE, Aiyer S, Skepper JN, et al. Significance of placental damage in vertical transmission of human immunodeficiency virus. J Med Virol [Internet] 1996 [cited 2019 May 7];50(3):237–43. Available from: <http://www.ncbi.nlm.nih.gov/pubmed/8923288>.

[80] Aagaard K, Ma J, Antony KM, Ganu R, Petrosino J, Versalovic J. The placenta harbors a unique microbiome. Sci Transl Med [Internet] 2014 [cited 2019 May 7];6(237):237ra65. Available from: <http://www.ncbi.nlm.nih.gov/pubmed/24848255>.

[81] Zheng J, Xiao X, Zhang Q, Mao L, Yu M, Xu J. The placental microbiome varies in association with low birth weight in full-term neonates. Nutrients [Internet] 2015 [cited 2019 May 7];7(8):6924–37. Available from: <http://www.ncbi.nlm.nih.gov/pubmed/26287241>.

[82] Leiby JS, McCormick K, Sherrill-Mix S, Clarke EL, Kessler LR, Taylor LJ, et al. Lack of detection of a human placenta microbiome in samples from preterm and term deliveries. Microbiome [Internet] 2018 [cited 2019 May 7];6(1):196. Available from: <https://microbiomejournal.biomedcentral.com/articles/10.1186/s40168-018-0575-4>.

[83] Lauder AP, Roche AM, Sherrill-Mix S, Bailey A, Laughlin AL, Bittinger K, et al. Comparison of placenta samples with contamination controls does not provide evidence for a distinct placenta microbiota. Microbiome [Internet] 2016 [cited 2019 May 7];4(1):29. Available from: <http://www.ncbi.nlm.nih.gov/pubmed/27338728>.

[84] Amarasekara R, Jayasekara RW, Senanayake H, Dissanayake VHW. Microbiome of the placenta in pre-eclampsia supports the role of bacteria in the multifactorial cause of pre-eclampsia. J Obstet Gynaecol Res [Internet] 2015 [cited 2019 May 7];41(5):662–9. Available from: <http://www.ncbi.nlm.nih.gov/pubmed/25492799>.

[85] Figueroa R, Bracero LA, Aguero-Rosenfeld M, Beneck D, Coleman J, Schwartz I. Confirmation of *Borrelia burgdorferi* spirochetes by polymerase chain reaction in placentas of women with reactive serology for Lyme antibodies. Gynecol Obstet Invest [Internet] 1996 [cited 2019 May 15];41(4):240–3. Available from: <http://www.ncbi.nlm.nih.gov/pubmed/8793493>.

[86] Li S, Fuhler GM, Nahush BN, Jose T, Bruno MJ, Peppelenbosch MP, et al. Pancreatic cyst fluid harbors a unique microbiome. Microbiome 2017;5(1).

[87] Perez-Muñoz ME, Arrieta M-C, Ramer-Tait AE, Walter J. A critical assessment of the "sterile womb" and "in utero colonization" hypotheses: implications for research on the pioneer infant microbiome. Microbiome [Internet] 2017 [cited 2019 May 17];5 (1):48. Available from: <http://www.ncbi.nlm.nih.gov/pubmed/28454555>.

[88] Collado MC, Rautava S, Aakko J, Isolauri E, Salminen S. Human gut colonisation may be initiated in utero by distinct microbial communities in the placenta and amniotic fluid. Sci Rep [Internet] 2016 [cited 2019 May 7];6(1):23129. Available from: <http://www.nature.com/articles/srep23129>.

[89] Malmuthuge N, Griebel PJ. Fetal environment and fetal intestine are sterile during the third trimester of pregnancy. Vet Immunol Immunopathol [Internet] 2018 [cited 2019 May 17];204:59–64. Available from: <https://www.sciencedirect.com/science/article/pii/S0165242718302927?via%3Dihub>.

[90] Jiménez E, Fernández L, Marín ML, Martín R, Odriozola JM, Nueno-Palop C, et al. Isolation of commensal bacteria from umbilical cord blood of healthy neonates born by cesarean section. Curr Microbiol [Internet] 2005 [cited 2020 Mar 12];51(4):270–4. Available from: <http://www.ncbi.nlm.nih.gov/pubmed/16187156>.

[91] Rautava S, Collado MC, Salminen S, Isolauri E. Probiotics modulate host-microbe interaction in the placenta and fetal gut: a randomized, double-blind, placebo-controlled trial. Neonatology [Internet] 2012 [cited 2019 May 17];102(3):178–84. Available from: <http://www.ncbi.nlm.nih.gov/pubmed/22776980>.

[92] Goldenberg RL, Culhane JF, Iams JD, Romero R. Epidemiology and causes of preterm birth. Lancet [Internet] 2008 [cited 2019 May 17];371(9606):75–84. Available from: <https://www.sciencedirect.com/science/article/pii/S0140673608600744?via%3Dihub#bib67>.

[93] Jones HE, Harris KA, Azizia M, Bank L, Carpenter B, Hartley JC, et al. Differing prevalence and diversity of bacterial species in fetal membranes from very preterm and term labor. Middleton P, editor. PLoS One [Internet] 2009 Dec 8 [cited 2019 May 17];4 (12):e8205. Available from: <https://dx.plos.org/10.1371/journal.pone.0008205>.

[94] Doyle RM, Alber DG, Jones HE, Harris K, Fitzgerald F, Peebles D, et al. Term and preterm labour are associated with distinct microbial community structures in placental membranes which are independent of mode of delivery. Placenta [Internet] 2014 [cited 2019 May 17];35(12):1099–101. Available from: <https://www.sciencedirect.com/science/article/pii/S0143400414008194?via%3Dihub#bib2>.

[95] Waites KB, Katz B, Schelonka RL. Mycoplasmas and ureaplasmas as neonatal pathogens. Clin Microbiol Rev [Internet] 2005 [cited 2019 May 17];18(4):757–89. Available from: <http://www.ncbi.nlm.nih.gov/pubmed/16223956>.

[96] Bianchi-Jassir F, Seale AC, Kohli-Lynch M, Lawn JE, Baker CJ, Bartlett L, et al. Preterm birth associated with Group B *Streptococcus* maternal colonization worldwide: systematic review and meta-analyses. Clin Infect Dis [Internet] 2017 [cited 2019 May 17];65(Suppl. 2):S133–42. Available from: <http://www.ncbi.nlm.nih.gov/pubmed/29117329>.

[97] Mendz GL, Kaakoush NO, Quinlivan JA. Bacterial aetiological agents of intra-amniotic infections and preterm birth in pregnant women. Front Cell Infect Microbiol [Internet] 2013 [cited 2019 May 17];3:58. Available from: <http://www.ncbi.nlm.nih.gov/pubmed/24137568>.

[98] Lappas M, Permezel M, Georgiou HM, Rice GE. Type II phospholipase A2in preterm human gestational tissues. Placenta [Internet] 2001 [cited 2019 May 17];22(1):64–9. Available from: <http://www.ncbi.nlm.nih.gov/pubmed/11162354>.

[99] Bejar R, Curbelo V, Davis C, Gluck L. Premature labor. II. Bacterial sources of phospholipase. Obstet Gynecol [Internet] 1981 [cited 2019 May 17];57(4):479–82. Available from: <http://www.ncbi.nlm.nih.gov/pubmed/7017516>.

[100] Doster RS, Sutton JA, Rogers LM, Aronoff DM, Gaddy JA. *Streptococcus agalactiae* induces placental macrophages to release extracellular traps loaded with tissue remodeling enzymes via an oxidative burstdependent mechanism. Torres VJ, editor. MBio [Internet] 2018 Nov 20 [cited 2019 May 17];9(6). Available from: <http://www.ncbi.nlm.nih.gov/pubmed/30459195>.

[101] Surve MV, Anil A, Kamath KG, Bhutda S, Sthanam LK, Pradhan A, et al. Membrane vesicles of Group B *Streptococcus* disrupt feto-maternal barrier leading to preterm birth. Sullam PM, editor. PLoS Pathog [Internet] 2016 Sep 1 [cited 2019 May 17];12(9): e1005816. Available from: <http://www.ncbi.nlm.nih.gov/pubmed/27583406>.

[102] García-Velasco JA, Menabrito M, Catalán IB. What fertility specialists should know about the vaginal microbiome: a review. Reprod Biomed Online [Internet] 2017 [cited 2019 May 17];35(1):103–12. Available from: <https://www.sciencedirect.com/science/article/pii/S1472648317301876?via%3Dihub>.

[103] Ravel J, Gajer P, Abdo Z, Schneider GM, Koenig SSK, McCulle SL, et al. Vaginal microbiome of reproductive-age women. Proc Natl Acad Sci U S A [Internet] 2011 [cited 2019 May 17];108(Suppl. 1):4680–7. Available from: <http://www.ncbi.nlm.nih.gov/pubmed/20534435>.

[104] Larsen B, Monif GRG. Understanding the bacterial flora of the female genital tract [Internet]. Clin Infect Dis 2001 [cited 2019 May 17];. Available from: <https://pdfs.semanticscholar.org/a0d0/ab47e6d7a9673767121379fb89e8856afffd.pdf>.

[105] Bitew A, Abebaw Y, Bekele D, Mihret A. Prevalence of bacterial vaginosis and associated risk factors among women complaining of genital tract infection. Int J Microbiol [Internet] 2017 [cited 2019 May 24];2017:1–8. Available from: <https://www.hindawi.com/journals/ijmicro/2017/4919404/>.

[106] Abdullateef RM, Ijaiya MA, Abayomi F, Adeniran AS, Idris H. Bacterial vaginosis: Prevalence and associated risk factors among non-pregnant women of reproductive age attending a Nigerian tertiary hospital. Malawi Med J [Internet] 2018 [cited 2019 May 24];29(4):290. Available from: <http://www.ncbi.nlm.nih.gov/pubmed/29963282>.

[107] Donders G, Van Calsteren K, Bellen G, Reybrouck R, Van den Bosch T, Riphagen I, et al. Predictive value for preterm birth of abnormal vaginal flora, bacterial vaginosis and aerobic vaginitis during the first trimester of pregnancy. BJOG Int J Obstet Gynaecol [Internet] 2009 [cited 2019 May 24];116(10):1315–24. Available from: <http://www.ncbi.nlm.nih.gov/pubmed/19538417>.

[108] Koumans EH, Sternberg M, Bruce C, McQuillan G, Kendrick J, Sutton M, et al. The prevalence of bacterial vaginosis in the United States, 2001–2004; Associations with symptoms, sexual behaviors, and reproductive health. Sex Transm Dis [Internet] 2007 [cited 2019 May 17];34(11):864–9. Available from: <https://insights.ovid.com/crossref?an = 00007435-200711000-00006>.

[109] Fredricks DN, Fiedler TL, Marrazzo JM. Molecular identification of bacteria associated with bacterial vaginosis. N Engl J Med [Internet] 2005 [cited 2019 May 17];353(18):1899–911. Available from: <http://www.nejm.org/doi/abs/10.1056/NEJMoa043802>.

[110] van de Wijgert JHHM, Borgdorff H, Verhelst R, et al. H, et al. The vaginal microbiota: what have we learned after a decade of molecular characterization. PLoS One [Internet] 2014 [cited 2019 May 17];9(8):e105998. Available from: <http://www.ncbi.nlm.nih.gov/pubmed/25148517>.

[111] Zhou X, Brown CJ, Abdo Z, Davis CC, Hansmann MA, Joyce P, et al. Differences in the composition of vaginal microbial communities found in healthy Caucasian and black women. ISME J [Internet] 2007 [cited 2019 May 24];1(2):121–33. Available from: <http://www.ncbi.nlm.nih.gov/pubmed/18043622>.

[112] Borgdorff H, van der Veer C, van Houdt R, Alberts CJ, de Vries HJ, Bruisten SM, et al. The association between ethnicity and vaginal microbiota composition in Amsterdam, the Netherlands. Fredricks DN, editor. PLoS One [Internet] 2017 Jul 11 [cited 2019 May 28];12(7):e0181135. Available from: <http://www.ncbi.nlm.nih.gov/pubmed/28700747>.

[113] Leclair CM, Hart AE, Goetsch MF, Carpentier H, Jensen JT, et al. Group B streptococcus: prevalence in a non-obstetric population. Low Genit Tract Dis 2010 [cited 2019 May 28];14(3):162–6. Available from: <http://www.ncbi.nlm.nih.gov/pubmed/20592549>.

[114] Rumyantseva T, Khayrullina G, Guschin A, Donders G. Prevalence of *Ureaplasma* spp. and *Mycoplasma hominis* in healthy women and patients with flora alterations. Diagn Microbiol Infect Dis [Internet] 2019 [cited 2019 May 17];93(3):227–31. Available from: <http://www.ncbi.nlm.nih.gov/pubmed/30344067>.

[115] Breugelmans M, Vancutsem E, Naessens A, Laubach M, Foulon W. Association of abnormal vaginal flora and *Ureaplasma* species as risk factors for preterm birth: a cohort study. Acta Obstet Gynecol Scand [Internet] 2010 [cited 2019 May 17];89(2):256–60. Available from: <http://www.ncbi.nlm.nih.gov/pubmed/19943819>.

[116] Ma'ayeh M, Rood K, Walker H, Oliver E, Gee S, Iams J. Vaginal progesterone is associated with decreased Group B *Streptococcus* colonisation at term: a retrospective cohort study. BJOG Int J Obstet Gynaecol [Internet] 2019 [cited 2019 May 28]; 1471–0528.15801. Available from: <http://www.ncbi.nlm.nih.gov/pubmed/31094064>.

[117] Freitas AC, Chaban B, Bocking A, Rocco M, Yang S, Hill JE, et al. The vaginal microbiome of pregnant women is less rich and diverse, with lower prevalence of Mollicutes, compared to non-pregnant women. Sci Rep [Internet] 2017 [cited 2019 May 17];7(1):9212. Available from: <http://www.ncbi.nlm.nih.gov/pubmed/28835692>.

[118] Romero R, Hassan SS, Gajer P, Tarca AL, Fadrosh DW, Nikita L, et al. The composition and stability of the vaginal microbiota of normal pregnant women is different from that of non-pregnant women. Microbiome [Internet] 2014 [cited 2019 May 28];2(1):4. Available from: <http://www.ncbi.nlm.nih.gov/pubmed/24484853>.

[119] Aagaard K, Riehle K, Ma J, Segata N, Mistretta T-A, Coarfa C, et al. A metagenomic approach to characterization of the vaginal microbiome signature in pregnancy. Ratner AJ, editor. PLoS One [Internet] 2012 Jun 13 [cited 2019 May 28];7(6):e36466. Available from: <http://www.ncbi.nlm.nih.gov/pubmed/22719832>.

[120] Wheeler S, Pryor K, Antczak B, Truong T, Murtha A, Seed P. The relationship of cervical microbiota diversity with race and disparities in preterm birth. J Neonatal Perinatal Med [Internet] 2018 [cited 2019 May 17];11(3):305–10. Available from: <http://www.medra.org/servlet/aliasResolver?alias = iospress&doi = 10.3233/NPM-17111>.

[121] Elovitz MA, Gajer P, Riis V, Brown AG, Humphrys MS, Holm JB, et al. Cervicovaginal microbiota and local immune response modulate the risk of spontaneous preterm delivery. Nat Commun [Internet] 2019 [cited 2019 May 28];10(1):1305. Available from: <http://www.nature.com/articles/s41467-019-09285-9>.

[122] Cortez RV, Taddei CR, Sparvoli LG, Ângelo AGS, Padilha M, Mattar R, et al. Microbiome and its relation to gestational diabetes. Endocrine [Internet] 2019 [cited 2019 May 28];64(2):254–64. Available from: <http://link.springer.com/10.1007/s12020-018-1813-z>.

[123] Conde-Agudelo A, Villar J, Lindheimer M. Maternal infection and risk of preeclampsia: systematic review and metaanalysis. Am J Obstet Gynecol [Internet] 2008 [cited 2019 May 30];198(1):7–22. Available from: <http://www.ncbi.nlm.nih.gov/pubmed/18166297>.

[124] Vander Haar EL, So J, Gyamfi-Bannerman C, Han YW. *Fusobacterium nucleatum* and adverse pregnancy outcomes: epidemiological and mechanistic evidence. Anaerobe [Internet] 2018 [cited 2019 May 29];50:55–9. Available from: <https://www.sciencedirect.com/science/article/pii/S1075996418300192?via%3Dihub#bib45>.

[125] Han YW, Redline RW, Li M, Yin L, Hill GB, McCormick TS. *Fusobacterium nucleatum* induces premature and term stillbirths in pregnant mice: implication of oral bacteria in preterm birth. Infect Immun [Internet] 2004 [cited 2019 May 29];72(4):2272–9. Available from: <http://www.ncbi.nlm.nih.gov/pubmed/15039352>.

[126] Belizário JE, Faintuch J, Garay-Malpartida M. Gut microbiome dysbiosis and immunometabolism: new frontiers for treatment of metabolic diseases. Mediators Inflamm [Internet] 2018 [cited 2019 May 28];2018:2037838. Available from: <http://www.ncbi.nlm.nih.gov/pubmed/30622429>.

[127] Wikoff WR, Anfora AT, Liu J, Schultz PG, Lesley SA, Peters EC, et al. Metabolomics analysis reveals large effects of gut microflora on mammalian blood metabolites. Proc Natl Acad Sci [Internet] 2009 [cited 2019 May 28];106(10):3698–703. Available from: <http://www.pnas.org/cgi/doi/10.1073/pnas.0812874106>.

[128] Turnbaugh PJ, Ley RE, Mahowald MA, Magrini V, Mardis ER, Gordon JI. An obesity-associated gut microbiome with increased capacity for energy harvest. Nature [Internet] 2006 [cited 2019 May 29];444(7122):1027–31. Available from: <http://www.ncbi.nlm.nih.gov/pubmed/17183312>.

[129] Thaiss CA, Itav S, Rothschild D, Meijer MT, Levy M, Moresi C, et al. Persistent microbiome alterations modulate the rate of post-dieting weight regain. Nature [Internet] 2016 [cited 2019 May 29];540(7634):544–51. Available from: <http://www.nature.com/articles/nature20796>.

[130] Khan MT, Nieuwdorp M, Bäckhed F. Microbial modulation of insulin sensitivity. Cell Metab [Internet] 2014 [cited 2019 May 29];20(5):753–60. Available from: <http://www.ncbi.nlm.nih.gov/pubmed/25176147>.

[131] Balan P, Chong YS, Umashankar S, Swarup S, Loke WM, Lopez V, et al. Keystone species in pregnancy gingivitis: a snapshot of oral microbiome during pregnancy and postpartum period. Front Microbiol [Internet] 2018 [cited 2019 May 29];9:2360. Available from: <https://www.frontiersin.org/article/10.3389/fmicb.2018.02360/full>.

[132] Onigbinde O, Sorunke M, Braimoh M, Adeniyi A. Periodontal status and some variables among pregnant women in a Nigeria tertiary institution. Ann Med Health Sci Res [Internet] 2014 [cited 2019 May 29];4(6):852–7. Available from: <http://www.ncbi.nlm.nih.gov/pubmed/25506475>.

[133] Han YW. Oral health and adverse pregnancy outcomes - what's next. J Dent Res [Internet] 2011 [cited 2019 May 29];90(3):289–93. Available from: <http://www.ncbi.nlm.nih.gov/pubmed/21041548>.

[134] Wandera M, Åstrøm AN, Okullo I, Tumwine JK. Determinants of periodontal health in pregnant women and association with infants' anthropometric status: a prospective cohort study from Eastern Uganda. BMC Pregnancy Childbirth [Internet] 2012 [cited 2019 May 29];12(1):90. Available from: <http://bmcpregnancychildbirth.biomedcentral.com/articles/10.1186/1471-2393-12-90>.

[135] Barak S, Oettinger-Barak O, Machtei EE, Sprecher H, Ohel G. Evidence of periopathogenic microorganisms in placentas of women with preeclampsia. J Periodontol [Internet] 2007 [cited 2019 May 29];78(4):670–6. Available from: <http://www.ncbi.nlm.nih.gov/pubmed/17397314>.

[136] Swati P, Thomas B, Vahab SA, Kapaettu S, Kushtagi P, Kushtagi P. Simultaneous detection of periodontal pathogens in subgingival plaque and placenta of women with hypertension in pregnancy. Arch Gynecol Obstet [Internet] 2012 [cited 2019 May 29];285(3):613–19. Available from: <http://www.ncbi.nlm.nih.gov/pubmed/21830010>.

[137] Wang J, Zheng J, Shi W, Du N, Xu X, Zhang Y, et al. Dysbiosis of maternal and neonatal microbiota associated with gestational diabetes mellitus. Gut [Internet] 2018 [cited 2019 May 28];67(9):1614–25. Available from: <http://www.ncbi.nlm.nih.gov/pubmed/29760169>.

[138] Nuriel-Ohayon M, Neuman H, Koren O. Microbial changes during pregnancy, birth, and infancy. Front Microbiol [Internet] 2016 [cited 2018 Dec 4];7:1031. Available from: <http://www.ncbi.nlm.nih.gov/pubmed/27471494>.

[139] Koren O, Goodrich JK, Cullender TC, Spor A, Laitinen K, Bäckhed HK, et al. Host remodeling of the gut microbiome and metabolic changes during pregnancy. Cell [Internet] 2012 [cited 2016 Aug 26];150(3):470–80. Available from: <http://www.ncbi.nlm.nih.gov/pubmed/22863002>.

[140] Progesterone increases *Bifidobacterium* relative abundance during late pregnancy. Cell Rep [Internet] 2019 Apr 16 [cited 2019 May 30];27(3):730–36.e3. Available from: <https://www.sciencedirect.com/science/article/pii/S221112471930405X>.

[141] Dahl C, Stanislawski M, Iszatt N, Mandal S, Lozupone C, Clemente JC, et al. Gut microbiome of mothers delivering prematurely shows reduced diversity and lower relative abundance of *Bifidobacterium* and *Streptococcus*. Abdo Z, editor. PLoS One [Internet] 2017 Oct 25 [cited 2019 May 30];12(10):e0184336. Available from: <http://www.ncbi.nlm.nih.gov/pubmed/29069100>.

[142] Shiozaki A, Yoneda S, Yoneda N, Yonezawa R, Matsubayashi T, Seo G, et al. Intestinal microbiota is different in women with preterm birth: results from terminal restriction fragment length polymorphism analysis. Zenclussen AC, editor. PLoS One [Internet] 2014 Nov 5 [cited 2019 May 30];9(11):e111374. Available from: <http://dx.plos.org/10.1371/journal.pone.0111374>.

[143] Makino H, Kushiro A, Ishikawa E, Kubota H, Gawad A, Sakai T, et al. Mother-to-infant transmission of intestinal bifidobacterial strains has an impact on the early development of vaginally delivered infant's microbiota. Sanz Y, editor. PLoS One [Internet] 2013 Nov 14 [cited 2019 May 30];8(11):e78331. Available from: <https://dx.plos.org/10.1371/journal.pone.0078331>.

[144] Yassour M, Jason E, Hogstrom LJ, Arthur TD, Tripathi S, Siljander H, et al. Strain-level analysis of mother-to-child bacterial transmission during the first few months of life. Cell Host Microbe [Internet] 2018 [cited 2019 May 30];24(1):146–154.e4. Available from: <http://www.ncbi.nlm.nih.gov/pubmed/30001517>.

[145] Fouhy F, Watkins C, Hill CJ, O'Shea C-A, Nagle B, Dempsey EM, et al. Perinatal factors affect the gut microbiota up to four years after birth. Nat Commun [Internet] 2019 [cited 2019 May 30];10(1):1517. Available from: <http://www.nature.com/articles/s41467-019-09252-4>.

[146] Ferretti P, Pasolli E, Tett A, Asnicar F, Gorfer V, Fedi S, et al. Mother-to-infant microbial transmission from different body sites shapes the developing infant gut microbiome. Cell Host Microbe [Internet] 2018 [cited 2018 Dec 4];24(1):133–145.e5. Available from: <http://www.ncbi.nlm.nih.gov/pubmed/30001516>.

[147] Duranti S, Lugli GA, Mancabelli L, Armanini F, Turroni F, James K, et al. Maternal inheritance of bifidobacterial communities and bifidophages in infants through vertical transmission. Microbiome [Internet] 2017 [cited 2019 May 30];5(1):66. Available from: <http://www.ncbi.nlm.nih.gov/pubmed/28651630>.

[148] Soh MC, Nelson-Piercy C. Biomarkers for adverse pregnancy outcomes in rheumatic diseases. Rheum Dis Clin North Am [Internet] 2017 [cited 2019 May 30];43(2):201–14. Available from: <http://www.ncbi.nlm.nih.gov/pubmed/28390563>.

[149] Moroni G, Ponticelli C. Pregnancy in women with systemic lupus erythematosus (SLE). Eur J Intern Med [Internet] 2016 [cited 2019 May 30];32:7–12. Available from: <http://www.ncbi.nlm.nih.gov/pubmed/27142327>.

[150] Cornish J, Tan E, Teare J, Teoh TG, Rai R, Clark SK, et al. A meta-analysis on the influence of inflammatory bowel disease on pregnancy. Gut [Internet] 2007 [cited 2017 Jan 30];56(6):830–7. Available from: <http://www.ncbi.nlm.nih.gov/pubmed/17185356>.

[151] Parikh K, Zhou L, Somasundaram R, Fuhler GM, Deuring JJ, Blokzijl T, et al. Suppression of p21Rac signaling and increased innate immunity mediate remission in Crohn's disease. Sci Transl Med [Internet] 2014;6:233ra53. Available from: <http://www.ncbi.nlm.nih.gov/pubmed/24760188>.

[152] Somasundaram R, Deuring JJ, Van Der Woude CJ, Peppelenbosch MP, Fuhler GM. Linking risk conferring mutations in NCF4 to functional consequences in Crohn's disease. Gut 2012;61(7).

[153] Somasundaram R, Nuij VJAA, Van Der Woude CJ, Kuipers EJ, Peppelenbosch MP, Fuhler GM. Peripheral neutrophil functions and cell signalling in crohn's disease. PLoS One 2013;8.

[154] Frank DN, St Amand AL, Feldman RA, Boedeker EC, Harpaz N, Pace NR. Molecular-phylogenetic characterization of microbial community imbalances in human inflammatory bowel diseases. Proc Natl Acad Sci U S A [Internet] 2007 [cited 2019 May 30];104(34):13780–5. Available from: <http://www.ncbi.nlm.nih.gov/pubmed/17699621>.

[155] Pascal V, Pozuelo M, Borruel N, Casellas F, Campos D, Santiago A, et al. A microbial signature for Crohn's disease. Gut [Internet] 2017 [cited 2018 Apr 20];66(5):813–22. Available from: <http://www.ncbi.nlm.nih.gov/pubmed/28179361>.

[156] Deuring JJ, Fuhler GM, Konstantinov SR, Peppelenbosch MP, Kuipers EJ, Haar CD, et al. Genomic ATG16L1 risk allele-restricted Paneth cell ER stress in quiescent Crohn's disease. Gut 2014;63(7).

[157] Kammerlander H, Nielsen J, Kjeldsen J, Knudsen T, Friedman S, Nørgård B. The Effect of disease activity on birth outcomes in a nationwide cohort of women with moderate to severe inflammatory bowel disease. Inflamm Bowel Dis [Internet] 2017 [cited 2019 May 30];23(6):1011–18. Available from: <http://www.ncbi.nlm.nih.gov/pubmed/28346274>.

[158] de Lima-Karagiannis A, Zelinkova-Detkova Z, van der Woude CJ. The effects of active IBD during pregnancy in the era of novel IBD therapies. Am J Gastroenterol [Internet] 2016 [cited 2019 May 30];111(9):1305–12. Available from: <http://www.ncbi.nlm.nih.gov/pubmed/27349339>.

[159] van der Giessen J, Huang VW, van der Woude CJ, Fuhler GM. Modulatory effects of pregnancy on inflammatory bowel disease. Clin Transl Gastroenterol [Internet] 2019 [cited 2019 May 6];10(3):e00009. Available from: <http://www.ncbi.nlm.nih.gov/pubmed/30908306>.

[160] Konstantinov SR, van der Woude CJ, Peppelenbosch MP. Do pregnancy-related changes in the microbiome stimulate innate immunity? Trends Mol Med 2013;454–9.

[161] Torres J, Hu J, Seki A, Eisele C, Nair N, Huang R, et al. Infants born to mothers with IBD present with altered gut microbiome that transfers abnormalities of the adaptive immune system to germ-free mice. Gut [Internet] 2019 [cited 2019 May 30]; gutjnl-2018-317855. Available from: <http://www.ncbi.nlm.nih.gov/pubmed/31036757>.

[162] Ferrocino I, Ponzo V, Gambino R, Zarovska A, Leone F, Monzeglio C, et al. Changes in the gut microbiota composition during pregnancy in patients with gestational diabetes mellitus (GDM). Sci Rep [Internet] 2018 [cited 2019 May 30];8(1):12216. Available from: <http://www.ncbi.nlm.nih.gov/pubmed/30111822>.

[163] Li J, Zhao F, Wang Y, Chen J, Tao J, Tian G, et al. Gut microbiota dysbiosis contributes to the development of hypertension. Microbiome [Internet] 2017 [cited 2019 May 29];5(1):14. Available from: <http://www.ncbi.nlm.nih.gov/pubmed/28143587>.

[164] Kim S, Goel R, Kumar A, Qi Y, Lobaton G, Hosaka K, et al. Imbalance of gut microbiome and intestinal epithelial barrier dysfunction in patients with high blood pressure. Clin Sci (Lond) [Internet] 2018 [cited 2019 May 29];132(6):701–18. Available from: <http://www.ncbi.nlm.nih.gov/pubmed/29507058>.

[165] Beale AL, Kaye DM, Marques FZ. The role of the gut microbiome in sex differences in arterial pressure. Biol Sex Differ 2019 [cited 2019 May 29];10:22. Available from: <https://doi.org/10.1186/s13293-019-0236-8>.

[166] Liu J, Yang H, Yin Z, Jiang X, Zhong H, Qiu D, et al. Remodeling of the gut microbiota and structural shifts in Preeclampsia patients in South China. Eur J Clin Microbiol Infect Dis [Internet] 2017 [cited 2019 May 29];36(4):713–19. Available from: <http://link.springer.com/10.1007/s10096-016-2853-z>.

[167] Gomez-Arango LF, Barrett HL, McIntyre HD, Callaway LK, Morrison M, Dekker Nitert M. Increased systolic and diastolic blood pressure is associated with altered gut microbiota composition and butyrate production in early pregnancy. Hypertension [Internet] 2016 [cited 2019 May 30];68(4):974–81. Available from: <https://www.ahajournals.org/doi/10.1161/HYPERTENSIONAHA.116.07910>.

[168] Ponzo V, Fedele D, Goitre I, Leone F, Lezo A, Monzeglio C, et al. Diet-gut microbiota interactions and gestational diabetes mellitus (GDM). 2019 [cited 2019 May 30]. Available from: <www.mdpi.com/journal/nutrients>.

[169] Mokkala K, Houttu N, Vahlberg T, Munukka E, Rönnemaa T, Laitinen K. Gut microbiota aberrations precede diagnosis of gestational diabetes mellitus. Acta Diabetol [Internet] 2017 [cited 2019 May 30];54(12):1147–9. Available from: <http://link.springer.com/10.1007/s00592-017-1056-0>.

[170] Kuang Y-S, Lu J-H, Li S-H, Li J-H, Yuan M-Y, He J-R, et al. Connections between the human gut microbiome and gestational diabetes mellitus. Gigascience [Internet] 2017 [cited 2019 May 30];6(8):1–12. Available from: <http://www.ncbi.nlm.nih.gov/pubmed/28873967>.

[171] Ganesan K, Chung SK, Vanamala J, Xu B. Causal relationship between diet-induced gut microbiota changes and diabetes: a novel strategy to transplant *Faecalibacterium prausnitzii* in preventing diabetes. Int J Mol Sci [Internet] 2018 [cited 2019 May 30];19 (12):3720. Available from: <http://www.ncbi.nlm.nih.gov/pubmed/30467295>.

[172] Grev J, Berg M, Soll R. Maternal probiotic supplementation for prevention of morbidity and mortality in preterm infants. Cochrane Database Syst Rev [Internet] 2018 [cited 2019 May 30];12:CD012519. Available from: <http://www.ncbi.nlm.nih.gov/pubmed/30548483>.

[173] Jarde A, Lewis-Mikhael A-M, Moayyedi P, Stearns JC, Collins SM, Beyene J, et al. Pregnancy outcomes in women taking probiotics or prebiotics: a systematic review and meta-analysis. BMC Pregnancy Childbirth [Internet] 2018 [cited 2019 May 30];18(1):14. Available from: <http://www.ncbi.nlm.nih.gov/pubmed/29310610>.

[174] Sugawara J, Ochi D, Yamashita R, Yamauchi T, Saigusa D, Wagata M, et al. Maternity Log study: a longitudinal lifelog monitoring and multiomics analysis for the early prediction of complicated pregnancy. BMJ Open [Internet] 2019 [cited 2019 May 2];9(2):e025939. Available from: <http://www.ncbi.nlm.nih.gov/pubmed/30782942>.

[175] Brummaier T, Syed Ahamed Kabeer B, Lindow S, Konje JC, Pukrittayaamee S, Utzinger J, et al. A prospective cohort for the investigation of alteration in temporal transcriptional and microbiome trajectories preceding preterm birth: a study protocol. BMJ Open [Internet] 2019 [cited 2019 May 2];9(1):e023417. Available from: <http://bmjopen.bmj.com/lookup/doi/10.1136/bmjopen-2018-023417>.

[176] Barthow C, Wickens K, Stanley T, Mitchell EA, Maude R, Abels P, et al. The Probiotics in Pregnancy Study (PiP Study): rationale and design of a double-blind randomised controlled trial to improve maternal health during pregnancy and prevent infant eczema and allergy. BMC Pregnancy Childbirth [Internet] 2016 [cited 2019 May 30];16(1):133. Available from: <http://bmcpregnancychildbirth.biomedcentral.com/articles/10.1186/s12884-016-0923-y>.

CHAPTER

3

Microbial signatures of preterm birth

David A. MacIntyre[1,2] *and Phillip R. Bennett*[1,2]

[1]Parturition Research Group, Institute of Reproductive and Developmental Biology, Imperial College London, London, United Kingdom, [2]March of Dimes European Preterm Birth Research Centre, Imperial College London, London, United Kingdom

Introduction

Preterm birth, defined as delivery before 37 weeks gestational age, remains a major global health issue. Despite improvements in neonatal care leading to significantly better survival rates in the developed world, preterm birth and its associated complications are the primary worldwide cause of death of children under the age of 5 years [1]. Almost one million babies die of prematurity each year and of those who survive, many millions more will suffer life-long morbidities such as cerebral palsy, blindness, deafness, and impaired respiratory function. This clearly has a major emotional and economic impact on society, yet robust prediction and effective prevention of preterm birth remain elusive. One reason for this is that preterm birth is a syndrome with multiple etiologies; however, it is estimated that around 30%–40% of all preterm birth cases have an infection or inflammation-based etiology [2]. While a relationship between infection and preterm birth has long been recognized, recent applications of molecular-based approaches have provided new insights into the microbial contribution toward preterm birth. This chapter will explore the evidence for microbial signatures of preterm birth and the mechanisms by which they may lead to the premature onset of labor and preterm birth.

The Human Microbiome in Early Life
DOI: https://doi.org/10.1016/B978-0-12-818097-6.00003-1

Infection and preterm birth: a historical perspective

François Mauriceau, one of the leading obstetricians of the 17th century, recorded the case of a woman in 1694, who in the sixth month of her pregnancy, "aborted on the ninth day of her pneumonia, and died on the thirteenth" [3]. Due to the historical differences in medical terminology, it is difficult to establish if this is indeed the first recorded infection-associated preterm birth; however, by the mid-1900s, systemic maternal infections such as pneumonia, typhoid fever, and malaria had been firmly associated with preterm labor and delivery [4–7]. Since then, advances in medicine, including the development of antibiotics, have led to a sharp reduction in the proportion of preterm birth cases attributable to systemic maternal infection, at least in higher income settings. However, intrauterine infection is still responsible for a disproportionate number of preterm births, particularly at earlier gestational ages [8,9].

There are a number of ways by which microbes can reach the uterine cavity including hematogenous spread, retrograde seeding through the fallopian tubes, or during invasive procedures such as amniocentesis or chorionic villus sampling [10–12]. However, the most common pathway of intrauterine infection is considered to be ascension from the vagina, through the cervix, and into the uterus [11,13]. There are a number of lines of evidence to support this. First, bacteria found in the placental and fetal membranes (amnion and chorion) of preterm birth cases are similar to those found in the vagina [14,15]. Second, histological chorioamnionitis (inflammation and neutrophil invasion of the fetal membranes) is observed most frequently and severely at the site of membrane rupture, which usually occurs in the lower part of the uterus close to the cervix [12]. Third, in twin pregnancies, histological chorioamnionitis and microbial invasion of the amniotic cavity are most commonly observed in the present and first born twin [14,16]. Finally, a substantial number of animal studies have been able to reproduce the process of ascending vaginal infection leading to preterm birth [17–22].

The relationship between reproductive tract infection and preterm birth was possibly first alluded toward the end of the 18th century. Then puerperal fever and sepsis were a major cause of peripartum mortality and usually followed an ascending infection of the decidua and/or placenta [23]. It is highly likely that a proportion of these puerperal fever cases was subsequent to infection-induced premature onset of labor and birth, which was then described as "abortion" when delivery was estimated to occur before 6 months of gestation. Initially, puerperal fever was attributed to "vapors in the air" that were seemingly transferred from one woman to the next [23]. In the 1790s during an epidemic of puerperal fever, naval surgeon Alexander Gordon noted that as a physician attending to pregnant women, he appeared to be propagating the fever between patients and accordingly suggested the burning of woman's bedclothes and fumigation of doctors and nurses involved in their care. His suggestions were ignored as were those of Hungarian physician, Ignaz Semmelweis who almost 50 years later showed that puerperal fever could be drastically reduced by hand disinfection using chlorinated lime solution before attending women in labor [24]. The pathogen responsible for puerperal sepsis was eventually isolated and identified as hemolytic *Streptococcus* by Louis Pasteur in 1879, and in 1936 Leonard Colebrook performed the first ever clinical trial of an antimicrobial drug at Queen Charlotte's Hospital, London where he successfully introduced sulfanilamide for the treatment of puerperal sepsis [25].

Work on puerperal sepsis was thus key in establishing a link between pathogen infection of the lower reproductive tract and poor pregnancy outcomes. As part of related investigations in 1892, Albert Döderlein isolated Döderlein's bacillus (now known as *Lactobacillus acidophilus* or a closely related species [26,27]) from vaginal secretions [28]. He noted that the vaginal bacterial communities of pregnant women could be stratified into normal (*Lactobacillus* spp. dominated) or abnormal (reduced *Lactobacillus* spp. with the presence of other organisms such as streptococci or staphylococci) and highlighted the capacity of the vaginal bacillus to inhibit the growth of staphylococci, which he attributed mainly to lactic acid production of these bacteria. Many of the clinical features associated with Döderlein's "abnormal" vaginal classification would now be recognized as symptomatic bacterial vaginosis (BV) [29].

Bacterial vaginosis and preterm birth

BV is a common clinical condition often accompanied by a thin vaginal discharge and malodor that has a major global burden on women's health [30,31]. Microbiologically, it is characterized by reduced levels of *Lactobacillus* spp. and increased diversity and relative abundance of facultative and strictly anaerobic bacteria including *Gardnerella, Prevotella, Mycoplasma,* and *Atopobium* species [32]. The Amsel criteria is the main method used clinically for BV diagnosis and is based upon a combined assessment of symptomatic (e.g., fishy odor, pH, and white discharge) and microscopic features of vaginal secretions (e.g., presence of vaginal epithelial cells coated by gram-variable bacteria or clue cells) [33], yet it is recognized that many affected women are asymptomatic [34]. The Nugent criteria is often employed in research setting for BV diagnosis and uses microscopic evaluation and grading for the presence or absence of three bacterial morphotypes (*Lactobacillus*, *Gardnerella vaginalis*, and *Mobiluncus* types) [35].

A large body of evidence supports an association between BV and increased risk of poor pregnancy outcome including preterm delivery [36–38], preterm prelabor rupture of the fetal membranes [39], and histological chorioamnionitis [40–42]. BV is estimated to increase the risk of preterm birth by twofold [43], irrespective of whether women are symptomatic or asymptomatic [44]. A meta-analysis in 2003 reported that preterm birth risk associates even more strongly with BV, the earlier it is screened for up to sevenfold increased risk if detected before 16 weeks of gestation [45]. A number of studies have also highlighted that BV-associated preterm birth risk can be mediated by gene–environment interactions. For example, women with both BV and a TNF-α-308 polymorphism are at an enhanced risk of preterm birth compared to women with BV or the TNF-α-308 polymorphism alone [46]. Similarly, synergistic interaction effects have also been described for BV and polymorphisms within the PRKCA gene, which can mediate both inflammatory and uterine contractile pathways [47].

Given the abundance of evidence linking BV to increased preterm birth risk, it is understandable that many efforts have been made to examine the effectiveness of antibiotic treatment of BV for the prevention of preterm birth. Unfortunately, the results of these trials have proven highly inconsistent [48,49]. This is partly attributable to heterogeneity between trial designs including differences in the timing of treatment, types of antibiotic used, administration routes, patient cohorts selected, and erroneous indications [50,51]. Some of these issues were addressed in the recently reported large, double-blinded PREMEVA trial of oral clindamycin in early pregnancy [52], suggested by some to be the preferable antibiotic for BV. In total, almost 85,000 women were screened, and 1904 BV-positive participants with low background risk for preterm birth (out of 5630 diagnosed) were assigned to clindamycin, whereas 956 participants received the placebo (at a pretrial defined 2:1 treatment:placebo ratio). No difference in the primary outcome of the composite of late miscarriage (16–21 weeks) or spontaneous very preterm birth (22–32 weeks) was detected between the two groups. As with all clinical trials, there are a number of limitations to consider including the low number of primary outcomes reported (1.2%, 22/1904 pregnancies) and the fact that almost 20% of women did not complete their assigned treatment course. The trial has also faced criticism for the method by which BV was diagnosed [50] and for not assessing the effectiveness of the treatment to resolve BV [53]. However, if these data hold true, they collectively highlight the ineffectiveness of oral clindamycin to prevent BV-associated preterm birth in low-risk women. Yet data from both the PREMEVA trial and the most recent Cochrane review on antibiotic use in pregnancy have led some experts to suggest that a randomized, placebo-controlled trial of screening and treating BV in women with a history of preterm birth is still warranted [49].

As detailed elsewhere, there are a number of additional reasons why antibiotic treatment of BV may not provide protection against preterm birth [43,54]. First, treatment of women with existing BV may mean that associated ascending infection or inflammatory pathway activation has already commenced. Second, the use of bactericidal agents such as metronidazole, as opposed to bacteriostatic agents, may lead to the lysing of bacteria and release of endotoxins and lipopolysaccharides [55], which are strong activators of uterine inflammation and preterm birth [56–58]. Third, some antibiotic treatments may actually be particularly effective against commensal *Lactobacillus* species, but not against pathogens associated with BV leading to inadvertent enrichment of these species in the vaginal niche [54,59,60]. There is also evidence that BV-associated bacteria have high rates of antibiotic resistance [61,62].

Finally, it is important to acknowledge that a large proportion of women with BV in pregnancy do not deliver preterm, which highlights the vital importance of individual-level host response to specific microbes in the reproductive tract. The ability to examine microbial composition in the reproductive tract in detail is thus critical to understanding the relationship between microbiome–host interactions in pregnancy and preterm birth risk.

Molecular-based characterization of the vaginal microbiome

Until relatively recently, investigation of vaginal microbes was limited to microscopy and culture-based approaches. These methods often failed to detect fastidious bacterial species and could not assess relative abundance of species within a community. The application of culture-independent, molecular-based approaches was first applied in 2002 and used a combination of polymerase chain reaction and denaturing gradient gel electrophoresis to assess the vaginal microbiome composition in a group of postmenopausal women [63]. Since then, phylogenetic-based analyses, using mainly 16S rRNA gene sequencing, have become widely used for the assessment of reproductive tract microbiome composition and structure, particularly in BV [64–66]. These metataxonomic studies have both confirmed that BV is characterized by a loss of lactobacilli and the enrichment of other facultative anaerobic bacteria but importantly, also led to the identification of other vaginal bacteria previously undetected by culture including *Lactobacillus iners, Atopobium vaginae,* and three species from the *Lachnospiraceae* family, BV-associated bacteria 1, 2, and 3 (BVAB1, 2, and 3) [65].

In 2011, a key study by Ravel et al. using 16S ribosomal RNA gene sequencing showed that in nonpregnant, asymptomatic women of reproductive age, the vaginal microbiome structure can be classified into five community state types (CSTs), four of which are characterized by almost complete dominance of either *L. crispatus* (CST I), *L. gasseri* (CST II), *L. iners* (CST III), or *L. jensenii* (CST V), whereas CST IV is characterized by low levels of *Lactobacillus* species and increased diversity and richness of anaerobic bacteria often associated with BV [67]. Interestingly, consistent with previous studies using culture and sequencing [68], CST IV was found to be particularly prevalent in Black (40%) and Hispanic (38%) women compared to Asians (20%) and Whites (10%). This observation has since been replicated in other cohorts [69,70], highlighting ethnicity as an intrinsic host factor that influences vaginal microbiome composition. Longitudinal studies have pointed toward the dynamic nature of these CSTs, which in some women readily flip and shift in response to menstruation, sexual intercourse, or hygiene practices such as douching, while in other women remain stable over long period of time [71]. It should be noted that additional CSTs, also referred to as "vaginal microbiome groups" or "vagitypes," have since been described, yet a clear consensus on their total number and functional role in health and disease has not been fully established [69,72–74].

Vaginal microbiome composition in term and preterm birth

Molecular-based bacterial compositional profiling in both cross-sectional and longitudinal pregnant patient populations has confirmed that the vaginal microbiome in healthy pregnancy is reduced in both overall diversity and richness and is typically associated with more stable, *Lactobacillus* species dominant CSTs [75–77]. Suggested reasons for this include the lack of cyclic hormonal variations and the high estrogenic state of pregnancy [78,79] that promote the deposition of glycogen in the vaginal epithelia. Glycogen can be broken down by maternal host α-amylase into complex sugars and preferentially be utilized by *Lactobacillus* spp. [80,81]. Consistent with this, soon after delivery, when maternal estrogen levels fall around 1000-fold [82], vaginal *Lactobacillus* spp. dominance decreases, and there is a shift toward a high diversity, CST IV-type community structure [76]. It has since been reported that this high-diversity state persists in some women for up to 1 year post-delivery [83,84]. A large study recently undertaken as part of the National Institutes of Health Human Microbiome Project's second initiative has largely confirmed these findings but also shown that the shift toward *Lactobacillus* spp. dominance and reduced diversity in normal pregnancy is most significant in women of African ancestry and occurs early in pregnancy [85].

Vaginal microbial signatures of preterm birth

In an attempt to better elucidate the relationship between vaginal microbiome signatures and preterm birth, numerous molecular-based studies have now been undertaken. As described in a recent systematic review that ultimately included nine studies published before November 2017, early work in this area failed to reach a consensus on the association between preterm birth risk and vaginal bacterial composition [86]. While some studies reported a positive association between increased vaginal microbial diversity and increased risk of preterm birth [83,87,88], others did not [89,90]. Inconsistencies between these studies can be partly attributed to differences in study design, methodologies used for microbiome compositional profiling, patient indications for preterm birth, and a lack of power. Given the additional heterogeneity of preterm birth etiology, it is not surprising that high variability between studies was observed.

In comparison, more recent studies have started to provide some clarity for the field. Generally, these studies have focused on larger patient populations and/or more careful characterization and consideration of preterm birth etiology [59,90–98]. A summary of studies examining the vaginal microbiome in women with preterm birth from January 2017 until July 2019 is provided in Table 3.1. In 2017 our laboratory reported data from a cohort of 161 women of whom 34 ultimately delivered preterm, which indicated that *L. crispatus* dominance of the vaginal microbiome at 16 weeks of pregnancy appeared to confer protection against subsequent preterm birth [96]. Despite the methodological differences, this finding has since been replicated by other groups [92,94,98], and many more have highlighted protection against preterm birth afforded by *Lactobacillus* spp. dominance more generally [59,91,93,97]. There is also a relatively high concordance between recent studies indicating a link between increased preterm birth risk and increased diversity, and the enrichment of species is often associated with BV, including *G. vaginalis*, *Sneathia* spp., *Prevotella* spp., and *Mollicutes* spp. [59,91–95,97].

A number of these studies have also highlighted the importance of ethnicity as a potential confounder of studies examining the vaginal microbiome in preterm and term pregnancies [85,92,94]. Such findings raise caution to the interpretation of mixed ethnic populations but also

TABLE 3.1 Summary of the studies examining the relationship between vaginal microbiome composition and preterm birth (PTB) from January 2017 through July 2019.

Year	Author [reference]	Control (*n*)	Cases (*n*)	Type and gestation of sampling	Background PTB risk (high/low), predominant ethnicity[a] (% total cohort)	Target for molecular characterization	Vaginal microbiome as a modifier of PTB risk	
							Microbiome features positively associated with increased PTB risk	Microbiome features positively associated with decreased PTB risk
2017	Kindinger et al. [88]	127	34	Cross-sectional (16 weeks)	High risk White (65%)	V1–V3 region of 16S rRNA gene	*L. iners*	*L. crispatus*
2017	Stout et al. [90]	53	24	Longitudinal (6–13, 14–26, and 27–40 weeks)	Mixed background risk Black (69%)	V1–V3 and V3–V5 regions of 16S rRNA gene	Decreased diversity and richness between first and second trimesters	—
2017	Callahan et al. [92]	30	9	Longitudinal (self-collected, weekly)	Low White	V4 region of 16S rRNA gene	Increased diversity *G. vaginalis*	*L. crispatus*
		55	41	Longitudinal (self-collected, weekly)	High Black		Decreased relative abundance of *L. jensenii* and *L. gasseri*	*L. crispatus*

2017	Stafford et al. [98]	107	26	Cross-sectional (20–36 weeks)	Mixed risk White (85%)	V1–V3 region of 16S rRNA gene	*L. jensenii*	*L. crispatus* *L. gasseri*
2018	Frietas et al. [95]	170	46	Cross-sectional (11–16 weeks)	Low White (61%)	549–567 bp of *cpn60* gene Targeted PCR	Increased diversity and richness *Mollicutes* spp.	Increased levels of eight rare taxa (<1% total sequence reads)
2018	Brown et al. [59]	20	15	Cross-sectional (28–34 weeks) and longitudinal after PPROM	Mixed risk Mixed ethnicity	V1–V2 region of 16S rRNA gene	*Bacteroides*, *Fusobacterium*, and *Clostridium* spp.	*Lactobacillus* spp. dominance
2019	Tabatabaei et al. [97]	356	94	Cross-sectional (self-collected, 8–13 weeks)	Background risk unreported White (71%)	V4 region of 16S rRNA gene	Increased diversity for early PTB (<34 weeks) but not late PTB (34–37weeks)	*Lactobacillus* spp. dominance
2019	Brown et al. [91]	36	60	Longitudinal (12–17, 18–23, 24–29, 30–36 weeks)	Mixed risk Mixed ethnicity	V1–V2 region of 16S rRNA gene	Increased diversity and richness	*Lactobacillus* spp. dominance

(*Continued*)

TABLE 3.1 (Continued)

Year	Author [reference]	Control (*n*)	Cases (*n*)	Type and gestation of sampling	Background PTB risk (high/low), predominant ethnicity[a] (% total cohort)	Target for molecular characterization	Vaginal microbiome as a modifier of PTB risk	
							Microbiome features positively associated with increased PTB risk	**Microbiome features positively associated with decreased PTB risk**
2019	Elovitz et al. [93]	432	107	Longitudinal (16–20, 20–24, 24–28 weeks)	Mixed risk Black (75%)	V3–V4 region of 16S rRNA gene	Increased diversity and richness in white women only Seven specific taxa including *Mobiluncus*, *Sneathia*, *Megasphaera*, and *Atopobium* spp.	*Lactobacillus* spp. dominance
2019	Fettweiss et al. [94]	90	45	Longitudinal (from 18 weeks and average 7 samples/ patient)	Low Black (78%)	V1–V3 region of 16S rRNA gene	BV-associated bacteria including Lachnospiraceae BVAB1, *Prevotella* spp., *Sneathia amnii*, and TM7-H1	*L. crispatus*

[a]*Mixed ethnicity, where no single ethnic group accounts for >50% of total cohort* PPROM, *Preterm, prelabour rupture of the membranes.*

point toward potential novel explanations underpinning discrepancies in preterm birth rates between these groups.

Despite the methodological differences still existing between recent studies, their findings collectively support historical data and suggest that preterm birth risk is increased by *Lactobacillus* spp. depletion and enrichment of potential pathogenic bacteria, many of which have previously been linked to BV, whereas *Lactobacillus* spp. dominance, particularly by *L. crispatus,* confers protection against preterm birth. The identification of additional, specific rarer taxa that associate with preterm birth in some patient cohorts is of interest but requires validation in independent patient populations.

Mechanisms of vaginal microbiome-driven preterm birth

To understand how the vaginal microbiome may cause preterm birth, it is first important to consider the mechanisms involved in regulating the onset of normal birth at term. To facilitate the delivery of the mature neonate at term, the maternal gestational tissues must undergo major phenotypic changes including ripening and dilatation of the cervix, the initiation of coordinated uterine contractions, and the rupture of the fetal membranes. Each of these processes involves activation of inflammatory pathways; hence, the onset of human labor is widely considered to be an inflammatory event [99–101]. Inflammatory activation in the gestational tissues occurs before the onset of labor [102,103] and is intricately mediated via tissue-specific activation of inflammatory transcription factors such as nuclear factor kappa B (NFκB) and activator protein 1 (AP1) [104]. Once activated, these transcription factors drive the expression of pro-labor genes such as cyclooxygenase-2 (COX2), which plays a pivotal role in prostaglandin synthesis and promotion of cervical ripening and myometrial contractility [105], connexin-43 (CX43), a major gap-junction protein that promotes connectivity and contractility of uterine cells [106], metalloproteinases (e.g., MMPs-1, -3, -8, and -9) involved in fetal membrane and cervical remodeling [104,107] as well as a wide array of other pro-inflammatory cytokines and chemokines such as IL8, IL1β, and TNFα [108–110].

While the precise mechanisms regulating inflammatory activation at term are yet to be fully elucidated, one widely held view is that during late pregnancy, circulating maternal immune cells are recruited into the uterus and cervix where they promote development of the pro-inflammatory state that initiates the phenotypic changes required for labor and birth [99]. This hypothesis highlights the importance of the maternal immune system for successful pregnancy outcome and provides the context by which vaginal microbiome composition may mediate preterm birth risk in some women. During pregnancy, the maternal immune system undergoes remarkable adaption that permits tolerance of the semiallogeneic fetus yet maintains the capacity for activation of innate and adaptive immune systems in response to pathogen challenge [100]. In the lower reproductive tract, this involves carefully mediated immune cell trafficking [111,112], spatially regulated expression of pattern recognition receptors (PRRs) throughout the reproductive tract [113,114], and secretion of antimicrobial peptides such as beta-defensins, elafin, and secretory leukocyte protease inhibitor [115–117]. Domination of vaginal microbial communities by *Lactobacillus* species appears to further contribute to suppression of inflammatory activation. For example, vaginal communities dominated by *L. crispatus* are generally characterized by low levels of pro-inflammatory cytokine expression compared to those colonized by BV-associated bacteria [118–123]. Acidification of the vaginal environment by *Lactobacillus* species through lactic acid production inhibits the growth of many potential pathogenic bacteria and viruses [124–126] and has an antiinflammatory action upon vaginal epithelial cells [125].

A shift away from vaginal dominance by *Lactobacillus* spp. in pregnancy could ultimately drive the development of a premature laboring phenotype in a number of ways (see Fig. 3.1). Formation of microbial communities with increased diversity enriched for pathogenic bacteria, of which some would have the capacity to ascend the reproductive tract, would be initially recognized by the maternal innate immune system through PRRs such as Toll-like receptors, C-type lectin receptors, and nucleotide-binding oligomerization domain-like (NOD-like) receptors [32]. These receptors recognize highly conserved pathogen-associated molecular patterns (e.g., lipopolysaccharides, flagellin, lipoteichoic acid, peptidoglycan, and dsRNA) presented by microbiome stimulating signaling pathways that eventually converge to activate inflammatory transcription factors such as NFκB and AP1. Subsequent increased expression of pro-inflammatory and pro-labor genes would lead to untimely initiation of cervical ripening, membrane rupture, and uterine contractility and eventually, preterm labor and birth. Evidence that this mechanism may involve ascension of vaginal pathogens, colonization of upper gestational tissues, and direct activation of inflammation in these tissues includes vaginal infection during murine pregnancy with group B *Streptococcus* (*Streptococcus agalactiae*) [22,127] or *Escherichia coli* [17]. It is also possible that vaginal microbiome may modulate upper reproductive tract inflammatory pathways without direct colonization of the upper gestational tissues. In mice, vaginal administration of bacterial lipopolysaccharide is sufficient to induce cervical remodeling, macrophage infiltration, and subsequent preterm birth [128].

Although substantial animal model data exists in support of vaginal microbiome-driven preterm birth, in-human evidence has largely been

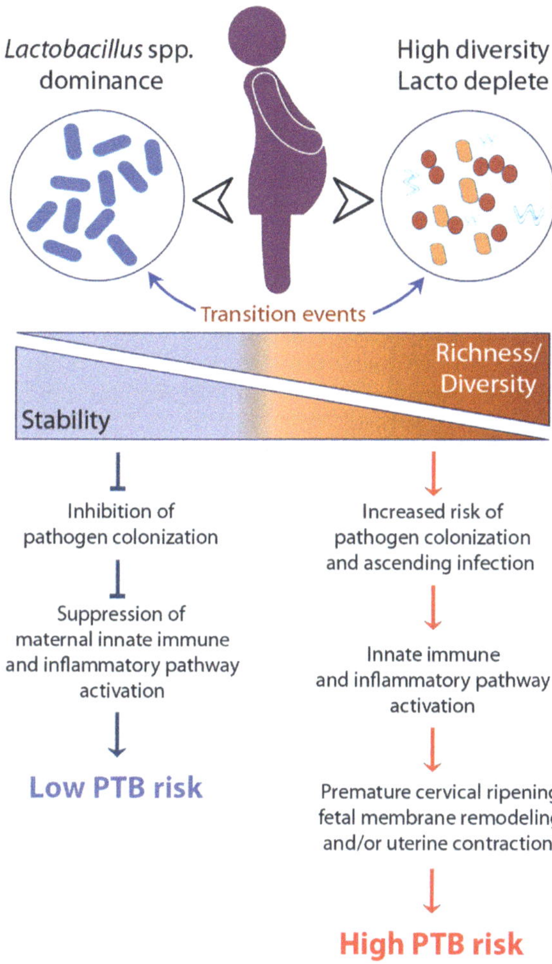

FIGURE 3.1 Pathways by which a shift in the vaginal microbiome composition from *Lactobacillus* spp. dominance toward a high diversity structure deplete of *Lactobacillus* spp. could modulate preterm birth (*PTB*) risk.

limited to associative studies. However, in a recent study performed by our laboratory, we showed that the use of multifilament suture material for cervical cerclage, where a stitch is placed around the cervix early in pregnancy to hold it closed in an attempt to delay the premature onset of labor, leads to a strong shift away from *Lactobacillus* spp. dominance toward high-diversity vaginal communities enriched for potential pathogens [88]. Importantly, we showed that this elicits a strong maternal innate immune response characterized by increased production of cervicovaginal pro-inflammatory chemokines and cytokines that leads to premature vascularization of the cervix, which is an early marker of cervical ripening and dilatation. No such changes in vaginal microbiome composition or inflammatory activation were observed in women who instead received a thin, inert, monofilament cerclage. These results provided a potential mechanistic explanation for retrospective data that reported significantly higher rates of intrauterine deaths (15% vs 5%, $P < .001$) and preterm births (28% vs 17%, $P < .001$) in women receiving a multifilament cerclage ($n = 337$) compared to women receiving a monofilament cerclage ($n = 344$). Consistent with these findings, data from the recently published integrative Human Microbiome Project also reported an increased expression of pro-inflammatory mediators in cervicovaginal fluid in women who experience a shift away from *Lactobacillus* spp. dominance toward "dysbiosis" and who experience increased rates of preterm birth [94].

Placental, oral, and gut signatures of preterm birth

Though strong evidence supports ascending vaginal microbiome as a major contributor to intrauterine infection and preterm birth, numerous studies have proposed hematogenous spread of bacteria and subsequent colonization of the placenta as a key secondary route of invasion and infection [129]. A commonly purported origin of these bacteria is the oral cavity [130], and oral-to-placental transmission has been demonstrated previously in murine models [131,132]. Taxa historically considered oral pathogens, such as *Fusobacterium nucleatum, Dialister* spp., *Prevotella* spp., and *Porphyromonas gingivalis,* have been identified in the placenta of women with periodontal disease [133,134], which itself is associated with increased risk of intrauterine infection and preterm birth [133,135,136]. In addition, a study asserting the existence and characterization of a core placental microbiome in healthy pregnancy suggested that taxonomically, the placental microbiome resembles more closely the oral, rather than the vaginal microbiome [137]. It is important to note, however, that placental samples used in this study were compared to oral and vaginal samples collected from unrelated, nonpregnant individuals enrolled in the Human Microbiome Project. A number of subsequent studies that have attempted to account for the low biomass of the placenta and high risk of confounding by contamination have failed to replicate the identification of a resident placental microbiome in healthy pregnancy [138–141]. Moreover, interventions that improve periodontal health during pregnancy do not prevent preterm birth [142–144]. Despite these findings, genuine colonization of the placenta and an accompanying molecular signature are observable in some preterm birth cases [138,145]; oral seeding seems unlikely to account for a large proportion of these cases.

An alternative source of placental seeding has been suggested to be the gastrointestinal tract [146,147]. Breakdown of the gut epithelia or active trafficking of gut bacteria sequestered in dendritic cells are the possible modes of translocation of microbes via the bloodstream to the placenta; however, emerging evidence also supports a role for the gut microbiome in the regulation of the host immune response, which may potentially influence inflammatory pathways that overlap the preterm birth phenotype [148,149]. Koren et al. reported a decrease in microbial diversity between the first and third trimesters, which was associated with increased pro-inflammatory cytokine concentrations in stool [149]. Transfer of third trimester maternal microbiome to germ-free mice induced a mild inflammatory response, with higher IL1β and lipocalin levels in stool. To date, only a couple of studies have investigated the relationship between preterm birth and gut microbial composition and have reported inconsistent findings. While DiGiulio et al. did not observe any relationship between maternal gut microbiome and preterm birth risk [83], a 2014 study by Shiozaki et al. reported significantly lower levels of *Clostridium* spp. and *Bacteroides*, as well as higher levels of Lactobacillales in stool of women subsequently experiencing preterm birth compared to term controls. While further research is clearly required, it seems plausible that maternal gut microbiome remodeling may fine-tune immunomodulatory and inflammatory pathways important for the onset of parturition [150].

Potential contribution of the virome and mycobiome to preterm birth

The majority of studies investigating microbial signatures and their association with preterm birth has been focused toward maternal bacteria. However, it is highly likely that other microbial constituents of the microbiome, that is, viruses and fungi, collectively referred as the virome and the mycobiome, respectively, play an important role in mediating preterm risk. Viral infections such as human immunodeficiency virus [151,152], herpes simplex virus [153,154], and Hepatitis B [155] are associated with preterm labor, and rates of chorioamnionitis are higher in preterm adenovirus-positive placentas compared to preterm adenovirus-negative placentas and term adenovirus-positive placentas [156]. There is also an increasing evidence that interaction between viral- and bacterial-driven maternal immune responses may accentuate inflammatory pathway activation leading to preterm birth. For example, in mice, priming of the maternal immune system with viral products or viral infection before bacterial inflammatory challenge leads to a significantly higher rate of preterm birth compared to viral or bacterial infection alone [157–160]. To date only one study has examined the relationship between the vaginal virome and preterm birth risk and reported an association between higher viral richness and preterm delivery, with highest risk observed in women with both high bacterial and viral diversity [161]. This points toward interactions between the viral and bacterial components of the vaginal microbiome and potential synergistic mediation of inflammatory pathways. Further evidence for such interactions includes the association between high-diversity cervicovaginal bacterial communities and human papillomavirus (HPV) infection [162] as well as increased risk of HIV infection [163,164]. The vaginal microbiome may also play a more distant role in the etiology of preterm labor through HPV infection-mediated cervical intraepithelial neoplasia (CIN) [165]. CIN is treated surgically through removal of the transformation zone of the cervix, either by cold knife excision, or more commonly by large loop diathermy excision. These surgical procedures, especially if they remove a large amount of cervical tissue, can cause preterm birth through creation of cervical incompetence.

There is also a growing appreciation of the role that the mycobiome may have on shaping vaginal bacterial communities and vice versa [166]. For example, women with symptomatic vulvovaginal candidiasis have been reported to have lower levels of *Lactobacillus* species [167,168], whereas short-chain fatty acids and lactate produced by *Lactobacillus* species inhibit the switch of nonpathogenic *Candida albicans* to pathogenic hyphae [169]. Vulvovaginal candidiasis is common in pregnancy, affecting up to 20% of all women, and while early studies did not detect any association between candidiasis and preterm birth [170,171], more recent retrospective analyses of women receiving treatment for candidiasis have suggested a significant reduction in preterm rates with treatment [172,173]. A recent open label pilot randomized controlled trial of clotrimazole also suggested a tendency toward reduction in preterm birth rates in treated women with asymptomatic candidiasis [174]. Collectively, these results indicate that further research into the dynamic relationship between the fungal, viral, and bacterial components of the vaginal microbiome in the context of preterm birth is warranted.

Conclusion

There is a strong evidence to support the concept of vaginal microbial signatures that are associated with increased risk of preterm birth. Depletion in *Lactobacillus* spp. leading to high diversity communities enriched for potentially pathogenic bacteria appears to drive local immune responses that lead to increased inflammatory pathway activation that can contribute to a host–microbe interaction etiology of preterm birth. This has led to a number of major research efforts designed to detect specific microbial signatures that can be broadly used for preterm birth prediction. However, the majority of these studies have focused on the application of DNA sequencing-based techniques, which while providing useful information regarding composition and relative abundance, fail to capture the functional interactions occurring in between the microbiome and the maternal host, which ultimately determine pregnancy outcome. Those studies that have examined maternal response through assessment of local innate immune activation have highlighted the fact that host–microbiome interactions are often patient group specific and can be even more highly individualized. Thus methods permitting simultaneous and rapid assessment of both bacterial composition and host responses may ultimately prove critical for exploiting microbial signatures for the prediction of preterm birth as well as the direction of personalized treatment strategies that could be integrated into patient care during pregnancy to optimize health outcomes. The development of strategies to encourage an "optimal" vaginal microbiome has the potential to improve multiple aspects of women's health, as well as reducing the rates of preterm birth, but will require better understanding of factors shaping community structure and stability.

References

[1] Liu L, et al. Global, regional, and national causes of child mortality in 2000-13, with projections to inform post-2015 priorities: an updated systematic analysis. Lancet 2015;385(9966):430–40.

[2] Agrawal V, Hirsch E. Intrauterine infection and preterm labor. Semin Fetal Neonatal Med 2012;17(1):12–19.

[3] Oxorn H. The changing aspects of pneumonia complicating pregnancy. Am J Obstet Gynecol 1955;70(5):1057–63.

[4] Finland M, Dublin TD. Pneumococcic pneumonias complicating pregnancy and the puerperium. JAMA 1939;112(11):1027–32.

[5] Wing ES, Troppoli DV. The intrauterine transmission of typhoid. JAMA 1930;95 (6):405–6.

[6] Stevenson CS, et al. Treatment of typhoid in pregnancy with chloramphenicol (chloromycetin). J Am Med Assoc 1951;146(13):1190–2.

[7] Gilles HM, et al. Malaria, anaemia and pregnancy. Ann Trop Med Parasitol 1969;63 (2):245–63.

[8] Hillier SL, et al. Microbiologic causes and neonatal outcomes associated with chorioamnion infection. Am J Obstet Gynecol 1991;165(4 Pt 1):955–61.

[9] Russell P. Inflammatory lesions of the human placenta. III. The histopathology of villitis of unknown aetiology. Placenta 1980;1(3):227–44.

[10] Blanc WA. Pathways of fetal and early neonatal infection. Viral placentitis, bacterial and fungal chorioamnionitis. J Pediatr 1961;59:473–96.

[11] Romero R, et al. The role of infection in preterm labour and delivery. Paediatr Perinat Epidemiol 2001;15(Suppl. 2):41–56.
[12] Romero R, et al. Infection and labor. V. Prevalence, microbiology, and clinical significance of intraamniotic infection in women with preterm labor and intact membranes. Am J Obstet Gynecol 1989;161(3):817–24.
[13] Goldenberg RL, et al. Epidemiology and causes of preterm birth. Lancet 2008;371 (9606):75–84.
[14] Benirschke K. Routes and types of infection in the fetus and the newborn. AMA J Dis Child 1960;99:714–21.
[15] Jones HE, et al. Differing prevalence and diversity of bacterial species in fetal membranes from very preterm and term labor. PLoS One 2009;4(12):e8205.
[16] Romero R, et al. Infection and labor. VI. Prevalence, microbiology, and clinical significance of intraamniotic infection in twin gestations with preterm labor. Am J Obstet Gynecol 1990;163(3):757–61.
[17] Suff N, et al. Ascending vaginal infection using bioluminescent bacteria evokes intrauterine inflammation, preterm birth, and neonatal brain injury in pregnant mice. Am J Pathol 2018;188(10):2164–76.
[18] Akgul Y, et al. Hyaluronan in cervical epithelia protects against infection-mediated preterm birth. J Clin Invest 2014;124(12):5481–9.
[19] Randis TM, et al. Group B *Streptococcus* beta-hemolysin/cytolysin breaches maternal-fetal barriers to cause preterm birth and intrauterine fetal demise in vivo. J Infect Dis 2014;210(2):265–73.
[20] Racicot K, et al. Viral infection of the pregnant cervix predisposes to ascending bacterial infection. J Immunol 2013;191(2):934–41.
[21] Vornhagen J, et al. Bacterial hyaluronidase promotes ascending GBS infection and preterm birth. MBio 2016;7(3).
[22] Vornhagen J, et al. Group B *Streptococcus* exploits vaginal epithelial exfoliation for ascending infection. J Clin Invest 2018;128(5):1985–99.
[23] Chamberlain G. British maternal mortality in the 19th and early 20th centuries. J R Soc Med 2006;99(11):559–63.
[24] Semmelweis I. The etiology, concept and prophylaxis of childbed ever (Die Ätiologie, der Begriff und die Prophylaxe des Kindbettfiebers). Madison, WI: The University of Wisconsin Press; 1983.
[25] Colebrook L, Purdie AW. Treatment of 106 cases of puerperal fever by sulphanilamide. Lancet 1937;230(5961):1237–42.
[26] Du Plessis EM, Dicks LM. Evaluation of random amplified polymorphic DNA (RAPD)-PCR as a method to differentiate *Lactobacillus acidophilus*, *Lactobacillus crispatus*, *Lactobacillus amylovorus*, *Lactobacillus gallinarum*, *Lactobacillus gasseri*, and *Lactobacillus johnsonii*. Curr Microbiol 1995;31(2):114–18.
[27] Thomas S. Döderlein's bacillus: *Lactobacillus acidophilus*. J Infect Dis 1928;43 (3):218–27.
[28] Döderlein A. Das Scheidensekret und seine Bedeutung für das Puerperalfieber. Leipzig: Verlag von Eduard Besold; 1892.
[29] Schroder R. Zur pathogenese und klinik des vaginalen fluors. Zentralb Gynakol 1921;38:1350–61.
[30] Peebles K, et al. High global burden and costs of bacterial vaginosis: a systematic review and meta-analysis. Sex Transm Dis 2019;46(5):304–11.
[31] van de Wijgert J, Jespers V. The global health impact of vaginal dysbiosis. Res Microbiol 2017;168(9–10):859–64.
[32] Hillier SL, et al. The normal vaginal flora, H2O2-producing lactobacilli, and bacterial vaginosis in pregnant women. Clin Infect Dis 1993;16(Suppl. 4):S273–81.

[33] Amsel R, et al. Nonspecific vaginitis. Diagnostic criteria and microbial and epidemiologic associations. Am J Med 1983;74(1):14–22.
[34] Klebanoff MA, et al. Vulvovaginal symptoms in women with bacterial vaginosis. Obstet Gynecol 2004;104(2):267–72.
[35] Nugent RP, Krohn MA, Hillier SL. Reliability of diagnosing bacterial vaginosis is improved by a standardized method of gram stain interpretation. J Clin Microbiol 1991;29(2):297–301.
[36] Gravett MG, et al. Independent associations of bacterial vaginosis and *Chlamydia trachomatis* infection with adverse pregnancy outcome. JAMA 1986;256(14):1899–903.
[37] Hay PE, et al. Abnormal bacterial colonisation of the genital tract and subsequent preterm delivery and late miscarriage. BMJ 1994;308(6924):295–8.
[38] Hillier SL, et al. Association between bacterial vaginosis and preterm delivery of a low-birth-weight infant. The vaginal infections and Prematurity Study Group. N Engl J Med 1995;333(26):1737–42.
[39] McGregor JA, French JI, Seo K. Premature rupture of membranes and bacterial vaginosis. Am J Obstet Gynecol 1993;169(2 Pt 2):463–6.
[40] Gibbs RS. Chorioamnionitis and bacterial vaginosis. Am J Obstet Gynecol 1993;169(2 Pt 2):460–2.
[41] Martius J, Eschenbach DA. The role of bacterial vaginosis as a cause of amniotic fluid infection, chorioamnionitis and prematurity--a review. Arch Gynecol Obstet 1990;247(1):1–13.
[42] Takei H, Ruiz B. Shift in vaginal flora (bacterial vaginosis) and the frequency of chorioamnionitis in a high-risk population. Acta Cytol 2006;50(4):410–14.
[43] Manns-James L. Bacterial vaginosis and preterm birth. J Midwifery Womens Health 2011;56(6):575–83.
[44] Leitich H, Kiss H. Asymptomatic bacterial vaginosis and intermediate flora as risk factors for adverse pregnancy outcome. Best Pract Res Clin Obstet Gynaecol 2007;21(3):375–90.
[45] Leitich H, et al. Bacterial vaginosis as a risk factor for preterm delivery: a meta-analysis. Am J Obstet Gynecol 2003;189(1):139–47.
[46] Macones GA, et al. A polymorphism in the promoter region of TNF and bacterial vaginosis: preliminary evidence of gene-environment interaction in the etiology of spontaneous preterm birth. Am J Obstet Gynecol 2004;190(6):1504–8. discussion 3A.
[47] Gomez LM, et al. Evidence of a gene-environment interaction that predisposes to spontaneous preterm birth: a role for asymptomatic bacterial vaginosis and DNA variants in genes that control the inflammatory response. Am J Obstet Gynecol 2010;202(4):386. e1-6.
[48] Lamont RF, et al. Treatment of abnormal vaginal flora in early pregnancy with clindamycin for the prevention of spontaneous preterm birth: a systematic review and metaanalysis. Am J Obstet Gynecol 2011;205(3):177–90.
[49] Brocklehurst P, et al. Antibiotics for treating bacterial vaginosis in pregnancy. Cochrane Database Syst Rev 2013;(1):CD000262.
[50] Lamont RF. Spontaneous preterm labour that leads to preterm birth: An update and personal reflection. Placenta 2019;79:21–9.
[51] Lamont RF, et al. The treatment of bacterial vaginosis in pregnancy with clindamycin to reduce the risk of infection-related preterm birth: a response to the Danish Society of Obstetrics and Gynecology guideline group's clinical recommendations. Acta Obstet Gynecol Scand 2017;96(2):139–43.
[52] Subtil D, et al. Early clindamycin for bacterial vaginosis in pregnancy (PREMEVA): a multicentre, double-blind, randomised controlled trial. Lancet 2018;392(10160):2171–9.

[53] Klebanoff MA, Brotman RM. Treatment of bacterial vaginosis to prevent preterm birth. Lancet 2018;392(10160):2141–2.
[54] Lamont RF. Advances in the prevention of infection-related preterm birth. Front Immunol 2015;6:566.
[55] Morency AM, Bujold E. The effect of second-trimester antibiotic therapy on the rate of preterm birth. J Obstet Gynaecol Can 2007;29(1):35–44.
[56] Migale R, et al. Specific lipopolysaccharide serotypes induce differential maternal and neonatal inflammatory responses in a murine model of preterm labor. Am J Pathol 2015;185(9):2390–401.
[57] MacIntyre DA, et al. Activator protein 1 is a key terminal mediator of inflammation-induced preterm labor in mice. FASEB J 2014;28(5):2358–68.
[58] MacIntyre DA, et al. Prevention of preterm labour via the modulation of inflammatory pathways. J Matern Fetal Neonatal Med 2012;25(Suppl. 1:):17–20.
[59] Brown RG, et al. Vaginal dysbiosis increases risk of preterm fetal membrane rupture, neonatal sepsis and is exacerbated by erythromycin. BMC Med 2018;16(1):9.
[60] Ferris MJ, et al. Association of *Atopobium vaginae,* a recently described metronidazole resistant anaerobe, with bacterial vaginosis. BMC Infect Dis 2004;4:5.
[61] Bostwick DG, et al. Antimicrobial resistance genes and modelling of treatment failure in bacterial vaginosis: clinical study of 289 symptomatic women. J Med Microbiol 2016;65(5):377–86.
[62] Beigi RH, et al. Antimicrobial resistance associated with the treatment of bacterial vaginosis. Am J Obstet Gynecol 2004;191(4):1124–9.
[63] Burton JP, Reid G. Evaluation of the bacterial vaginal flora of 20 postmenopausal women by direct (Nugent score) and molecular (polymerase chain reaction and denaturing gradient gel electrophoresis) techniques. J Infect Dis 2002;186(12):1770–80.
[64] Srinivasan S, Fredricks DN. The human vaginal bacterial biota and bacterial vaginosis. Interdiscip Perspect Infect Dis 2008;2008:750479.
[65] van de Wijgert JH, et al. The vaginal microbiota: what have we learned after a decade of molecular characterization? PLoS One 2014;9(8):e105998.
[66] Fredricks DN, Fiedler TL, Marrazzo JM. Molecular identification of bacteria associated with bacterial vaginosis. N Engl J Med 2005;353(18):1899–911.
[67] Ravel J, et al. Vaginal microbiome of reproductive-age women. Proc Natl Acad Sci U S A 2011;108.
[68] Zhou X, et al. The vaginal bacterial communities of Japanese women resemble those of women in other racial groups. FEMS Immunol Med Microbiol 2010;58(2):169–81.
[69] Borgdorff H, et al. The association between ethnicity and vaginal microbiota composition in Amsterdam, the Netherlands. PLoS One 2017;12(7):e0181135.
[70] Fettweis JM, et al. Differences in vaginal microbiome in African American women versus women of European ancestry. Microbiology 2014;160(Pt 10):2272–82.
[71] Gajer P, Brotman RM, Bai G, Sakamoto J, Schütte UME, Zhong X, et al. Temporal dynamics of the human vaginal microbiota. Sci Transl Med 2012;4:132ra52.
[72] Martin DH, Marrazzo JM. The vaginal microbiome: current understanding and future directions. J Infect Dis 2016;214(Suppl. 1):S36–41.
[73] MacIntyre DA, Sykes L, Bennett PR. The human female urogenital microbiome: complexity in normality. Emerg Top Life Sci 2017;1(4):363–72.
[74] Fettweis JM, et al. A new era of the vaginal microbiome: advances using next-generation sequencing. Chem Biodivers 2012;9(5):965–76.
[75] Aagaard K, et al. A metagenomic approach to characterization of the vaginal microbiome signature in pregnancy. PLoS One 2012;7(6):e36466.
[76] MacIntyre DA, et al. The vaginal microbiome during pregnancy and the postpartum period in a European population. Sci Rep 2015;5:8988.

[77] Romero R, et al. The composition and stability of the vaginal microbiota of normal pregnant women is different from that of non-pregnant women. Microbiome 2014;2 (1):4.

[78] Roy EJ, Mackay R. The concentration of oestrogens in blood during pregnancy. J Obstet Gynaecol Br Emp 1962;69:13–17.

[79] Roy EJ, MacKay R. The concentration of oestrogens in blood during pregnancy. BJOG 1962;69:13–17.

[80] Spear GT, et al. Human alpha-amylase present in lower-genital-tract mucosal fluid processes glycogen to support vaginal colonization by Lactobacillus. J Infect Dis 2014;210(7):1019–28.

[81] Ayre WB. The glycogen-estrogen relationship in the vaginal tract. J Clin Endocrinol Metab 1951;11(1):103–10.

[82] Nott PN, et al. Hormonal changes and mood in the puerperium. Br J Psychiatry 1976;128:379–83.

[83] DiGiulio DB, et al. Temporal and spatial variation of the human microbiota during pregnancy. Proc Natl Acad Sci U S A 2015;112(35):11060–5.

[84] Doyle R, et al. A *Lactobacillus*-deficient vaginal microbiota dominate post-partum women in rural Malawi. Appl Environ Microbiol 2018;84.

[85] Serrano MG, et al. Racioethnic diversity in the dynamics of the vaginal microbiome during pregnancy. Nat Med 2019;25(6):1001–11.

[86] Peelen MJ, et al. The influence of the vaginal microbiota on preterm birth: A systematic review and recommendations for a minimum dataset for future research. Placenta 2019;79:30–9.

[87] Hyman RW, et al. Diversity of the vaginal microbiome correlates with preterm birth. Reprod Sci 2014;21(1):32–40.

[88] Kindinger LM, et al. Relationship between vaginal microbial dysbiosis, inflammation, and pregnancy outcomes in cervical cerclage. Sci Transl Med 2016;8(350):350ra102.

[89] Romero R, et al. The vaginal microbiota of pregnant women who subsequently have spontaneous preterm labor and delivery and those with a normal delivery at term. Microbiome 2014;2:18.

[90] Stout MJ, et al. Early pregnancy vaginal microbiome trends and preterm birth. Am J Obstet Gynecol 2017;217(3):356.e1–356.e18.

[91] Brown RG, et al. Establishment of vaginal microbiota composition in early pregnancy and its association with subsequent preterm prelabor rupture of the fetal membranes. Transl Res 2019;207:30–43.

[92] Callahan BJ, et al. Replication and refinement of a vaginal microbial signature of preterm birth in two racially distinct cohorts of US women. Proc Natl Acad Sci U S A 2017;114(37):9966–71.

[93] Elovitz MA, et al. Cervicovaginal microbiota and local immune response modulate the risk of spontaneous preterm delivery. Nat Commun 2019;10(1):1305.

[94] Fettweis JM, et al. The vaginal microbiome and preterm birth. Nat Med 2019;25 (6):1012–21.

[95] Freitas AC, et al. Increased richness and diversity of the vaginal microbiota and spontaneous preterm birth. Microbiome 2018;6(1):117.

[96] Kindinger LM, et al. The interaction between vaginal microbiota, cervical length, and vaginal progesterone treatment for preterm birth risk. Microbiome 2017;5(1):6.

[97] Tabatabaei N, et al. Vaginal microbiome in early pregnancy and subsequent risk of spontaneous preterm birth: a case-control study. BJOG 2019;126(3):349–58.

[98] Stafford GP, et al. Spontaneous preterm birth is associated with differential expression of vaginal metabolites by *Lactobacilli*-dominated microflora. Front Physiol 2017;8:615.

[99] Gomez-Lopez N, et al. Immune cells in term and preterm labor. Cell Mol Immunol 2014;11(6):571–81.
[100] Sykes L, et al. The Th1:th2 dichotomy of pregnancy and preterm labour. Mediators Inflamm 2012;2012:967629.
[101] Keelan JA. Intrauterine inflammatory activation, functional progesterone withdrawal, and the timing of term and preterm birth. J Reprod Immunol 2018;125:89–99.
[102] Lim S, et al. Nuclear factor kappa B activation occurs in the amnion prior to labour onset and modulates the expression of numerous labour associated genes. PLoS One 2012;7(4):e34707.
[103] Migale R, et al. Modeling hormonal and inflammatory contributions to preterm and term labor using uterine temporal transcriptomics. BMC Med 2016;14(1):86.
[104] Lindstrom TM, Bennett PR. The role of nuclear factor kappa B in human labour. Reproduction 2005;130(5):569–81.
[105] Allport VC, et al. Human labour is associated with nuclear factor-kappaB activity which mediates cyclo-oxygenase-2 expression and is involved with the 'functional progesterone withdrawal'. Mol Hum Reprod 2001;7(6):581–6.
[106] Piersanti M, Lye SJ. Increase in messenger ribonucleic acid encoding the myometrial gap junction protein, connexin-43, requires protein synthesis and is associated with increased expression of the activator protein-1, c-fos. Endocrinology 1995;136 (8):3571–8.
[107] Osmers R, et al. Origin of cervical collagenase during parturition. Am J Obstet Gynecol 1992;166(5):1455–60.
[108] Khanjani S, et al. NFkappaB and AP-1 drive human myometrial IL8 expression. Mediators Inflamm 2012;2012:504952.
[109] Belt AR, et al. The nuclear transcription factor NF-kappaB mediates interleukin-1beta-induced expression of cyclooxygenase-2 in human myometrial cells. Am J Obstet Gynecol 1999;181(2):359–66.
[110] Kniss DA, et al. Blockade NF-kappaB activation prohibits TNF-alpha-induced cyclooxygenase-2 gene expression in ED27 trophoblast-like cells. Placenta 2001;22 (1):80–9.
[111] Hafner LM, Cunningham K, Beagley KW. Ovarian steroid hormones: effects on immune responses and *Chlamydia trachomatis* infections of the female genital tract. Mucosal Immunol 2013;6(5):859–75.
[112] Yeaman GR, et al. CD8+ T cells in human uterine endometrial lymphoid aggregates: evidence for accumulation of cells by trafficking. Immunology 2001;102 (4):434–40.
[113] Aflatoonian R, Fazeli A. Toll-like receptors in female reproductive tract and their menstrual cycle dependent expression. J Reprod Immunol 2008;77(1):7–13.
[114] Hart KM, et al. Functional expression of pattern recognition receptors in tissues of the human female reproductive tract. J Reprod Immunol 2009;80(1–2):33–40.
[115] King AE, Critchley HO, Kelly RW. Innate immune defences in the human endometrium. Reprod Biol Endocrinol 2003;1:116.
[116] King AE, et al. Expression of secretory leukocyte protease inhibitor and elafin in human fallopian tube and in an in-vitro model of *Chlamydia trachomatis* infection. Hum Reprod 2009;24(3):679–86.
[117] Wira C, et al. Effect of the menstrual cycle on immunological parameters in the human female reproductive tract. J Acquir Immune Defic Syndr 2005;38(Suppl. 1): S34–6.
[118] Anahtar MN, et al. Cervicovaginal bacteria are a major modulator of host inflammatory responses in the female genital tract. Immunity 2015;42(5):965–76.

[119] Hedges SR, et al. Local and systemic cytokine levels in relation to changes in vaginal flora. J Infect Dis 2006;193(4):556–62.
[120] Mattsby-Baltzer I, et al. IL-1 beta, IL-6, TNF alpha, fetal fibronectin, and endotoxin in the lower genital tract of pregnant women with bacterial vaginosis. Acta Obstet Gynecol Scand 1998;77(7):701–6.
[121] Platz-Christensen JJ, et al. Endotoxin and interleukin-1α in the cervical mucus and vaginal fluid of pregnant women with bacterial vaginosis. Am J Obstet Gynecol 1993;169(5):1161–6.
[122] Spandorfer SD, et al. Relationship of abnormal vaginal flora, proinflammatory cytokines and idiopathic infertility in women undergoing IVF. J Reprod Med 2001;46 (9):806–10.
[123] Wennerholm UB, et al. Interleukin-1 alpha, interleukin-6 and interleukin-8 in cervico/vaginal secretion for screening of preterm birth in twin gestation. Acta Obstet Gynecol Scand 1998;77(5):508–14.
[124] Becher N, et al. The cervical mucus plug: structured review of the literature. Acta Obstet Gynecol Scand 2009;88(5):502–13.
[125] Hearps AC, et al. Vaginal lactic acid elicits an anti-inflammatory response from human cervicovaginal epithelial cells and inhibits production of pro-inflammatory mediators associated with HIV acquisition. Mucosal Immunol 2017;10(6):1480–90.
[126] Witkin SS, Linhares IM. HIV inhibition by lactobacilli: easier in a test tube than in real life. MBio 2015;6(5). e01485-15.
[127] Seale AC, et al. Estimates of the burden of group B streptococcal disease worldwide for pregnant women, stillbirths, and children. Clin Infect Dis 2017;65(Suppl. 2): S200–19.
[128] Gonzalez JM, et al. Complement activation triggers metalloproteinases release inducing cervical remodeling and preterm birth in mice. Am J Pathol 2011;179 (2):838–49.
[129] Mendz GL, Kaakoush NO, Quinlivan JA. Bacterial aetiological agents of intra-amniotic infections and preterm birth in pregnant women. Front Cell Infect Microbiol 2013;3:58.
[130] Cobb CM, et al. The oral microbiome and adverse pregnancy outcomes. Int J Womens Health 2017;9:551–9.
[131] Han YW, et al. *Fusobacterium nucleatum* induces premature and term stillbirths in pregnant mice: implication of oral bacteria in preterm birth. Infect Immun 2004;72 (4):2272–9.
[132] Fardini Y, et al. Transmission of diverse oral bacteria to murine placenta: evidence for the oral microbiome as a potential source of intrauterine infection. Infect Immun 2010;78(4):1789–96.
[133] Chaim W, Mazor M. Intraamniotic infection with fusobacteria. Arch Gynecol Obstet 1992;251(1):1–7.
[134] Barak S, et al. Evidence of periopathogenic microorganisms in placentas of women with preeclampsia. J Periodontol 2007;78(4):670–6.
[135] Offenbacher S, et al. Periodontal infection as a possible risk factor for preterm low birth weight. J Periodontol 1996;67(10 Suppl.):1103–13.
[136] Pattanashetti JI, Nagathan VM, Rao SM. Evaluation of periodontitis as a risk for preterm birth among preeclamptic and non-preeclamptic pregnant women - a case control study. J Clin Diagn Res 2013;7(8):1776–8.
[137] Aagaard K, et al. The placenta harbors a unique microbiome. Sci Transl Med 2014;6 (237):237ra65.
[138] Leiby JS, et al. Lack of detection of a human placenta microbiome in samples from preterm and term deliveries. Microbiome 2018;6(1):196.

[139] Lauder AP, et al. Comparison of placenta samples with contamination controls does not provide evidence for a distinct placenta microbiota. Microbiome 2016;4(1):29.
[140] Theis KR, et al. Does the human placenta delivered at term have a microbiota? Results of cultivation, quantitative real-time PCR, 16S rRNA gene sequencing, and metagenomics. Am J Obstet Gynecol 2019;220(3):267. e1-39.
[141] de Goffau MC, et al. Human placenta has no microbiome but can contain potential pathogens. Nature 2019;572(7769):329–34.
[142] Michalowicz BS, et al. Treatment of periodontal disease and the risk of preterm birth. N Engl J Med 2006;355(18):1885–94.
[143] Offenbacher S, et al. Effects of periodontal therapy on rate of preterm delivery: a randomized controlled trial. Obstet Gynecol 2009;114(3):551–9.
[144] Macones GA, et al. Treatment of localized periodontal disease in pregnancy does not reduce the occurrence of preterm birth: results from the Periodontal Infections and Prematurity Study (PIPS). Am J Obstet Gynecol 2010;202(2):147. e1-8.
[145] Leon LJ, et al. Enrichment of clinically relevant organisms in spontaneous preterm-delivered placentas and reagent contamination across all clinical groups in a large pregnancy cohort in the United Kingdom. Appl Environ Microbiol 2018;84(14).
[146] Perez PF, et al. Bacterial imprinting of the neonatal immune system: lessons from maternal cells? Pediatrics 2007;119(3):e724–32.
[147] Gomez-Arango LF, et al. Contributions of the maternal oral and gut microbiome to placental microbial colonization in overweight and obese pregnant women. Sci Rep 2017;7(1):2860.
[148] Round JL, Mazmanian SK. The gut microbiota shapes intestinal immune responses during health and disease. Nat Rev Immunol 2009;9(5):313–23.
[149] Koren O, et al. Host remodeling of the gut microbiome and metabolic changes during pregnancy. Cell 2012;150(3):470–80.
[150] Mor G, Aldo P, Alvero AB. The unique immunological and microbial aspects of pregnancy. Nat Rev Immunol 2017;17(8):469–82.
[151] Hardy JM, et al. The effect of Asian influenza on the outcome of pregnancy, Baltimore, 1957-1958. Am J Public Health Nations Health 1961;51:1182–8.
[152] Harris JW. Influenza occurring in pregnant women a statistical study of thirteen hundred and fifty cases. JAMA. 1919;72(14):978–80.
[153] Brown ZA, et al. Asymptomatic maternal shedding of herpes simplex virus at the onset of labor: relationship to preterm labor. Obstet Gynecol 1996;87(4):483–8.
[154] Srinivas SK, et al. Placental inflammation and viral infection are implicated in second trimester pregnancy loss. Am J Obstet Gynecol 2006;195(3):797–802.
[155] Sirilert S, et al. Pregnancy outcomes among chronic carriers of hepatitis B virus. Int J Gynaecol Obstet 2014;126(2):106–10.
[156] Tsekoura EA, et al. Adenovirus genome in the placenta: association with histological chorioamnionitis and preterm birth. J Med Virol 2010;82(8):1379–83.
[157] Cardenas I, et al. Viral infection of the placenta leads to fetal inflammation and sensitization to bacterial products predisposing to preterm labor. J Immunol 2010;185 (2):1248–57.
[158] Cardenas I, et al. Placental viral infection sensitizes to endotoxin-induced pre-term labor: a double hit hypothesis. Am J Reprod Immunol 2011;65(2):110–17.
[159] Cappelletti M, et al. Type I interferons regulate susceptibility to inflammation-induced preterm birth. JCI Insight 2017;2(5):e91288.
[160] Cross SN, et al. Viral infection sensitizes human fetal membranes to bacterial lipopolysaccharide by MERTK inhibition and inflammasome activation. J Immunol 2017;199(8):2885–95.

[161] Wylie KM, et al. The vaginal eukaryotic DNA virome and preterm birth. Am J Obstet Gynecol 2018;219(2):189. e1-19.
[162] Mitra A, et al. The vaginal microbiota, human papillomavirus infection and cervical intraepithelial neoplasia: what do we know and where are we going next? Microbiome 2016;4(1):58.
[163] Martin HL, et al. Vaginal lactobacilli, microbial flora, and risk of human immunodeficiency virus type 1 and sexually transmitted disease acquisition. J Infect Dis 1999;180(6):1863–8.
[164] Taha TE, et al. Bacterial vaginosis and disturbances of vaginal flora: association with increased acquisition of HIV. AIDS 1998;12(13):1699–706.
[165] Kyrgiou M, et al. Adverse obstetric outcomes after local treatment for cervical preinvasive and early invasive disease according to cone depth: systematic review and meta-analysis. BMJ 2016;354:i3633.
[166] Bradford LL, Ravel J. The vaginal mycobiome: a contemporary perspective on fungi in women's health and diseases. Virulence 2017;8(3):342–51.
[167] Peeters F, et al. Observations on candidal vaginitis. Vaginal pH, microbiology, and cytology. Am J Obstet Gynecol 1972;112(1):80–6.
[168] Auger P, Joly J. Microbial flora associated with *Candida albicans* vulvovaginitis. Obstet Gynecol 1980;55(3):397–401.
[169] Noverr MC, Huffnagle GB. Regulation of *Candida albicans* morphogenesis by fatty acid metabolites. Infect Immun 2004;72(11):6206–10.
[170] McGregor JA, et al. Prevention of premature birth by screening and treatment for common genital tract infections: results of a prospective controlled evaluation. Am J Obstet Gynecol 1995;173(1):157–67.
[171] Cotch MF, et al. Epidemiology and outcomes associated with moderate to heavy Candida colonization during pregnancy. Vaginal Infections and Prematurity Study Group. Am J Obstet Gynecol 1998;178(2):374–80.
[172] Czeizel AE, Rockenbauer M. A lower rate of preterm birth after clotrimazole therapy during pregnancy. Paediatr Perinat Epidemiol 1999;13(1):58–64.
[173] Czeizel AE, Fladung B, Vargha P. Preterm birth reduction after clotrimazole treatment during pregnancy. Eur J Obstet Gynecol Reprod Biol 2004;116(2):157–63.
[174] Roberts CL, et al. Treatment of asymptomatic vaginal candidiasis in pregnancy to prevent preterm birth: an open-label pilot randomized controlled trial. BMC Pregnancy Childbirth 2011;11:18.

CHAPTER

4

Prenatal origins of the infant gut microbiome

Michael D. Jochum[1], *Maxim D. Seferovic*[1] and *Kjersti M. Aagaard*[1,2,3,4]

[1]Division of Maternal-Fetal Medicine, Department of Obstetrics and Gynecology, Baylor College of Medicine, Texas Children's Hospital, Houston, TX, United States, [2]Center for Microbiome and Metagenomics Research, Baylor College of Medicine, Texas Children's Hospital, Houston, TX, United States, [3]Department of Molecular and Human Genetics, Baylor College of Medicine, Texas Children's Hospital, Houston, TX, United States, [4]Department of Molecular and Cell Biology, Baylor College of Medicine, Texas Children's Hospital, Houston, TX, United States

Key points

1. The intrauterine environment retains a low abundance, low-biomass microbiome that may be important for establishing tolerance to commensal organisms in utero.
2. Multiple perinatal factors (including the maternal diet) demonstrate a lasting impact on the developing human microbiome.
3. Cultivating these and other low abundance, low-biomass communities have a unique set of technical and biological challenges. Metagenomic sequencing remains an important mainstay of the current science.

Abbreviations

HMO Human milk oligosaccharide
NICU Neonatal intensive care unit

The Human Microbiome in Early Life
DOI: https://doi.org/10.1016/B978-0-12-818097-6.00004-3

When considered through an immunologist's lens, why would we presume that fetal development ought to occur in a "sterile" setting?

In their early descriptions of immunological tolerance in 1949, Burnet and Fenner [1] put forward the notion that self/nonself-discriminations were fundamental cornerstones of a functional mature immune system. Such discriminations, by their very nature, would occur in fetal and early life and inform our immune repertoire in such a manner as to enable a lifetime of immune tolerance and pathogen recognition. Although subsequent giants in the field (such as Medawar, Lederberg, Matzinger, Marrack and Kappler, von Boehmer, and Nossal, Gathings, and Unanue) would provide the supportive experimental evidence, Burnet's legacy of generating a testable immunological tolerance hypothesis stands as one of the greatest contributions to translational medicine in the last hundred years. The recent onslaught of metagenomics and microbiome science [2,3] has brought us to another crossroads of our understandings how mammalian immune systems differentiate to allow maintenance of functional integrity across a lifetime. How do we come to not only tolerate each of our commensal microbiome friends in their distinct ecological human body niches, but recognize their not too distant cousins as potential foes? How can a commensal pathobiont strain in one body niche (for example, Group B *Streptococcus* in the vagina of a pregnant mother about to give birth) become an invasive pathogen in another site at another life stage (neonatal invasive Group B streptococcal disease)? Put another way, in relatively short order following birth, the infant will populate.

Although parturition was traditionally assumed the initial contact point during which neonates are exposed to their mother's microbes [4], emerging evidence indicates that this is unlikely to be true. The presence of microbes and low-biomass microbial communities within the intrauterine space—the uterine decidua, the placenta, and the amnion and chorionic membranes and amniotic fluid—has now been consistently documented in a growing multitude of mammalian species [5–46]. Distinct microbial communities in the meconium among preterm and healthy-term neonates are detected and described within minutes to hours of birth, and these impressively expand in the first days to weeks of life to readily show discrete body niche communities long before that same infant will alter its diet or engage in meaningful contact with the outside world. Moreover, although the notion of an "unsterile womb" has been challenged by a few [47–52], there are several rather consistent observations that collectively support a functional role for these sparse microbial communities [5–46,53,54]. First, it appears that the source and sink of these early developmental communities (i.e., breast milk, placenta, amniotic fluid, meconium, newborn oral and skin communities) are general of low biomass and low abundance. Second, being truly "germ free" (or at least germ depleted) carries with it imminent and known risk of disease. Namely, gnotobiotic mice still carry pregnancies to term and can propagate relatively deplete or replete lineages, but are impressively prone to postnatal inflammation and sepsis from intestinal pathogens. Thus there appears to be a role for these microbes (or at least their antigens and DNA) in generating immune tolerance in the offspring, although what that role is remains poorly understood. Third, within a short span of time and devoid of extensive exposures, which might seed horizontal transfer, the microbiome will remarkably expand and speciate by body niche. Collectively, such consistency of observations suggests that these processes very likely play some role in later immune development and allergic and atopic responses, albeit imperfectly understood at present.

The challenge we are facing in bringing together these rather consistent observations with their deserving mechanistic explanations is one of sparsity and inverse proportionality. Reliably detecting these few and far islands of microbes at their strain level in a sea of host genetic material requires tremendously more metagenomic sequence generation and computational capacity than is typically performed (i.e., to reach 2X coverage of a 2 Mb bacterial genome at 1%–10% relative abundance, one would need to sequence single low-biomass samples to upwards of 20 Gb and as high as 150 Gb). Despite ours and other investigators' best efforts to control for sources of environmental or kit contamination [5–35], in the absence of any known true "sterile tissue," there is inherent risk of misidentification of bacterial signal in high- and low-biomass specimens alike. As we recently demonstrated in our publication on utilization of in situ hybridization for the detection of these sparse microbes in the human placenta, the equivalent of a placental tissue section's volume of breast milk (~0.5 μL) would contain only ~5 bacteria [24]. Stated differently, observed bacteria occupy a hundred millionth ($1/10^8$) of the relative placental volume [24]. Nevertheless, they are present, and with ardent efforts, they can be distinguished from potential artifact and contaminant controls [24].

Ultimately, understanding when, how, and why these initially sparse communities expand and diversify will enable not only a fuller appreciation of Burnet and Fenner's notions of immune modulation and tolerance, but also provide fundamental clues to when these processes go awry to render both communicable and noncommunicable disease. In this chapter, we will attempt to provide an objective overview of the evidence to date supporting the presence of a low biomass, low abundance community of microbes consistently observed in intrauterine and fetal environments. We will then discuss several practical and clinically relevant examples of where developmental exposures alter the early life microbiome, resulting in apparently lasting consequences in juvenile animals. We will then engage in a robust discussion of the technical and biological challenges confronting scientists working with these low-biomass communities. Finally, we will conclude with our proposal of several potential unifying hypotheses concerning what "rules of the road" might govern the establishment of these early life microbial communities.

Consistency of evidence to date supporting the presence of microbes during fetal development

Current estimates indicate that the number of microbes that inhabits the entire body roughly equals the total number of cells that comprises the human body [2]. The bulk of this biomass is found in the large and small intestines, though bacteria are known to inhabit nearly every niche throughout the body including the skin and the vagina [3]. Together, these bacteria comprise an individual's microbiota that encode for thousands of metabolic functions known in totality as the microbiome. The community genetic repertoire encoded by both human and microbe is collectively referred to as the metagenome. As detailed earlier, while the historic paradigm assumes that the neonate is born sterile and is colonized differently depending on mode of delivery (i.e., Cesarean vs vaginal) [4], emerging evidence suggests to the contrary [5–23,26,28,29,55–57].

Observations arising from healthy pregnancies and studies within relevant animal models have indicated that the fetus may be first exposed to bacteria during gestation [5,7,11,18–35]. Recent studies of the reproductive tract in comparably aged women demonstrate a continuum of microbiota from the vagina, cervix, endometrial lining, and fallopian tubes, indicative of a nonsterile intrauterine environment before and during the time of conception, implantation, and placental development [4,25–33]. Other studies using culture and PCR-based techniques have positively identified bacteria in the fetal membranes, cord blood, and possibly in the amniotic fluid of healthy, term pregnancies, suggesting that microbiota can inhabit the in utero environment without overtly affecting the pregnancy or the health of the infant [6–23,26,28,29,34,35,37,55–57]. In addition, the use of metagenomics sequencing technologies revealed the diversity of the low-biomass microbial community of the placenta parenchyma and chorionic villus [6], which was also historically considered a sterile tissue in the absence of disease. Across the 320 placentae examined in this initial metagenomics study, the most common bacterial taxa were Proteobacteria, such as *Escherichia coli*, and other microbiota common to the oral cavity, such as *Fusobacterium* and *Streptococcus* species [6]. This work has since galvanized efforts to reconsider our fundamental assumptions about when and whence we first begin to acquire our microbes in early life. Observations of mother–neonatal pairs by Dong et al. and Collado et al. demonstrated that the microbiota found within the placenta share significant similarity to that of the neonate's meconium, indicating that microbiota may be additionally transferred across the placenta at the maternal–fetal interface into the fetus, where it would thereafter be presumptively excreted into the amniotic fluid as fetal urine [11,38]. Of note, in the mid-gestation (17–20 weeks human gestation), amniotic fluid transitions from a placental-derived fluid to being comprised of fetal urine. Although *Lactobacillus* species have been detected in the placental membranes in healthy, term pregnancies by metagenomics sequencing methodologies, the overall diversity of commensal species found within the placental parenchyma, amniotic fluid, and neonatal meconium suggests that the vaginal microbiome is unlikely to be the only origin for the full gamut of microbial species found within the intrauterine space [6–23,26,28,29,33–46,53–58].

Although our lab has been one of several pioneers in this field, we absolutely concur with the raised concerns of other investigators that ours and others studies ought be interpreted with a note of caution [47–52]. Specifically, we agree that determinations of the womb being "sterile" [53] or "not so sterile" [27] cannot rely solely on human observational studies, since these are inherently prone to both overinterpretation and misclassification. As we recently stated [24], with respect to the conclusions that can be drawn by the state of scientific evidence, we remain agnostic as to whether this is a community that is live and colonizing, or rather a highly predictable and consistent collection of metagenomes that may or may not be alive (e.g., harboring a unique microbiome, viability unknown). We similarly make no definitive conclusions regarding neither the origin nor the fate of these presumptive microbiome constituents. Although we have published previously the degree of taxonomic similarity between the placental and oral metagenomic communities, these were reported as cross-sectional comparisons from different cohorts and not meant to do anything more than speculate on their source nor sink [6]. In this current chapter and in our primary research, we have consistently iterated that the placental metagenomics community is of low biomass and low abundance, and its diversity does not nearly approximate that of the oral or any other relatively higher biomass community [6,24]. However, given the weighted evidence from dozens of laboratories employing multiple and varied techniques, including metagenomics and targeted PCR sequencing, cultivation, microscopy, and cross-validation with human and animal models warrant ongoing consideration and experimental testing.

To this end, it is worthwhile noting the animal model evidence similarly supporting early exposure to low abundance, low-biomass communities occurs in utero. In an initial and now landmark study, Jimenez et al. orally administered genetically labeled *Enterococcus faecium* to pregnant mice and sterilely delivered their pups 1 day ahead of anticipated delivery. Interestingly, they could culture and identify the labeled bacterium from the fetal intestine, indicating that microbiota can be transferred from mother to offspring even before delivery occurs [54]. However, the precise route of transmission was not examined in this study. Subsequent work by Han et al. and Fardini et al. has put forth a hematogenous model of placental colonization that potentially explains these observations [41,42]. In the former study, *Fusobacterium nucleatum* was given to pregnant mice intravenously during late gestation (embryonic day 16–17). While peripheral organs cleared *F. nucleatum* within 24 hours, this bacterial species persisted in the placenta and could be detected in the amniotic fluid and fetus at 72 hours postinfection [42]. In the latter study, the authors intravenously administered commensal bacteria typical of the human oral cavity to pregnant mice late in gestation, and found that they could selectively detect many of these administered microbiota in the placental tissues by PCR [42]. However, the fetal tissues were not specifically examined in this study, and thus a hematogenous route of placental and subsequently fetal colonization remains speculative without more definitive evidence. In sheep, a series of elegant studies from a collaborative team at the University of Florida has shown that identical strain isolates can be recovered from the ewe, its placenta, and the fetal tissues (including brain) [45]. Most recently this team has shown that they can inject GFP-, RFP-, and FRFP-tagged *Staphylococcus aureus* into the maternal blood stream at very low levels (100 CFU) in late gestation, and detect and recover these same tagged microbes in fetal and placenta tissue. Of interest, they could not recover live bacteria, and many of the bacteria were now intracellular [46]. Other mammalian species that have similarly been shown to harbor microbes in the intrauterine environment include other rodents and ungulates, bovine, and primates—to name a few. Indeed, if humans were not to harbor intrauterine microbes, they would be the sole known exception.

In summary, while the "womb" has traditionally been considered "sterile," this notion is no longer uniformly accepted. Bacteria or their metagenomes have been found in not only the reproductive tissues before pregnancy, but also in the intrauterine tissues of pregnancy [5–57]. When combined with several recent studies demonstrating the maturation of the fetal gut immune repertoire in this same mid-gestational interval, it seems reasonable that the development of the human microbiome might first begin in utero [59–66]. Further evidence for this probability arises from data suggesting that maternal exposures have a lasting impact on the offspring microbiome; this is discussed in the next section.

Is there evidence for a role of maternal exposures on her offspring's microbiome?

Impact of maternal nutrition in pregnancy on offspring gut microbiome

We have published extensively our findings arising from our nonhuman primate model of maternal high-fat diet feeding, showing that offspring exposed to a high-fat diet in pregnancy demonstrates increased anxiety-like behaviors, reduced thyroid hormone production, hepatic circadian gene expression, and nonalcoholic fatty liver disease [67–71]. Changes to the fetal epigenome is a likely driver of these observed phenotypes, as we have extensively documented changes to histone acetylation and promotor occupancy around key genes, as well as altered expression of the deacetylase Sirtuin 1 (SIRT1) [67,71–75].

However, diet is also known to be a potent modifier of the gut microbiome, favoring a microbiome capable of metabolizing the available substrates [76]. We have recently demonstrated maternal diet to have a long-term impact on the offspring gut microbiome, which may independently contribute to the phenotypes seen or help induce the epigenetic changes previously documented [77]. In this study, dams were either provided a high-fat diet or a control diet before and during pregnancy and lactation. As with humans, a high-fat diet caused significant weight gain and induced corresponding shifts in the nonpregnant gut microbiome [76,77]. To isolate the effects of maternal diet, the offspring of dams provided both with a high-fat and control diet were weaned onto a control diet at 6–7 months of age. Interestingly, at 1-year of age, the offspring gut microbiome could be discriminated based on whether their mothers consumed a high-fat or control diet, despite the offspring consuming a control diet for several months [77]. Specifically, a high-fat diet appeared to persistently diminish the relative abundance of commensal *Campylobacter* species in the offspring gut, indicating that maternal diet may play a significant role in shaping the transmission of commensal microbiota that can persist beyond infancy and may extend into adulthood [77].

The effect of maternal diet appears to extend beyond gestation. Research has shown that a high-fat diet leads to increased milk fat concentration and content when compared to a high-carbohydrate diet [78,79]. However, no differences in milk production or quantity of milk were observed. Therefore, neonates consuming breast milk from mothers with a high-fat diet consume higher energy intake, which can have potential effects on the development of their microbiome. Although it has not been studied, differences in the properties of the breast milk likely affect which bacteria flourish in the neonatal microbiome, but further studies are needed to confirm this hypothesis. Along these same lines, the maternal diet may be associated with alterations in the breast milk microbiome. Unfortunately, currently there are no studies published evaluating this association. Investigations in our lab are currently exploring this hypothesis and will help us understand any substantial impact the maternal diet has on the breast milk microbiome.

Limited evidence to date has similarly indicated that maternal gestational diet may influence offspring adiposity in early life by altering offspring gut microbiota. Independent studies have implicated certain bacterial species, including *E. coli*, as a major modifying factor of this phenotype in the mouse [80], though it is uncertain if *E. coli* has a similar impact on infant adiposity and growth trajectories in humans or primates. Intriguingly, mitigating the effects of a high-fat diet on the maternal gut microbiota with a prebiotic supplement has been reported to attenuate the impact of maternal diet on the offspring's propensity for adiposity [30], though our recent work in a nonhuman primate model of a maternal high-fat diet has demonstrated that probiotics do not alter gut microbiome structure, nor last persistently in the gut microbiome [31]. Nevertheless, the development of obesity is an incredibly complex pathophysiological process that may be first programmed in fetal life, but is likely sustained in postnatal life by continued environmental exposure to high-density dietary intake or aberrant microbiota.

In addition to obesity and immunity, recent data suggest that maternal diet may have an impact on offspring behavior by modulating gut microbiota. Bidirectional communication between the brain and the enteric nervous system has long been recognized, but only within the last few years has the impact of gut microbiota been explored in greater detail [81]. By producing neurotransmitters in the gut, such as serotonin or GABA, gut microbiota are hypothesized to contribute to a number of neurological and behavioral disorders, including anxiety, depression, and autism, by activating or depressing neural pathways in the enteric and central nervous system [81]. Recent work in the mouse by Buffington et al. has indicated that maternal gestational diet may modify offspring behavior by altering offspring gut microbiota in early life [82]. In this study, offspring whose mothers consumed a high-fat diet in pregnancy exhibited profound social deficits associated with significant changes to oxytocin levels in the brain and specific alterations of the offspring gut microbiota [82]. Intriguingly, postnatal reintroduction of *Lactobacillus reuteri*, which was depleted as a result of a maternal high-fat diet, to the affected offspring was found to ameliorate the deficient social behavior and enhance oxytocin levels in the brain, indicating a causal linkage between maternal diet, gut microbiota, and neurologic development [82]. Thus future studies examining the impact of maternal gestational diet on offspring gut microbiota will likely continue to refine our understanding of which microbiota are important to neurological development, how these microbiota are capable of modulating the gut–brain axis, and when these interactions are required.

Preterm birth

Being born preterm, and/or the underlying factors leading to a preterm birth, have lasting effects on the short- and long-term outcomes of neonates [83]. Preterm neonates are at higher risk for infection and intestinal problems, among other illnesses, owing to the lack of sufficient development of host tissues and immaturity of immune regulation at birth. The neonatal microbiome differs in preterm as compared to term infants, and it has been hypothesized that some of the disease morbidity association with prematurity is due to, or exacerbated by, the changes in the neonatal microbiome that result from the gestational age at delivery, the underlying predisposing factors, and/or the extensive postnatal environmental exposures necessitated for preterm neonatal care [84,85].

Compared to term infants, the gut microbiome of preterm infants tends to be much more sparsely populated. One group of investigators from Spain evaluated 21 premature neonates' intestinal microbiota during the first 3 months of life and compared it to term, exclusively breastfed, vaginally delivered neonates. Preterm neonates had increased levels of facultative anaerobic microorganisms and decreased levels of strict anaerobes such as *Bifidobacterium*, *Bacteroides*, and *Atopobium* [86]. However, it is difficult to assess if the changes they found are due to the lack of exclusive human milk feeding (all preterm infants included in this study received mixed feeding) or other associations with hospitalization and/or premature birth itself such as antibiotics, for example. Furthering the idea that the microbiome is different among premature neonates when compared to term neonates, other investigators have shown that very low birthweight neonates (birthweights <1500 g) have decreased diversity of their microbiota, which may be due to living in an neonatal intensive care unit (NICU) itself, alongside generally continuous antibiotic therapy, sterile isolators, and receipt of parenteral feeds [87–91].

Not only do premature neonates have a delay in the colonization of "healthy" commensal bacteria such as *Bifidobacterium*, but also the premature neonate's microbiota contain higher quantities of pathogenic bacteria. *Klebsiella, Weissella, Clostridium,* Enterobacteriaceae, Enterococcaceae, Streptococcaceae, and Staphylococcaceae have all been found more commonly in premature neonates microbiota than in neonates born at term [32]. Concurrent with these results, other investigators found increased levels of *Klebsiella pneumonia* in the preterm infant microbiota, and *Clostridium difficile* was detected exclusively in the preterm infants [84,92]. These observations may be a result of the types of bacteria that tend to exist in the immediate environment, as well as the relatively sparse microbiome that allows for opportunistic colonization. Premature neonates are typically treated in a NICU, which has been demonstrated to harbor a wide range of bacteria, many of which are known opportunistic pathogens that contain antibiotic resistance genes [93]. Interestingly, many of the neonatal gut microbes can be traced to bacteria found on NICU surfaces, indicating that the environment, in this case, plays a large role in seeding the preterm gut microbiome. Despite this, however, long-term observations of the preterm gut microbiome have demonstrated that by 1–3 years of age, the preterm microbiome develops similar complexity in term infants, indicating the resilient potential of the gut microbiome [93]. Nevertheless, considering that early life interactions between the host and its microbes are likely critical for crucial patterning, this has the potential to drastically impact a child's long-term health.

In summary, premature neonates are more prone to foster and harbor pathogenic (or at least pathobionts) bacteria rather than beneficial commensals, and the diversity and richness of their microbial communities that are first seen at birth simplify days–weeks later. However, this follows periods of often intense interventions and isolation, as well as antimicrobial therapy. Thus it is nearly impossible to gage whether the parsing down of the microbiome postbirth in premature neonates is a cause or consequence of these necessary interventions. Harboring pathogenic bacteria with less commensal, protective bacteria may be one key reason why this age group has a higher likelihood of necrotizing enterocolitis and other infectious maladies than term neonates.

Mode of delivery

A woman can give birth one of two ways: vaginally or via Cesarean delivery. Although one is often considered the more "traditional" path and the other is surgical, there is a concern that a lack of exposure to the vaginal microbiome may lead to higher rates of certain diseases occurring later in life such as atopic conditions, inflammatory bowel disease, type I and II diabetes mellitus, and asthma [94–97]. Given such challenges in establishing a causal relationship between Cesarean and later disease, one option is to alternately provide a mechanistic link between the exposure (Cesarean) and the outcome (later in life atopic disease, for example). Attempts in doing so have led researchers to postulate that absence of vaginal microbes in cases of neonates born via Cesarean delivery may be the malefactor. But how well warranted is this concern?

If the postulate that Cesarean-born infants fail to be colonized in their gut (or skin or mouth) by vaginally derived microbes is true, then microbes living in the maternal vaginal niche should (as a rule) establish stable, long-term communities in the infant. However, several lines of evidence have shown this not to be true. First, one of the key tenets of the Human Microbiome Project was the observation of unique body niche speciation. In the vagina, the ecology is dominated by *Lactobacillus* spp. which are highly adept at living at low pH; these same microbes do not dominate the neonatal gut nor other body niches. Second, experimental manipulations have not yet provided strong evidence suggesting that vaginal microbes establish long-term community dominance outside the vagina [98]. The reasons to perform a Cesarean delivery are often due to clear and evident medical conditions. For instance, conditions such as placenta previa (where the placenta overlies the cervix) or vasa previa (where vessels overlie the cervix) are absolute indications for performing Cesarean deliveries because if a vaginal birth occurred, the fetus would likely exsanguinate from severe blood loss. Another underlying medical indication or reason to perform a Cesarean delivery is evidence of significant fetal distress, thus necessitating emergent delivery or otherwise risk intrapartum fetal demise. Thus, the underlying reasons for performing a Cesarean delivery with any given pregnancy are many, and will thus provide a distinctly different medical landscape from that which accompanies a vaginal delivery.

Does a lack of exposure to the vaginal microbiome confer risks to the neonate as he or she grows and develops into adulthood? Some studies have found an association between delivery via Cesarean and increased rates of atopic disorders, food allergies, metabolic syndrome, and obesity later in life [94–97]. For instance, a recent large prospective study conducted over 16 years with more than 22,000 participants found that Cesarean-delivered infants had a 13% increased risk of obesity later in life [94]. Many often-cited studies have not fully elucidated if it is the mode of delivery versus the underlying indication for delivery that is the culprit [4]. Our team published one of the first large, prospectively followed cohort of women that includes the maternal indications for delivery along with mode of delivery in evaluating the effects of the development of the neonatal microbiome across multiple body sites [98]. While neonates born via Cesarean delivery have distinct microbiome profiles in the first week of life, this effect did not last past 2 months of age [98,99]. Interestingly, the neonatal microbiome differentiates at specific body sites by 6–8 weeks postnatally [98]. Immediately after birth, we found that the mode of delivery was associated with differences in the neonatal microbiome within the nares, skin, and oral cavity. However, the neonatal meconium microbiome clustered separately among Cesarean born, suggesting a distinct maternal origin that separated it from the other body sites studied. By 6 weeks of age, the infant microbiota had diversified with body site specificity, and, most remarkably, no differences were identified between infants born via vaginal delivery and those born via Cesarean in any and all body sites examined. When these were further parsed by indication for Cesarean, or by labored versus unlabored, Cesarean-born neonates without labor were the most dissimilar. In addition, when controlling for the indication for delivery and other clinical factors, only maternal diet and formula feeding appear to have a lasting effect on the neonatal microbiome at 6 weeks of age, and the maternal indication for delivery plays a significant role in this development. Along these lines, in preterm neonates, other investigators have shown that tracheal aspirates have been evaluated for the evaluation of the microbiome, and there are no appreciable differences between those premature neonates born via Cesarean delivery versus those born vaginally [100]. The recent TEDDY cohort studies further failed to delineate a fundamental role for Cesarean delivery in shaping the early developmental microbiome.

In summary, while others have previously observed that neonates born vaginally or vaginally seeded demonstrated increased *Lactobacillus* and *Bacteroides*. Interestingly, Chu et al. found low levels of *Bacteroides* and/or high levels of *Lactobacillus* to be equally probable in vaginally and Cesarean-delivered neonates and infant, but their variation was dependent upon several other collinear factors such as the percent fat in the maternal diet, formula feeding, labored versus unlabored delivery, and gestational age at delivery. These observations were consistent with much earlier studies showing that vaginal delivery rarely transfers *Lactobacillus* to the neonate and hence have been recapitulated by multiple groups that collectively fail to demonstrate a significant nor lasting distinction in the gut microbiota of vaginal versus Cesarean-delivered infants or children. Based on the emerging data in well-designed, prospective, and population-based cohorts, we conclude that it is highly probable that the maternal indication for Cesarean delivery plays a larger role than the mode of delivery itself. This renders the alternate view that a lack of vaginal "seeding" [101,102] by the maternal vaginal microbiome during parturition is unlikely to be the key culprit in any subsequent association between Cesarean birth and late in life disease. Rather, the origin of the risk lies in the maternal "soil" from which the neonate originated from (i.e., the maternal and intrauterine milieu). While these may or may not be the indications that led to the decision to perform a Cesarean delivery, parsing cause from surgery is of fundamental importance since the nature and timing of any corrective interventions would be widely different.

Why is it difficult to cultivate sparse microbes, such as in the amniotic fluid or placenta?

Since the first, albeit accidental, culture of bacteria "back in the day" to the now highly scientific process of optimization and discovery, we have increasingly expanded on the techniques of microbial culture [103]. Microbiologists have always understood that we must develop and design our culture environments and techniques to account for the near unlimited variety in environmental requirements within the bacterial domain, whose members are so varied that their diversity cannot be accurately accounted for [104,105]. Medical microbiologists have optimized culture techniques of a very few and selected number that are currently known to be medically relevant to human infection through pragmatic refinement and study of their predictive and diagnostic reliability. Because of this, cultivation is historically considered the "gold standard" for the identification of clinical pathogenic infections [106,107]. As a result, some would suggest that if we cannot culture or cultivate bacteria, they must not be there. This is demonstrably false and led to the current advent of metagenomics in not only microbiome research, but emerging and under development clinical applications (e.g., Karius diagnostic tools). Thus it is obvious to many but recognized by few that we are quite limited in the scope of biologically relevant commensal and pathogenic human microbiota amenable to culture [106,108].

The era of first 16S sequencing and now metagenomics has made this all too apparent [109]. The rise of DNA sequencing has revealed that, *at the species level*, there are as many as 80% of bacterial taxa estimated to be uncultivable [110]. At the subspecies level this problem magnifies. For example, in a study that estimated the levels of *F. nucleatum*, there was a gross underestimation of levels by culture compared to molecular techniques, suggesting that a significant portion of the strains of this species, a common oral and sometimes intrauterine pathogen considered eminently culturable, are in fact quite uncultivable [111]. Therefore, some strains are more amenable to culture than others, and through the medical necessity of pathogen diagnoses, we may have inadvertently directed our refinement and culture development toward the pathogenic strains that are most correlative in specificity and sensitivity to clinical disease. A lack of diversity of strains, as may happen with infectious or commensal growth, therefore diminishes the opportunity for culture-based detection. Scarce bacteria are intrinsically limited in their diversity. Thus culture-based detection of rare bacteria in low-biomass communities is even further challenging.

Even within strains, some bacteria that are thought to be cultivatable will often not be. It is a significant and recognized problem for public health that strains, some even pathogenic (e.g., *E. coli*), are viable but nonculturable (VBNC), and therefore they evade enumeration or detection [110]. The mechanisms that allow strains or single bacteria to become dormant, viable but not proliferating, are increasingly understood [112]. Indeed, culturable strains can be induced to be uncultivable through metabolic stress, or antibiotics in a mechanism independent of selection or genetic resistance (bacterial persisters) [113,114]. Bacterial cytokines, notably resuscitation-promoting factor, present in both gram positive and negative bacteria, signals resurgence from dormancy induced by metabolic stress. Bacteria remain VBNC as an adaptive mechanism to survive, and presumably transit, in inhospitable environments. It has been demonstrated in many instances that they even temporarily lose their virulence [112]. Proliferating, infectious bacteria by definition are not dormant and therefore more amenable to culture-based detection. The corollary is that rare viable, but non proliferating bacteria are likely to be dormant and less amenable to culture-based detection as may be the case in low-biomass samples.

It thus ought not be surprising that there is discordance in many cases between what translational science reveals by microbial DNA, or other molecular, or microscopy-based techniques, and what can be determined as present by ex vivo cultivation alone. The prediction of amniotic fluid microbes in clinically evident chorioamnionitis, for example, has been demonstrated to have greater sensitivity using molecular techniques than diagnosis by either cultivation alone or by clinical criteria with and without cultivation. A study of preterm deliveries with intact membranes revealed colonization in association with inflammation and microbial invasion of the amniotic cavity found significant discordance between 16S and culture-based testing, and a greater association with inflammation rather than bacterial colonization [115]. 16S rRNA positive amniotic fluid at time of amniocentesis was associated with decreased gestational age despite largely culture-negative findings [116]. Molecular techniques correlated with preterm delivery, despite nearly 90% of samples testing negative by culture [117]. A false negative culture rate of 40% was estimated for mycoplasma infection of the amniotic fluid based on comparison to PCR techniques in studies of both preterm prolonged rupture of membranes and spontaneous preterm birth patients [118,119]. Greater diversity of taxa and accurate prediction of preterm birth was observed by employing PCR or molecular diagnostic rather than culture-based techniques, suggesting a lack of cultivability of many of the molecular-identified and presumptive causally-related strains [120]. Discordance between culture-based findings and other methods of bacterial observation has been shown in the placenta as well. As many as 25% of 174 indicated Cesarean-delivered preterm birth placentas were culture positive [121]. Histological studies in the absence of clinical infection have identified placental or membrane bacteria [35,122–124]. The identification of bacteria has been shown to be in the absence of histological inflammation [35,123].

The placental conduit

The placenta has largely been considered as a barrier used to protect the fetus from bloodborne pathogens and in preventing rejection from the mother [53,125]. That being said, it is well demonstrated that this "barrier" is imperfect, and that some bacteria (e.g., syphilis), protozoa (e.g., toxoplasmosis), and viruses (e.g., herpes simplex virus, cytomegalovirus, varicella, and Zika) can indeed be transmitted across the placenta [125–127]. This notion of an imperfect barrier suggests that the placenta may serve not just as passive barrier, but rather a conduit used for interfacing with the fetus, functioning in the normal growth and development through exposing the fetus to a low abundance microbiome of vertically transmitted beneficial symbionts, while simultaneously protecting against pathogenic infection [6].

If we accept this role as a conduit rather than a barrier, then perhaps the placenta is also used in the transmission of select microbiota that play a critical role in normal growth and development in utero through the exposure of beneficial microbes in low abundance. This hypothesis is further supported with several other examples outside of human perinatal biology that indicate vertical transmission of the gut microbiota. For example, using the murine model, Moeller et al. demonstrated vertical transmission of a microbiota profile in two wild-type mouse lineages over a 3-year period [128]. Although the specific mechanisms of transmission were not studied, most microbiota were confirmed to be vertically inherited and remained distinct after over 10 generations [128].

The plant kingdom also provides multiple examples of seedborne vertical transmission of mutualistic endophytic bacteria [129,130] and fungi [131]. Through natural selection, the process of vertical transmission provides selective pressure for these microbes to decrease in pathogenicity and increase in their ability to support plant growth and development [129], therein increasing both host and microbial survival and expansion [132]. Examples of endophytic seedborne vertical transmission of bacteria in rice *Oryza sativa* and wheat *Triticum aestivum* include *Bacillus* spp. [133], *Enterobacter* spp. [134,135], *Ochrobactrum* spp. [136], *Stenotrophomonas* spp. [135,136], *Paenibacillus* spp. [133,136], *Pseudomonas* spp. [134,136], as well as multiple other genera [129]. These microbes have been demonstrated to aid in host growth and development, disease suppression, and abiotic tolerance through mechanisms such as the production/modification of phytohormones, alteration of root/shoot system architecture, manipulation of reactive oxygen species, production of long-chain extracellular polysaccharides, and transcriptional regulation of host stress response genes [137–146]. Given the multiple examples of vertical transmission in animals [128,147], insects [147,148], and plants [136], it would appear that if microbes were not *present* in utero in human pregnancies, it would truly be the exception across the plant and animal kingdoms.

Rules of the road?

Although the precise mechanism and extent to which microbes can be vertically transmitted in humans are currently unknown, there do seem to be some overarching rules that might be reasonably assumed and subsequently govern vertical transmission. *We will thus conclude this chapter with our proposal of several potential unifying hypotheses concerning what "rules of the road" might govern the establishment of these early life microbial communities.* These are not meant to be taken as statements of fact, but rather hypotheses with varying degrees of data supporting or refuting. They are inherently testable hypotheses, and likely worth testing with well-designed experiments.

Possible "rule of the road" number 1: The introduction and transmission of microbes in utero occur in low abundance and in low-biomass concentrations, perhaps in order to avoid the endosymbionts from overtaking the host during the initial developmental stages. The evidence both for and against this notion is extensively detailed throughout this chapter.

Possible "rule of the road" number 2: There also seem to be a number of small molecule intermediate candidates that may play a critical role in recruitment and/or establishment of the selected mutualists in this perinatal setting (e.g., human milk oligosaccharides) [149,150]. Similarly, in plants, nodulation factors are lipochitooligosaccharides produced by N2 fixing bacteria collectively known as rhizobia. These rhizobia initiate root nodule formation in legumes through the recognition binding from lysin motif (LysM) receptor kinases [151]. In addition, the presence of human milk oligosaccharides (HMOs) in breast milk may also be a specialized source of nutrients dedicated to beneficial bacteria, therein allowing them to competitive exclude non-HMO-metabolizing bacteria [152]. Similarly, it has also been demonstrated that in a coniferous forest, hub trees, also known as "mother trees," are able to send carbon to younger seedlings through a mycorrhizal network in a deliberate and kin-specific manner, therein aiding in seedling growth and development (Fig. 4.1). In addition, these hub trees have been shown to reduce their own root growth in an effort to allow their offspring to grow without competition, and in a wounding event, will release defense signals to their offspring, therein priming the seedling's defense and increasing resistance to future stressors [153–156].

Possible "rule of the road" number 3: The permissiveness of vertical transmission may be less important for colonization permission than it is for colonization resistance. Colonization resistance refers to the capacity (largely functional) for some microbes to prevent other microbes from colonizing or flourishing through the niche exclusion. There are several examples of colonization resistance biocontrol strategies being implemented (with mixed results) in commercial agriculture. For example, application of the biocontrol agent *Agrobacterium radiobacter* strain K84 has been commercially used in

FIGURE 4.1 Illustration showing the vertical transmission of bacteria in humans and plants leading to the normal growth and development of offspring (*left*). Illustration demonstrating the transfer of nutrients through either breastfeeding in humans (*upper right*) or the mycorrhizal network in plants (*bottom right*).

preventing crown gall formation from the pathogen *Agrobacterium tumefaciens* in grapevine root grafting procedures by niche exclusion and bacteriocin production of agrocin 84 [157].

In summary, there are still more questions than answers when it comes to in utero microbiomics. For instance, it is yet unclear as to whether beneficial microbes are truly being vertically transferred through the placenta, or simply creating tolerance through the exposure of the immune system to microbial DNA. Future research should focus on elucidating the structure and function of small molecule intermediates such as HMOs in order to better understand their role in seeding the infant gut microbiome, and the role that transferal of microbes in utero might play in metabolism.

All-in-all, while some key questions have been answered regarding the development of the human microbiome and its impact on human disease, much still remains unknown. Physicians and scientists alike must seek answers to these vital questions in order to elucidate potential etiologies for disease states, including those which arise due to aberrant development of the human microbiome and thus can be amenable to therapies early in life. Moreover, continuing to operate under age-old paradigms of "purity at birth" that ignore mounting evidence for low-biomass metagenomes (be they of sparse abundance or not) in favor of a priori assumptions will limit progress in our field. While our laboratory presently remains agnostic as to the issue of whether these consistently observed low-biomass communities actually are alive and colonize the fetus, or alternately enable later colonization through a process of immune tolerance, we look forward to the future and bold science that will help elucidate one of the most significant questions of our time. In the interval until that occurs, we must remain both fairly open minded and constructively critical of the available evidence.

Financial disclosure

We have nothing to disclose

Conflicts of interest

None

Clinical trial registration

Not applicable

References

[1] Burnet FM, Fenner F. The production of antibodies. 2nd ed. Melbourne: Macmillan; 1949.

[2] Sender R, Fuchs S, Milo R. Revised estimates for the number of human and bacteria cells in the body. PLoS Biol 2016;14(8):e1002533.

[3] Huttenhower C, Gevers D, Knight R, et al. Structure, function and diversity of the healthy human microbiome. Nature 2012;486(7402):207.

[4] Dominguez-Bello MG, Costello EK, Contreras M, et al. Delivery mode shapes the acquisition and structure of the initial microbiota across multiple body habitats in newborns. Proc Natl Acad Sci 2010;107(26):11971–5.

[5] Zhu L, Luo F, Hu W, et al. Bacterial communities in the womb during healthy pregnancy. Front Microbiol 2018;9:2163.

[6] Aagaard K, Ma J, Antony KM, Ganu R, Petrosino J, Versalovic J. The placenta harbors a unique microbiome. Sci Transl Med 2014;6(237):237–65.

[7] Doyle RM, Alber DG, Jones HE, et al. Term and preterm labour are associated with distinct microbial community structures in placental membranes which are independent of mode of delivery. Placenta 2014;35(12):1099–101.

[8] Antony KM, Ma J, Mitchell KB, Racusin DA, Versalovic J, Aagaard K. The preterm placental microbiome varies in association with excess maternal gestational weight gain. Am J Obstet Gynecol 2015;212(5) 653 e651-616.

[9] Zheng J, Xiao X, Zhang Q, Mao L, Yu M, Xu J. The placental microbiome varies in association with low birth weight in fullterm neonates. Nutrients 2015;7 (8):6924–37.

[10] Bassols J, Serino M, Carreras-Badosa G, et al. Gestational diabetes is associated with changes in placental microbiota and microbiome. Pediatr Res 2016;80 (6):777–84.

[11] Collado MC, Rautava S, Aakko J, Isolauri E, Salminen S. Human gut colonisation may be initiated in utero by distinct microbial communities in the placenta and amniotic fluid. Sci Rep 2016;6:23129.

[12] Prince AL, Ma J, Kannan PS, et al. The placental membrane microbiome is altered among subjects with spontaneous preterm birth with and without chorioamnionitis. Am J Obstet Gynecol 2016;214(5) 627 e621-627 e616.

[13] Doyle RM, Harris K, Kamiza S, et al. Bacterial communities found in placental tissues are associated with severe chorioamnionitis and adverse birth outcomes. PLoS One 2017;12(7):e0180167.

[14] Parnell LA, Briggs CM, Cao B, Delannoy-Bruno O, Schrieffer AE, Mysorekar IU. Microbial communities in placentas from term normal pregnancy exhibit spatially variable profiles. Sci Rep 2017;7(1):11200.

[15] Gomez-Arango LF, Barrett HL, McIntyre HD, Callaway LK, Morrison M, Nitert MD. Contributions of the maternal oral and gut microbiome to placental microbial colonization in overweight and obese pregnant women. Sci Rep 2017;7(1):2860.

[16] Zheng J, Xiao XH, Zhang Q, et al. Correlation of placental microbiota with fetal macrosomia and clinical characteristics in mothers and newborns. Oncotarget 2017;8 (47):82314–25.

[17] Leon LJ, Doyle R, Diez-Benavente E, et al. Enrichment of clinically relevant organisms in spontaneous preterm delivered placenta and reagent contamination across all clinical groups in a large UK pregnancy cohort. Appl Environ Microbiol 2018;84(14): e00483-18.

[18] Mitchell CM, Haick A, Nkwopara E, et al. Colonization of the upper genital tract by vaginal bacterial species in nonpregnant women. Am J Obstet Gynecol 2015;212 (5):611 e611–19.

[19] Franasiak JM, Werner MD, Juneau CR, et al. Endometrial microbiome at the time of embryo transfer: next-generation sequencing of the 16S ribosomal subunit. J Assist Reprod Genet 2016;33(1):129–36.
[20] Moreno I, Codoner FM, Vilella F, et al. Evidence that the endometrial microbiota has an effect on implantation success or failure. Am J Obstet Gynecol 2016;215 (6):684–703.
[21] Verstraelen H, Vilchez-Vargas R, Desimpel F, et al. Characterisation of the human uterine microbiome in non-pregnant women through deep sequencing of the V1-2 region of the 16S rRNA gene. PeerJ 2016;4:e1602.
[22] Chen C, Song X, Wei W, et al. The microbiota continuum along the female reproductive tract and its relation to uterine-related diseases. Nat Commun 2017;8 (1):875.
[23] Kyono K, Hashimoto T, Nagai Y, Sakuraba Y. Analysis of endometrial microbiota by 16S ribosomal RNA gene sequencing among infertile patients: a single-center pilot study. Reprod Med Biol 2018;17(3):297–306.
[24] Seferovic MD, Pace RM, Caroll M, et al. Visualization of microbes by 16S in situ hybridization in term and preterm placentae without intraamniotic infection. Am J Obstet Gynecol 2019;221(2):146.e1–146.e23.
[25] Prince AL, Pace RM, Dean T, et al. The development and ecology of the Japanese macaque gut microbiome from weaning to early adolescence in association with diet. Am J Primatol 2019;81(10-11):e22980.
[26] Giudice LC. Challenging dogma: the endometrium has a microbiome with functional consequences!. Am J Obstet Gynecol 2016;215(6):682–3.
[27] Stinson LF, Boyce MC, Payne MS, Keelan JA. The not-so-sterile womb: evidence that the human fetus is exposed to bacteria prior to birth. Front Microbiol 2019;10:1124.
[28] Pelzer E, Gomez-Arango LF, Barrett HL, Nitert MD. Review: maternal health and the placental microbiome. Placenta 2017;54:30–7.
[29] Benner M, Ferwerda G, Joosten I, Van der Molen RG. How uterine microbiota might be responsible for a receptive, fertile endometrium. Hum Reprod Update 2018;24 (4):393–415.
[30] Paul HA, Bomhof MR, Vogel HJ, Reimer RA. Diet-induced changes in maternal gut microbiota and metabolomic profiles influence programming of offspring obesity risk in rats. Sci Rep 2016;6:20683.
[31] Pace RM, Prince AL, Ma J, et al. Modulations in the offspring gut microbiome are refractory to postnatal synbiotic supplementation among juvenile primates. BMC Microbiol 2018;18(1):28.
[32] Baker JM, Chase DM, Herbst-Kralovetz MM. Uterine microbiota: residents, tourists, or invaders? Front Immunol 2018;9:208.
[33] Seo SS, Arokiyaraj S, Kim MK, et al. High prevalence of *Leptotrichia amnionii, Atopobium vaginae, Sneathia sanguinegens*, and factor 1 microbes and association of spontaneous abortion among Korean women. BioMed Res Int 2017;2017:11.
[34] Dong Y, St Clair PJ, Ramzy I, Kagan-Hallet KS, Gibbs RS. A microbiologic and clinical study of placental inflammation at term. Obstet Gynecol 1987;70(2):175–82.
[35] Cao B, Mysorekar IU. Intracellular bacteria in placental basal plate localize to extravillous trophoblasts. Placenta 2014;35(2):139–42.
[36] Blaser MJ, Dominguez-Bello MG. The human microbiome before birth. Cell Host Microbe 2016;20(5):558–60.
[37] Mysorekar IU, Cao B. Microbiome in parturition and preterm birth. Semin Reprod Med 2014;32(1):50–5.
[38] Dong XD, Li XR, Luan JJ, et al. Bacterial communities in neonatal feces are similar to mothers' placentae. Can J Infect Dis Med Microbiol J canadien des maladies infectieuses et de la microbiologie medicale 2015;26(2):90–4.

[39] Ding T, Schloss PD. Dynamics and associations of microbial community types across the human body. Nature 2014;509(7500):357.
[40] Pelzer ES, Allan JA, Waterhouse MA, Ross T, Beagley KW, Knox CL. Microorganisms within human follicular fluid: effects on IVF. PLoS One 2013;8(3): e59062.
[41] Han YW, Redline RW, Li M, Yin L, Hill GB, McCormick TS. *Fusobacterium nucleatum* induces premature and term stillbirths in pregnant mice: implication of oral bacteria in preterm birth. Infect Immun 2004;72(4):2272–9.
[42] Fardini Y, Chung P, Dumm R, Joshi N, Han YW. Transmission of diverse oral bacteria to murine placenta: evidence for the oral microbiome as a potential source of intrauterine infection. Infect Immun 2010;78(4):1789–96.
[43] Zarate MA, Rodriguez MD, Chang EI, et al. Post-hypoxia Invasion of the fetal brain by multidrug resistant *Staphylococcus*. Sci Rep 2017;7(1):6458.
[44] Yu K, Rodriguez MD, Paul Z, et al. Proof of principle: physiological transfer of small numbers of bacteria from mother to fetus in late-gestation pregnant sheep. PLoS one 2019;14(6):e0217211.
[45] Aagaard K, Riehle K, Ma J, et al. A metagenomic approach to characterization of the vaginal microbiome signature in pregnancy. PLoS one 2012;7(6):e36466.
[46] Walther-António MR, Jeraldo P, Miller MEB, et al. Pregnancy's stronghold on the vaginal microbiome. PLoS One 2014;9(6):e98514.
[47] Theis KR, Romero R, Winters AD, et al. Does the human placenta delivered at term have a microbiota? Results of cultivation, quantitative real-time PCR, 16S rRNA gene sequencing, and metagenomics. Am J Obstet Gynecol 2019;220(3) 267. e261–267. e239.
[48] Lauder AP, Roche AM, Sherrill-Mix S, et al. Comparison of placenta samples with contamination controls does not provide evidence for a distinct placenta microbiota. Microbiome 2016;4(1):29.
[49] Kim D, Hofstaedter CE, Zhao C, et al. Optimizing methods and dodging pitfalls in microbiome research. Microbiome 2017;5(1):52.
[50] Bushman F. De-discovery of the placenta microbiome. Am J Obstet Gynecol 2019;220 (3):213–14.
[51] Leiby JS, McCormick K, Sherrill-Mix S, et al. Lack of detection of a human placenta microbiome in samples from preterm and term deliveries. Microbiome 2018;6(1):196.
[52] Glassing A, Dowd SE, Galandiuk S, Davis B, Chiodini RJ. Inherent bacterial DNA contamination of extraction and sequencing reagents may affect interpretation of microbiota in low bacterial biomass samples. Gut Pathog 2016;8(1):24.
[53] Perez-Munoz ME, Arrieta MC, Ramer-Tait AE, Walter J. A critical assessment of the "sterile womb" and "in utero colonization" hypotheses: implications for research on the pioneer infant microbiome. Microbiome 2017;5(1):48.
[54] Jimenez E, Marin ML, Martin R, et al. Is meconium from healthy newborns actually sterile? Res Microbiol 2008;159(3):187–93.
[55] Prince AL, Chu DM, Seferovic MD, Antony KM, Ma J, Aagaard KM. The perinatal microbiome and pregnancy: moving beyond the vaginal microbiome. Cold Spring Harb Perspect Med 2015;5(6).
[56] Prince AL, Antony KM, Ma J, Aagaard KM. The microbiome and development: a mother's perspective. Semin Reprod Med 2014;32(1):14–22.
[57] Moreno I, Franasiak JM. Endometrial microbiota-new player in town. Fertil Steril 2017;108(1):32–9.
[58] MacIntyre DA, Chandiramani M, Lee YS, et al. The vaginal microbiome during pregnancy and the postpartum period in a European population. Sci Rep 2015;5:8988.
[59] Backhed F, Roswall J, Peng Y, et al. Dynamics and stabilization of the human gut microbiome during the first year of life. Cell Host Microbe 2015;17(6):852.

[60] Matamoros S, Gras-Leguen C, Le Vacon F, Potel G, de La Cochetiere MF. Development of intestinal microbiota in infants and its impact on health. Trends Microbiol 2013;21(4):167–73.
[61] Mackie RI, Sghir A, Gaskins HR. Developmental microbial ecology of the neonatal gastrointestinal tract. Am J Clin Nutr 1999;69(5):1035S–45S.
[62] Stewart CJ, Ajami NJ, O'Brien JL, et al. Temporal development of the gut microbiome in early childhood from the TEDDY study. Nature 2018;562(7728):583–8.
[63] Stewart CJ, Embleton ND, Clements E, et al. Cesarean or vaginal birth does not impact the longitudinal development of the gut microbiome in a cohort of exclusively preterm infants. Front Microbiol 2017;8:1008.
[64] Stewart CJ, Embleton ND, Marrs EC, et al. Temporal bacterial and metabolic development of the preterm gut reveals specific signatures in health and disease. Microbiome 2016;4(1):67.
[65] Stewart CJ, Embleton ND, Marrs ECL, et al. Longitudinal development of the gut microbiome and metabolome in preterm neonates with late onset sepsis and healthy controls. Microbiome 2017;5(1):75.
[66] Vebo HC, Sekelja M, Nestestog R, et al. Temporal development of the infant gut microbiota in immunoglobulin E-sensitized and nonsensitized children determined by the GA-map infant array. Clin Vaccine Immunol 2011;18(8):1326–35.
[67] Suter M, Bocock P, Showalter L, et al. Epigenomics: maternal high-fat diet exposure in utero disrupts peripheral circadian gene expression in nonhuman primates. FASEB J 2011;25(2):714–26.
[68] Suter MA, Aagaard-Tillery KM. Environmental influences on epigenetic profiles. Semin Reprod Med 2009;27(5):380–90.
[69] Suter MA, Chen A, Burdine MS, et al. A maternal high-fat diet modulates fetal SIRT1 histone and protein deacetylase activity in nonhuman primates. FASEB J 2012;26 (12):5106–14.
[70] Suter MA, Sangi-Haghpeykar H, Showalter L, et al. Maternal high-fat diet modulates the fetal thyroid axis and thyroid gene expression in a nonhuman primate model. Mol Endocrinol 2012;26(12):2071–80.
[71] Suter MA, Takahashi D, Grove KL, Aagaard KM. Postweaning exposure to a high-fat diet is associated with alterations to the hepatic histone code in Japanese macaques. Pediatr Res 2013;74(3):252–8.
[72] Suter M, Abramovici A, Aagaard-Tillery K. Genetic and epigenetic influences associated with intrauterine growth restriction due to in utero tobacco exposure. Pediatr Endocrinol Rev: 2010;8(2):94–102.
[73] Sullivan EL, Grayson B, Takahashi D, et al. Chronic consumption of a high-fat diet during pregnancy causes perturbations in the serotonergic system and increased anxiety-like behavior in nonhuman primate offspring. J Neurosci 2010;30(10):3826–30.
[74] Thorn SR, Baquero KC, Newsom SA, et al. Early life exposure to maternal insulin resistance has persistent effects on hepatic NAFLD in juvenile nonhuman primates. Diabetes 2014;63(8):2702–13.
[75] Sullivan EL, Smith MS, Grove KL. Perinatal exposure to high-fat diet programs energy balance, metabolism and behavior in adulthood. Neuroendocrinology 2011;93 (1):1–8.
[76] David LA, Maurice CF, Carmody RN, et al. Diet rapidly and reproducibly alters the human gut microbiome. Nature 2014;505(7484):559–63.
[77] Ma J, Prince AL, Bader D, et al. High-fat maternal diet during pregnancy persistently alters the offspring microbiome in a primate model. Nat Commun 2014;5:3889.
[78] Mohammad MA, Sunehag AL, Haymond MW. Effect of dietary macronutrient composition under moderate hypocaloric intake on maternal adaptation during lactation. Am J Clin Nutr 2009;89(6):1821–7.

[79] Munch EM, Harris RA, Mohammad M, et al. Transcriptome profiling of microRNA by Next-Gen deep sequencing reveals known and novel miRNA species in the lipid fraction of human breast milk. PLoS One 2013;8(2):e50564.
[80] Fak F, Karlsson CL, Ahrne S, Molin G, Westrom B. Effects of a high-fat diet during pregnancy and lactation are modulated by *E. coli* in rat offspring. Int J Obes 2012;36 (5):744–51.
[81] Foster JA, McVey Neufeld KA. Gut-brain axis: how the microbiome influences anxiety and depression. Trends Neurosci 2013;36(5):305–12.
[82] Buffington SA, Di Prisco GV, Auchtung TA, Ajami NJ, Petrosino JF, Costa-Mattioli M. Microbial reconstitution reverses maternal diet-induced social and synaptic deficits in offspring. Cell 2016;165(7):1762–75.
[83] Saigal S, Doyle LW. An overview of mortality and sequelae of preterm birth from infancy to adulthood. Lancet 2008;371(9608):261–9.
[84] Moles L, Gomez M, Jimenez E, et al. Preterm infant gut colonization in the neonatal ICU and complete restoration 2 years later. Clin Microbiol Infect 2015;21(10):936 e931–936 e 910.
[85] La Rosa PS, Warner BB, Zhou Y, et al. Patterned progression of bacterial populations in the premature infant gut. Proc Natl Acad Sci U S A 2014;111(34):12522–7.
[86] Arboleya S, Binetti A, Salazar N, et al. Establishment and development of intestinal microbiota in preterm neonates. FEMS Microbiol Ecol 2012;79(3):763–72.
[87] Rouge C, Goldenberg O, Ferraris L, et al. Investigation of the intestinal microbiota in preterm infants using different methods. Anaerobe 2010;16(4):362–70.
[88] Roudiere L, Jacquot A, Marchandin H, et al. Optimized PCR-temporal temperature gel electrophoresis compared to cultivation to assess diversity of gut microbiota in neonates. J Microbiol Methods 2009;79(2):156–65.
[89] Magne F, Abely M, Boyer F, Morville P, Pochart P, Suau A. Low species diversity and high interindividual variability in faeces of preterm infants as revealed by sequences of 16S rRNA genes and PCR-temporal temperature gradient gel electrophoresis profiles. FEMS Microbiol Ecol 2006;57(1):128–38.
[90] Schwartz S, Friedberg I, Ivanov IV, et al. A metagenomic study of diet-dependent interaction between gut microbiota and host in infants reveals differences in immune response. Genome Biol 2012;13(4):r32.
[91] Neu J. Intestinal microbiota studies in preterm infants. J Pediatr Gastroenterol Nutr 2016;62(2):193–4.
[92] Gibson MK, Wang B, Ahmadi S, et al. Developmental dynamics of the preterm infant gut microbiota and antibiotic resistome. Nat Microbiol 2016;1:16024.
[93] Brooks B, Firek BA, Miller CS, et al. Microbes in the neonatal intensive care unit resemble those found in the gut of premature infants. Microbiome 2014;2(1):1.
[94] Yuan C, Gaskins AJ, Blaine AI, et al. Association between cesarean birth and risk of obesity in offspring in childhood, adolescence, and early adulthood. JAMA Pediatr 2016;170(11):e162385.
[95] Cardwell CR, Stene LC, Joner G, et al. Caesarean section is associated with an increased risk of childhood-onset type 1 diabetes mellitus: a meta-analysis of observational studies. Diabetologia 2008;51(5):726–35.
[96] Kolokotroni O, Middleton N, Gavatha M, Lamnisos D, Priftis KN, Yiallouros PK. Asthma and atopy in children born by caesarean section: effect modification by family history of allergies - a population based cross-sectional study. BMC Pediatr 2012;12:179.
[97] Sevelsted A, Stokholm J, Bisgaard H. Risk of asthma from cesarean delivery depends on membrane rupture. J Pediatr 2016;171 38-42 e31-38-42 e34.
[98] Chu DM, Ma J, Prince AL, Antony KM, Seferovic MD, Aagaard KM. Maturation of the infant microbiome community structure and function across multiple body sites and in relation to mode of delivery. Nat Med 2017;23(3):314–26.

[99] Chu DM, Meyer KM, Prince AL, Aagaard KM. Impact of maternal nutrition in pregnancy and lactation on offspring gut microbial composition and function. Gut Microbes 2016;7(6):459–70.
[100] Lohmann P, Luna RA, Hollister EB, et al. The airway microbiome of intubated premature infants: characteristics and changes that predict the development of bronchopulmonary dysplasia. Pediatr Res 2014;76(3):294–301.
[101] Committee on Obstetric Practice. Committee opinion No. 725: vaginal seeding. Obstet Gynecol 2017;130(5):e274–8.
[102] Dominguez-Bello MG, De Jesus-Laboy KM, Shen N, et al. Partial restoration of the microbiota of cesarean-born infants via vaginal microbial transfer. Nat Med 2016;22 (3):250–3.
[103] Schaechter M. A brief history of bacterial growth physiology. Front Microbiol 2015;6:289.
[104] Stewart EJ. Growing unculturable bacteria. J Bacteriol 2012;194(16):4151–60.
[105] Woese CR, Kandler O, Wheelis ML. Towards a natural system of organisms: proposal for the domains Archaea, Bacteria, and Eucarya. Proc Natl Acad Sci U S A 1990;87(12):4576–9.
[106] Mancini N, Carletti S, Ghidoli N, Cichero P, Burioni R, Clementi M. The era of molecular and other non-culture-based methods in diagnosis of sepsis. Clin Microbiol Rev 2010;23(1):235–51.
[107] Opota O, Croxatto A, Prod'hom G, Greub G. Blood culture-based diagnosis of bacteraemia: state of the art. Clin Microbiol Infect 2015;21(4):313–22.
[108] Browne HP, Forster SC, Anonye BO, et al. Culturing of 'unculturable' human microbiota reveals novel taxa and extensive sporulation. Nature 2016;533(7604):543–6.
[109] Wade W. Unculturable bacteria--the uncharacterized organisms that cause oral infections. J R Soc Med 2002;95(2):81–3.
[110] Ward DM, Weller R, Bateson MM. 16S rRNA sequences reveal numerous uncultured microorganisms in a natural community. Nature 1990;345(6270):63–5.
[111] Dymock D, Weightman AJ, Scully C, Wade WG. Molecular analysis of microflora associated with dentoalveolar abscesses. J Clin Microbiol 1996;34(3):537–42.
[112] Ramamurthy T, Ghosh A, Pazhani GP, Shinoda S. Current perspectives on viable but non-culturable (VBNC) pathogenic bacteria. Front Public Health 2014;2:103.
[113] Balaban NQ, Merrin J, Chait R, Kowalik L, Leibler S. Bacterial persistence as a phenotypic switch. Science 2004;305(5690):1622–5.
[114] Ayrapetyan M, Williams TC, Oliver JD. Bridging the gap between viable but nonculturable and antibiotic persistent bacteria. Trends Microbiol 2015;23(1):7–13.
[115] Combs CA, Gravett M, Garite TJ, et al. Amniotic fluid infection, inflammation, and colonization in preterm labor with intact membranes. Am J Obstet Gynecol 2014;210 (2) 125 e121-125 e115.
[116] Markenson GR, Martin RK, Tillotson-Criss M, Foley KS, Stewart Jr. RS, Yancey M. The use of the polymerase chain reaction to detect bacteria in amniotic fluid in pregnancies complicated by preterm labor. Am J Obstet Gynecol 1997;177(6):1471–7.
[117] Oyarzun E, Yamamoto M, Kato S, Gomez R, Lizama L, Moenne A. Specific detection of 16 micro-organisms in amniotic fluid by polymerase chain reaction and its correlation with preterm delivery occurrence. Am J Obstet Gynecol 1998;179 (5):1115–19.
[118] Yoon BH, Romero R, Lim JH, et al. The clinical significance of detecting Ureaplasma urealyticum by the polymerase chain reaction in the amniotic fluid of patients with preterm labor. Am J Obstet Gynecol 2003;189(4):919–24.
[119] Yoon BH, Romero R, Kim M, et al. Clinical implications of detection of Ureaplasma urealyticum in the amniotic cavity with the polymerase chain reaction. Am J Obstet Gynecol 2000;183(5):1130–7.

[120] Han YW, Shen T, Chung P, Buhimschi IA, Buhimschi CS. Uncultivated bacteria as etiologic agents of intra-amniotic inflammation leading to preterm birth. J Clin Microbiol 2009;47(1):38–47.
[121] Onderdonk AB, Hecht JL, McElrath TF, et al. Colonization of second-trimester placenta parenchyma. Am J Obstet Gynecol 2008;199(1) 52 e51-52 e10.
[122] Fortner KB, Grotegut CA, Ransom CE, et al. Bacteria localization and chorion thinning among preterm premature rupture of membranes. PLoS One 2014;9(1):e83338.
[123] Stout MJ, Conlon B, Landeau M, et al. Identification of intracellular bacteria in the basal plate of the human placenta in term and preterm gestations. Am J Obstet Gynecol 2013;208(3):226 e221–226 e 227.
[124] Steel JH, Malatos S, Kennea N, et al. Bacteria and inflammatory cells in fetal membranes do not always cause preterm labor. Pediatr Res 2005;57(3):404–11.
[125] Gude NM, Roberts CT, Kalionis B, King RG. Growth and function of the normal human placenta. Thromb Res 2004;114(5–6):397–407.
[126] Soilleux EJ, Coleman N. Transplacental transmission of HIV: a potential role for HIV binding lectins. Int J Biochem Cell Biol 2003;35(3):283–7.
[127] Oliveira GM, Pascoal-Xavier MA, Moreira DR, et al. Detection of cytomegalovirus, herpes virus simplex, and parvovirus b19 in spontaneous abortion placentas. J Matern Fetal Neonatal Med 2019;32(5):768–75.
[128] Moeller AH, Suzuki TA, Phifer-Rixey M, Nachman MW. Transmission modes of the mammalian gut microbiota. Science 2018;362(6413):453–7.
[129] Shahzad R, Khan AL, Bilal S, Asaf S, Lee IJ. What is there in seeds? Vertically transmitted endophytic resources for sustainable improvement in plant growth. Front Plant Sci 2018;9(24):24.
[130] Frank AC, Saldierna Guzman JP, Shay JE. Transmission of bacterial endophytes. Microorganisms 2017;5(4):70.
[131] Schardl CL. Epichloe festucae and related mutualistic symbionts of grasses. Fungal Genet Biol 2001;33(2):69–82.
[132] Rudgers JA, Afkhami ME, Rúa MA, Davitt AJ, Hammer S, Huguet VM. A fungus among us: broad patterns of endophyte distribution in the grasses. Ecology 2009;90 (6):1531–9.
[133] Diaz Herrera S, Grossi C, Zawoznik M, Groppa MD. Wheat seeds harbour bacterial endophytes with potential as plant growth promoters and biocontrol agents of *Fusarium graminearum*. Microbiol Res 2016;186–187:37–43.
[134] Verma SK, Kingsley K, Irizarry I, Bergen M, Kharwar RN, White Jr. JF. Seed-vectored endophytic bacteria modulate development of rice seedlings. J Appl Microbiol 2017;122(6):1680–91.
[135] Walitang DI, Kim CG, Jeon S, Kang Y, Sa T. Conservation and transmission of seed bacterial endophytes across generations following crossbreeding and repeated inbreeding of rice at different geographic locations. Microbiologyopen 2019;8(3): e00662.
[136] Hardoim PR, Hardoim CC, van Overbeek LS, van Elsas JD. Dynamics of seed-borne rice endophytes on early plant growth stages. PLoS One 2012;7(2):e30438.
[137] Ngumbi E, Kloepper J. Bacterial-mediated drought tolerance: current and future prospects. Appl Soil Ecol 2016;105:109–25.
[138] Osakabe Y, Osakabe K, Shinozaki K, Tran LS. Response of plants to water stress. Front Plant Sci 2014;5(86):86.
[139] Timmusk S, Abd El-Daim IA, Copolovici L, et al. Drought-tolerance of wheat improved by rhizosphere bacteria from harsh environments: enhanced biomass production and reduced emissions of stress volatiles. PLoS One 2014;9(5):e96086.
[140] Forni C, Duca D, Glick BR. Mechanisms of plant response to salt and drought stress and their alteration by rhizobacteria. Plant Soil 2017;410(1–2):335–56.

[141] Barnawal D, Bharti N, Pandey SS, Pandey A, Chanotiya CS, Kalra A. Plant growth-promoting rhizobacteria enhance wheat salt and drought stress tolerance by altering endogenous phytohormone levels and TaCTR1/TaDREB2 expression. Physiol Plant 2017;161(4):502–14.
[142] Gontia-Mishra I, Sapre S, Sharma A, Tiwari S. Amelioration of drought tolerance in wheat by the interaction of plant growth-promoting rhizobacteria. Plant Biol (Stuttg) 2016;18(6):992–1000.
[143] Liu F, Xing S, Ma H, Du Z, Ma B. Cytokinin-producing, plant growth-promoting rhizobacteria that confer resistance to drought stress in *Platycladus orientalis* container seedlings. Appl Microbiol Biotechnol 2013;97(20):9155–64.
[144] Vurukonda SS, Vardharajula S, Shrivastava M, Sk ZA. Enhancement of drought stress tolerance in crops by plant growth promoting rhizobacteria. Microbiol Res 2016;184:13–24.
[145] Vacheron J, Desbrosses G, Bouffaud ML, et al. Plant growth-promoting rhizobacteria and root system functioning. Front Plant Sci 2013;4(356):356.
[146] Dimkpa C, Weinand T, Asch F. Plant-rhizobacteria interactions alleviate abiotic stress conditions. Plant Cell Env 2009;32(12):1682–94.
[147] Funkhouser LJ, Bordenstein SR. Mom knows best: the universality of maternal microbial transmission. PLoS Biol 2013;11(8):e1001631.
[148] Koga R, Meng XY, Tsuchida T, Fukatsu T. Cellular mechanism for selective vertical transmission of an obligate insect symbiont at the bacteriocyte-embryo interface. Proc Natl Acad Sci U S A 2012;109(20):E1230–7.
[149] Charbonneau MR, O'Donnell D, Blanton LV, et al. Sialylated milk oligosaccharides promote microbiota-dependent growth in models of infant undernutrition. Cell 2016;164(5):859–71.
[150] Ramani S, Stewart CJ, Laucirica DR, et al. Human milk oligosaccharides, milk microbiome and infant gut microbiome modulate neonatal rotavirus infection. Nat Commun 2018;9(1):5010.
[151] Gage DJ. Infection and invasion of roots by symbiotic, nitrogen-fixing rhizobia during nodulation of temperate legumes. Microbiol Mol Biol Rev 2004;68(2):280–300.
[152] Bode L. Human milk oligosaccharides in the prevention of necrotizing enterocolitis: a journey from in vitro and in vivo models to mother-infant cohort studies. Front Pediatr 2018;6:385.
[153] Asay AK. Mycorrhizal facilitation of kin recognition in interior Douglas-fir (Pseudotsuga menziesii var. glauca) [Text] [M.Sc. thesis]. Vancouver, Canada: University of British Columbia; 2013.
[154] Teste FP, Simard SW, Durall DM, Guy RD, Jones MD, Schoonmaker AL. Access to mycorrhizal networks and roots of trees: importance for seedling survival and resource transfer. Ecology 2009;90(10):2808–22.
[155] Gorzelak MA, Asay AK, Pickles BJ, Simard SW. Inter-plant communication through mycorrhizal networks mediates complex adaptive behaviour in plant communities. AoB Plants 2015;7:plv050.
[156] File AL, Klironomos J, Maherali H, Dudley SA. Plant kin recognition enhances abundance of symbiotic microbial partner. PLoS One 2012;7(9):e45648.
[157] Cook RJ. Making greater use of introduced microorganisms for biological control of plant pathogens. Annu Rev Phytopathol 1993;31(1):53–80.

SECTION II

Birth – entering the world dominated by microbes

CHAPTER

5

Impact of delivery mode in early life microbiome and risk of disease

Eugenie Martineau and Jose C. Clemente

[1]Department of Genetics and Genomic Sciences, Icahn Institute for Data Science and Genomic Technology, Immunology Institute, Icahn School of Medicine at Mount Sinai, New York, NY, United States

Introduction

Maternal and infant mortality used to be an important health problem, as pregnancy was often associated with complications, particularly during delivery. Cesarean section delivery (C-section delivery, CSD) notably improved mortality rates and health outcomes for both mothers and newborns [1]. The World Health Organization currently recommends CSD when the health or life of the mother/newborn is threatened by the use of vaginal delivery (VD). For instance, CSD will be favored in cases of placenta previa, when the fetus is not adequately positioned for birth, when the mother has uncontrolled high blood pressure, diabetes, obesity, or in case of multiple pregnancies. The use of CSD should not exceed 10%–15% of total deliveries, as lower rates significantly increase risk for mother and infant, and overtreatment is wasteful and potentially harmful [2]. However, the actual CSD rates vary significantly across countries, ranging from 6.0% in less-developed countries to 27.2% in developed regions [3]. Importantly, the global average CSD rate increased from 6.7% to 19.1% between 1990 and 2014, at an average annual rate of increase of 4.4% [3].

The Human Microbiome in Early Life
DOI: https://doi.org/10.1016/B978-0-12-818097-6.00005-5

This increase is, however, not without risks: a 14-year follow-up of more than two million Canadian women showed that CSD was associated with various complications, including infections, blood clots, and heart attack, compared with those who delivered vaginally [4]. In addition, mothers who delivered by C-section require more time to recover. Epidemiological studies also show that infants are at higher risk for neonatal infections, respiratory distress [5], and miss the surge of plasma cortisol and catecholamines that occurs during the passage through the birth canal [6]. This hormonal increase positively impacts postnatal glycemia levels, blood pressure, alveolar liquid absorption, and body temperature, and it contributes to neurological adaptation [6]. In addition, epidemiologic studies have shown that CSD increases the risk of several metabolic and chronic immune diseases later in life [7–9]. Although there are numerous factors involved in the pathogenesis of these conditions, it has been often suggested that their recent increase in prevalence is linked to perturbations in early-life microbiome. CSD interrupts the natural transfer of vaginal microbiome to the newborn, and therefore could partially explain this association. In this chapter, we will review recent findings regarding the importance of delivery mode on early-life microbiome, and explore how it impacts the developing immune system and contribute to immune programming and sensitization. Finally, we will investigate the existing and future solutions that aim to reduce the aberrations in the establishment of the microbiome ecosystem in CSD infants and decrease the risk for immune and metabolic diseases.

Delivery mode and dynamics of the microbiome in early life

Age is one of the main drivers of gut microbiome composition in early life [10]. Longitudinal sampling of infants has shown that there is a consistent succession of gut microbes during the first 3 years of life across different populations [11]. Although children younger than 3 years of age have significantly lower microbial diversity than adults, their interindividual variability is significantly higher [11]. This variability can originate from various external factors, such as diet, environment, medications, or delivery mode. In this section, we will present current knowledge on how CSD shapes early microbiome in comparison to VD, and how these differences persist or change throughout childhood. We have summarized the main results and conclusions from each of the studies described in Table 5.1.

TABLE 5.1 Summary of recent studies comparing the effect of delivery mode on the infant's microbiome.

Studies	Samples collection	Proportion of C-section (CSD/Total)	Perinatal antibiotics (VD/CSD)	Breast milk (VD/CSD)	Ethnicity (VD/CSD)				Median BMI or weight	Countries	Sample types	CSD microbiome		Conclusions
					Caucasian	Hispanic	Asian	African				Elevated abundance	Decreased abundance	
Azad et al. [12]	3/4 months after delivery	6/24	5/6	11/4	—				—	Canadian	Fecal	—	*Escherichia–Shigella* and an absence of *Bacteroides*	The microbial community is modified by mode of delivery and infant diet No strong differences by mode of delivery in the prevalence of *Clostridium difficile* or the relative abundances of *Bifidobacterium* or *Clostridium*
Bäckhed et al. [13]	Birth, 1 week, 4 months, 12 months	15/98	2/11	41/15	—				68.5 kg	Swedish	Fecal	Bacteria from skin, oral, and surrounding environment during delivery	*Bacteroides* and *Bifidobacterium*—*Escherichia/Shigella*	Mode of delivery and feeding patterns have major effects on gut microbiota assembly: mother–infant transmission was less frequent in CSD The difference between delivery modes gradually decreased at 4 months and 12 months of age as the infant's microbiome becomes more similar to their mothers
Biasucci et al. [14]	Newborn	23/46	0	46	46	0			—	Italy	Fecal	—	Bifidobacterial species	Early stage of the bacterial intestinal colonization in CSD infants is altered in comparison to VD infants, with no or little influence of the type of feeding Less diversity for CS newborn

(Continued)

TABLE 5.1 (Continued)

Studies	Samples collection	Proportion of C-section (CSD/Total)	Perinatal antibiotics (VD/CSD)	Breast milk (VD/CSD)	Ethnicity (VD/CSD)				Median BMI or weight	Countries	Sample types	CSD microbiome		Conclusions
					Caucasian	Hispanic	Asian	African				Elevated abundance	Decreased abundance	
Bokulich et al. [15]	24 h after delivery and every month during 2 years	19/43	12/13	20/11	28	4	8	2	24.4 ± 4.3	United States	*Baby*: Stool samples; *Mother*: stool samples, vaginal swabs, and rectal swabs (prepartum and postpartum)	—	Persistent decrease in *Bacteroides* populations (1 year)	Diversity was disturbed by delivery mode
Chu et al. [16]	Longitudinal: Day 1, Day 30–45 cross-sectional: Day 1	52/157	45/19	—	1/1	90/45	4/4	10/3	(27.7/27.4)	United States	Stool/oral/nasal/skin	—		No discernable effect of CSD on the early microbiota
Dominguez-Bello et al. [17]	<24 h after delivery	5/10	1/5	—	Mestizo (2/4) and Amerindians (2/1)				—	Venezuela	*Mother*: skin, oral mucosa, and vagina; *Baby*:skin, oral mucosa, and nasopharyngeal	Skin microbial community, (*Staphylococcus*, *Streptococcus*, Propionibacterineae, *Haemophilus*, and *Acinetobacter*)	*Lactobacillus*, *Prevotella*, or *Sneathia* spp.	Neonates bacterial communities undifferentiated across multiple body habitats, regardless of delivery mode VD infants: microbiome similar in composition to the vaginal communities of the mothers CSD infants lacked bacteria from the vaginal community but harbor bacterial communities similar to the skin communities of the mothers

Goedert et al. [18]	Adult	92/1097	248	—	995	16	38	10	24	United States	Fecal	Lachnospiraceae, Peptococcaceae, and Ruminococcaceae families of Clostridia (*Firmicutes*), *Dysgonomonas* (Bacteroidetes, Bacteroidales) and *Salmonella* and *Haemophilus* (Proteobacteria, Enterobacteriaceae)	*Coprobacillus* and *Holdemania* (Firmicutes) and of *Neisseria* (*Proteobacteria*)	Adults born by CSD appear to have a distinctly different composition of their fecal microbial population
Grönlund et al. [19]	3 to 5, 10, 30, 60, and 180 days of age	30/64	For all CSD	—	—				—	Finland	Fecal	—	*Bifidobacterium* and *Lactobacillus* colonization rates reached the rates of vaginally delivered infants at 1 month and 10 days, respectively/ low colonization of *Bacteroides*	The primary gut flora in infants born by cesarean delivery may be disturbed for up to 6 months after the birth
Hill et al. [20]	Week 1, 4, 8, 24	105/192	3/21	49/47	—				—	Ireland	Stool	*Clostridium*	*Bifidobacterium* and *Bacteroides* and *Lactobacillus*	Mode of delivery and gestational age at birth: strong influencers of early gut microbiota structure following birth Impact of mode of delivery and gestational age still apparent 24 weeks after birth, but decrease with age
Jakobsson et al. [21]	1 week and 1, 3, 6, 12, and 24 months after birth	9/24	4/2	0.83	—				—	Sweden	Fecal	*Enterococcus*	*Bacteroides*/circulating levels of Th1-associated chemokines during infancy	CSD was associated with a lower total microbial diversity, delayed colonization of the Bacteroidetes phylum, and reduced Th1 responses during the first 2 years of life

(Continued)

TABLE 5.1 (Continued)

Studies	Samples collection	Proportion of C-section (CSD/ Total)	Perinatal antibiotics (VD/CSD)	Breast milk (VD/ CSD)	Ethnicity (VD/CSD)				Median BMI or weight	Countries	Sample types	CSD microbiome		Conclusions
					Caucasian	Hispanic	Asian	African				Elevated abundance	Decreased abundance	
Martin et al. [22]	First defecation, day 2, 1 week, 1/3/6 months	28/108	0	102	—				—	Netherlands	fecal	—	*Bifidobacteria*, *Bacteroides*. *Lactobacillus* (until day 2)	CSD infants had a delayed colonization of several bacterial groups or species VD Infants had significantly higher bacterial counts in the meconium than those born by CS
McCann et al. [23]	After 1 year	10/20	—	8/7	—				—	Ireland	Stool	—	Anelloviridae: TTV	Birth mode results in distinctly different viral communities, with VD infants having greater viral and bacteriophage diversity
Mueller et al. [24]	Day 1, 7, 30, and 90	46/59	3/5	13/40	12/42	1/4			(22.4/ 25.4)	Brazil	Meconium/stool	—	Bacteroidetes, *Parabacteroides*, and *Clostridium*	Impact of delivery mode on the initial transition in neonatal intestinal bacterial structure, composition, and predicted metagenome
Mueller et al. [25]	Day 1	48/74	3/8	49/7	70	4			CSD <25:30 and >25:26. VD <25:13 and >25:5	Brazil	Neonatal stool	*Enterococcus*, *Bifidobacterium*	*Bacteroides* (delayed), Clostridiaceae	CSD: perturbation to the asssembly of neonatal intestinal microbiome shortly after birth

Nagpal et al. [26]	first 24 h of life	17/151	—	—	0	0	151	0	—	Japan	Meconium	—	*Lactobacillus*	The infant microbiota is strongly influenced by delivery mode and the differences are not noticeable at meconium stage and become prominent thereafter Delayed/lower colonization of *Lactobacilli* and relatively different proportions of facultative vs obligatory anaerobes in the meconium of CSD babies
Salminem et al. [27]	7 years old infant	31/60	—	—	—				—	Finland	Fecal	—	*Clostridial*	Abnormal development of the intestinal microbiota reported following cesarean section
Wampach et al. [28]	Day 1, 3, 5, 28, 150, and 365.	7/15	6/5	0/5	7/5	0	0	1/1	(21.8/26.8)	Luxembourg	Meconium/stool	*Haemophilus* spp., Streptococcus spp., *Enterobacter* spp., *Propionibacterium* spp., *Staphylococcus* spp., and the genus *Lactobacillus* over the first year of life—*Bifidobacterium* spp.	*Bacteroides*, *Parabacteroides*, *Methanosphaera* spp., *Dothideomycetes* spp., and Pezizomycotina	CSD infants: delay in colonization and succession affecting all three domains of life. First year: archaea and fungi were more often detected by qPCR in VD infants compared to CSD infants Major differences between VD and CSD infants were less apparent at 1 year of age
Wampach et al. [29]	≤24 h, 3 days and 5 days postpartum	18/33	2/10	15/16	—				(60/75) kg	Luxembourg	Stool (mothers and babies), vaginal swab	OTUs assigned to *Staphylococcus*	*Bacteroides*, *Parabacteroides*	Several functional pathways significantly underrepresented in CSD neonates
Yassour et al. [30]	(3, 6, 12, 18, 24, and 36 months	4/39	—	—	—				—	Finland	Fecal	*Bifidobacterium* spp.	*Bacteroides*	Delivery mode has a strong long-term effect on microbial diversity Low-*Bacteroides* signature in all CSD infant microbiome and 20% of vaginally born children

Effect on the neonatal microbiome (delivery day)

During VD, passage through the birth canal seeds the neonate with vaginal bacteria, particularly *Lactobacillus* spp. such as *Lactobacillus crispatus, Lactobacillus gasseri, Lactobacillus jensenii,* or *Lactobacillus iners.* Neonates born by CSD, on the other hand, lack this first critical microbial exposure. Accordingly, it is assumed that VD neonates are primarily colonized by vaginal bacteria, whereas CD infants are exposed to a mixture of environmental microbes [17]. Such differences should be most obvious in the neonatal skin and oral microbiomes. Interestingly, previous studies have reported that neonatal microbiome and its associated functional pathways are relatively homogeneous across all body sites (skin, oral, nasopharyngeal), with the notable exception of the neonatal meconium [17,31].

The meconium is the earliest stool of the newborn and is usually discharged within the first 24 hours postdelivery. It is composed of intestinal epithelial cells, mucus, amniotic fluids, and other substances ingested while the fetus is in the uterus. While the meconium was previously thought to be sterile, a number of recent studies have reported bacterial cells and DNA in meconium samples [32–34], although it is still unclear whether these bacteria originate from contaminants [35] or other sources. While microbiome diversity in the meconium is not significantly different between VD and CSD infants [24,26,31], the meconium of infants born by VD has significantly higher bacterial counts [22].

The skin and oral microbiome have reduced diversity in CSD neonates [31] and harbor bacterial communities similar to those found in the maternal skin, whereas VD neonates harbor bacterial taxa similar in composition to the vaginal communities of the mothers, which are absent in CSD neonate. Accordingly, the dominant taxa are reflective of delivery mode: *Propionibacterium, Staphylococcus* spp., and *Streptococcus* in CSD neonates, while *Lactobacillus, Prevotella, Atopobium,* or *Sneathia* spp. are associated with VD [17,31].

It is worth noting that neonatal samples, particularly meconium and skin, have very low microbial biomass. Because the amount of microbial DNA is lower than in, for example, stool samples, contaminant DNA and cross contamination can become important experimental confounders [36]. Contaminant DNA can originate from sample collection or during sample preparation: laboratory environment, researchers, plastic consumables, nucleic acid extraction kits, or laboratory reagents are often sources of contamination [35]. Cross contamination might occur when high biomass samples are processed simultaneously with low biomass samples [37]. To ensure the quality and robustness of studies using low biomass samples, Eisenhofer et al. provide a set of guidelines to minimize the effects of contamination, including experimental design reports, use of control samples, and detailed analysis of potential contaminant taxa [36].

Effect on infant microbiome (from birth to 6 months)

During the first months of life, the infant microbiota structure and function gradually changes and diversifies to adapt to different body sites. Therefore, differences in the initial communities (e.g., lack of vaginal exposure in CSD babies) may lead to variation in the microbial succession patterns in the gut and other body habitats that persist over time [17].

Accordingly, several studies highlighted that CSD infants experience a delay in colonization and succession by *Lactobacillus*, *Bifidobacterium*, and *Bacteroides* [22,28,38,39]. This could be caused by the delay in exposure to these bacteria, or due to the colonization by different microbial taxa, which could be less adapted to the human gastrointestinal tract and therefore have lower growth rates. Prokaryotic richness stabilizes between the first and fifth month of life [28], and the differences between delivery modes gradually decrease from 4 months [40]. However, another study still observed an effect of delivery mode in the microbiome 6 months after birth, albeit diminished [20].

Overall, several studies have reported an effect of delivery mode in microbiome composition of CSD infants. A notable exception is the work of Chu et al. in which no discernible effect of delivery mode in the microbiota was appreciated [31]. One possible explanation of this finding is the fact that this cohort was composed mostly of Hispanic women (over 85%), a population that has been previously shown to harbor low levels of vaginal *Lactobacillus* [41]. Importantly, no study has yet investigated the relevance of maternal vaginal *Lactobacillus* in the composition of the offspring microbiome, and additional work will be required to validate this hypothesis.

While the bacterial microbiome in early life has been well documented, less is known about the viral, fungal, and archeal components. A study by Lim et al. observed that while the gut eukaryotic virome and bacterial microbiome expand during the first 2 years of life, bacteriophage richness and diversity are higher in early life and decrease with age [42]. This study, however, did not present results on the effect of delivery mode in the virome. A more recent study following newborns for the first year of life found that archaea and fungi were more often detected in VD than in CSD infants [28]. Birth mode also results in distinct viral communities at 1 year of age, with VD infants having greater eukaryotic and bacterial viral diversity [23]. In addition to these community-wide differences, several taxa were also decreased in CSD infants, including viruses

(*Torque Teno Virus* [TTV]), archaea (*Methanosphaera* spp.), and fungi (*Dothideomycetes* spp. and Pezizomycotina) [23,28].

Effect on microbiome maturation (6 months and older)

After the introduction of solid food in the diet, approximately at 6 months of age, gut microbiome richness continues to gradually increase over time [15], and the differences between VD and CSD infants become less apparent as the ratio of Firmicutes/Bacteroides between CSD and VD infants converge with age [28]. It is however unclear whether this ratio represents a molecular marker of underlying biological processes or is a cohort-specific estimate, as some recent meta-analyses have found inconsistencies across various studies in the context of obesity [43,44]. Other studies have found that differences in specific bacterial taxa remain for prolonged periods, with CSD infants having significantly more abundant Clostridiales and Enterobacteriaceae at 4 and 12 months [40] and lower microbiome diversity at 8 months [15].

After 6 months, the microbiome of infants undergoes a gradual maturation, with many of the taxa characteristic of early childhood being replaced by a mixture of Firmicutes, primarily Clostridiales, and leading to an adultlike microbiome configuration. This change could be partially driven by the introduction of solid foods and subsequent exposure to new and diverse nutrients, which could modulate gut microbial content and decrease early differences in profiles caused by delivery mode or other maternal and neonatal characteristics [45,46]. Microbial diversity has also been reported to be lower in CSD children even at 2 years of age [21], that being said another group confirmed this results for all infants with low-*Bacteroides* signature regardless of their delivery mode [30]. At 7 years of age, Salminen et al. reported an abnormal development of the intestinal microbiota following CSD [27]. While the bifidobacterial levels in the feces at birth were comparable regardless of the delivery mode at birth, the numbers of *Clostridia* were significantly higher in VD children 7 years after birth.

The infant microbiome is not solely defined by delivery mode

Although several studies are consistent in that CSD infants experience a lower microbiome diversity and richness as well as a lower prevalence of *Bacteroides* during the first 6 months of life (see Table 5.1), conclusions are, however, less definitive after 6 months of life, as results tend to vary across studies. For instance, a study of 39 Finnish children found that the lack of *Bacteroides* is not restricted to CSD children, as 7 of 35 (20%) children delivered vaginally also had low *Bacteroides* abundance [30]. Various maternal and medical factors could influence how the early-life microbiome is shaped, which we will summarize in the following section.

Although most studies generally describe birthing as "cesarean" or "vaginal," the delivery procedure is significantly more complex and often associated with complications such as preterm birth, eclampsia, umbilical cord complications, or prolonged pregnancy that can impact the infant microbiome outcomes [47]. Moreover, the presence of labor during CSD causes changes in levels of endocrine, inflammatory, and contractile factors. These changes might influence the maternal microbiome or the establishment of the neonatal microbiome [6]. In addition, labor is often accompanied by rupture of the fetal membranes, exposing the fetus to maternal vaginal bacteria. Accordingly, previous studies have demonstrated that the neonatal microbiome richness and diversity can be drastically impacted by the presence/absence of labor during CSD [12]. As for VD, the procedure can also deeply impact the neonate microbiome diversity, as the presence/absence of fecal matter impacts the *Bifidobacterium* level [48], which could partially explain some of the differences observed between studies.

Antibiotic intake is another major confounder in microbiome studies. In the context of birthing, mothers delivering by CS are administered intrapartum antibiotic prophylaxis as standard of care, although the exact timing of administration can vary in different regions. It has been largely documented that maternal antibiotic intakes are a powerful disruptor as it affects the developing gut microbiome of the infant by decreasing the relative abundance of Actinobacteria, Bacteroidetes, *Bifidobacteria* spp., *Bacteroides* spp., *Parabacteroides* spp., *Atopobium* spp., and *Lactobacillus* spp., and there is an increase in the relative abundance of Proteobacteria, Firmicutes, Enterobacteriaceae, *Enterococcus* spp., and *Clostridium* spp. [49–53]. These perturbations of the neonatal gut microbiome directly overlap with those observed in CSD infants. Therefore, the use of antibiotics at late stages of pregnancy and before delivery should be registered and considered when assessing the effect of CSD in the offspring microbiome.

Maternal condition can also play an important role in delivery mode and, by extension, in how the infant's microbiome is shaped early in life. For instance, maternal obesity ($BMI > 25\ kg/m^2$) is a known risk factor for CSD, with an odds ratio (OR) between 2.01 and 2.36 compared with women of normal weight. Interestingly, a recent study published by Mueller et al. shows that high BMI in women is associated with drastic increase of *Bacteroides* in VD but not in CSD infants [25]. Similarly, Tun et al. reported that VD infants ($n = 708$) born to overweight or obese mothers had a statistically higher abundance of several genera (e.g., *Bacteroides* spp., *Megasphaera* spp., *Blautia* spp., and *Oscillospira* spp.) and lower abundance of others (e.g., *Haemophilus* and *Veillonella*) compared to those born from normal weight mothers [54]. A more recent study has also reported that maternal inflammatory bowel disease (IBD) status impacts bacterial diversity and richness in the infant stool at days 7, 14, and 90, although surprisingly did not find an association with delivery mode [55]. This could be partially due to the overall microbial community structure (beta diversity) that did not differ significantly by delivery mode [55]. Of note, CSD on itself has been associated with an increased risk of immune disorders later in life, including IBD (see next section).

Geographical factors can also impact the infant microbiome. For instance, Yatsunenko et al. reported a progressive adaptation of the gut microbiome in American, Venezuelan, and Malawian infants during the first years of life: as infants age, their bacterial communities gradually become more similar to those of adults in their population and differentiate from those of other populations [11]. Similarly, Vatanen et al. identified marked differences in the gut microbiome development of 222 infants from three populations in Finnish Karelia, Russian Karelia, and Estonia during their first 3 years of life. They show that infants in Finnish Karelia and Estonia have a distinct gut microbiome compared to those from Russian Karelia, with significantly lower abundance of *Bacteroides* species [39]. It is important to notice, however, that geographic origin can be often intertwined by other factors such as diet, which on itself can also impact microbial diversity and composition.

Overall, these results emphasize the need to carefully design studies investigating microbiome differences between VD and CSD, as many of the confounders can introduce important biases. In practice, this will require comprehensive characterization of the cohorts (e.g., exposure to medications, comorbidities, diet, and detailed description of delivery procedure) followed by analysis that rigorously adjusts for confounders.

Impact of delivery mode on health outcomes

In the first section, we reviewed the evidence linking delivery mode to changes in early-life microbiome. While this difference appears to be gradually reduced after weaning, its consequences may persist through its impact on the development of the immune system. In this section, we will describe potential mechanisms through which disruptions to the newborn's microbiota can lead to important metabolic and immune outcomes.

Interplay between the immune system and the early microbiome

The early-life microbiome modulates immune development and health outcomes

The composition and timing of early neonatal colonization are important factors influencing the microbial education of the developing immune system, potentially modifying systemic gene expression and increased risk of various diseases. The early microbiome can modulate the infant's developing immune phenotype by interacting with molecular receptors of gut immune cells and activating signaling cascades that lead to cell differentiation and control inflammatory status [56]. For instance, a study using germ-free (GF) mouse models of allergy has demonstrated that regulatory T cells (Treg) and invariant natural killer T (iNKT) cells can only be "educated" by microbial colonization during a specific postnatal time window to enable tolerance to later environmental exposures [57,58]. Signals from the intestinal microbiota are involved in several immune mechanisms, such as the maturation of the intestinal mucosa and its gut-associated lymphoid tissues, the completion of the Peyer's patches of the distal ileum, but also the development of nonlymphoid structures of the intestinal mucosa that contributes to the establishment of host–microbiota mutualism [59,60]. Overall, these results strongly suggest that early-life colonization plays a critical role in shaping immune development, which can subsequently impact disease risk.

Disrupted microbial colonization due to C-section birthing can have long-term health consequences through the impaired immune development. For instance, members of the *Bacteroides* genus are known to exert strong effects on the immune system. The surface polysaccharide of *Bacteroides fragilis* is recognized by toll-like receptor 2 (TLR2) in Treg cells, inducing the production of IL10 and other cytokines that affects the Th1/Th2 balance and promotes immunotolerance [61]. The delayed colonization of CSD infants with *Bacteroides* [22,28,38,39] could therefore explain the delayed activation of Th1-type immunity observed in these infants [21].

Remarkably, some studies have shown that although the microbial changes due to CSD are only temporal, their effect on the regulation of the immune system or metabolic profiles can persist in adulthood. For example, mice born by CSD have a distinct bacterial profile at weaning, but while these microbial differences gradually disappear, various regulatory immune cells (Foxp3 + regulatory T cells, tolerogenic CD103 +, dendritic cells (DC) in mesenteric lymph nodes, and spleens) are significantly lower than in VD mice at adulthood (Hansen et al., 2014). In another study in mice, Cox et al. found that antibiotic exposure (low-dose penicilin) during infancy transiently disturbs its microbiome. This perturbation, while limited to early life, leads to long-term metabolic effect in adulthood [62]. These findings in mouse models have also been observed in humans. A clinical study following infants with IgE-associated eczema during the first 2 years of life, found lower diversity of the total microbiota at 1 month and a lower diversity of the bacterial phylum Bacteroidetes and the genus *Bacteroides* at 1 month and the phylum Proteobacteria at 12 months of age [63].

Timing of colonization matters

Timing of colonization can have important consequences for bacterial engraftment and subsequent impact on immune development. In fact, varying order and timing of colonization by complex microbial communities and/or a cocktail of four bacterial strains, Martinez et al. have highlighted the importance of early colonization history in gut microbiota assembly [64]. Because CSD infants are initially colonized by a different set of bacteria than VD newborns, the dynamics of the gut microbiota follow notably different trajectories in these groups. This can delay the establishment of intestinal anaerobes most commonly associated with the adult gut, such as Bacteroidetes. Indeed, the timing of colonization of GF mice has been shown to significantly impact immune development. Sudo et al. have demonstrated that intestinal colonization in early life is required to induce systemic tolerance to oral antigen [ovalbumin (OVA)] exposure. Monocolonization of GF mice with *Bifidobacterium infantis* during infancy provided immune responsiveness to OVA. However, when *B. infantis* was introduced in adulthood, the GF phenotype persisted, and oral OVA did not induce systemic tolerance [65]. GF mice are also more susceptible to both ulcerative colitis and OVA-induced asthma than specific-pathogen-free mice [58]. This susceptibility is mediated by an increase in iNKT cells and chemokine ligand CXCL16. However, the susceptibility was diminished when the GF pups and their dams were colonized on the day of birth, but not when the mice were colonized at 5 weeks of age. The absence of colonization during the newborn period resulted in increased hypermethylation of the Cxcl16 gene, causing increased iNKT cell accumulation in the colon and lungs [58]. Overall, these studies strongly suggest that timing and order of bacterial colonization are important determinants of immune development, providing a potential mechanism to explain how CSD impacts health phenotypes later in life.

Delivery mode may have a lifelong impact on health outcomes

Immune diseases

As described earlier, aberrant neonatal microbiota composition following CSD is associated with an alteration of the immune system. Numerous clinical and epidemiological studies have reported an increased risk for immune disorders in children born by cesarean section. Specifically, CSD is associated with a moderate risk increase for asthma. For instance, a study of two million term children in Denmark (1977–2012) found that the 14% born by CSD had significantly higher risk of asthma (OR of 1.23) even when accounting for confounders [8]. The epidemiological work of Håkansson et al. and the meta-analyses of Bager et al. (26 studies in total) confirmed this association and found that the risk of asthma was increased by 31% and 18%, respectively for CSD children [9,66]. In addition, Bager et al. also found an association between CSD and food allergies (OR of 1.32). Further, among children whose mothers were allergic, CSD was associated with a sevenfold increased risk of parentally perceived reactions to egg, fish, or nuts and a fourfold increased risk of confirmed egg allergy [67]. This result is of particular interest, as it suggests that the combination of genetic predisposition and CSD-driven dysbiosis can result in an increased risk for immune diseases.

Although other associations with immunological diseases have been reported, results are less robust and often contradictory. For instance, Bager and Sevelsted both report an association of CSD with IBD, with OR of 1.29 and 1.20, respectively [8,68]. In contrast, other groups have not identified such association. Bruce et al. reported an OR of 1.00 [69], and Decker et al. found ORs of 1.14 for Crohn's disease and 0.87 for ulcerative colitis [70]. Celiac disease was found to be associated with CSD in this last study (OR of 1.8), although a different study found no such association [71]. Type 1 diabetes has also been associated with CSD: a meta-analysis of 21 studies found a 20% increase in the risk of childhood-onset type 1 diabetes after CSD that cannot be explained by other confounders [72]. Yet, Sevelsted et al. did not find such association in their Danish cohort (OR of 1.01) [8].

Overall, CSD appears to be a shared risk factor for several immune-related diseases, including asthma and food allergies. The association with IBD, celiac disease, or type 1 diabetes, however, appears less clear, and different studies are inconsistent. This heterogeneity could originate from differences in the populations as some of the aforementioned studies have been conducted in Europe, whereas others were performed in Canada or China. The inclusion of different confounders, or the use of different methods for data analysis, could also impact the robustness of the conclusions. An important factor that is often unclear in many of these studies is the presence/absence of labor during C-section. In fact, it has been suspected that nonlabored C-sections might have a stronger association with allergic outcomes than labored C-sections [66,73]. Finally, it is worth noting that immune diseases can also be a partial consequence of stress hormone induction in the fetus, which is lower in CSD infants and may affect immune maturation [74].

Metabolic disorders

Numerous studies have described an association between obesity and imbalances in gut microbiota. For instance, Ley et al. observed that genetically obese (ob/ob) mice have decreased Bacteroidetes/Firmicutes ratios compared with their lean (ob/ + and + / + wildtype) siblings [75]. Further, GF mice colonized with feces from obese mice gained more fat compared to mice colonized with feces from lean mice [76], and similar results have also been found when colonizing GF mice with microbiota from lean or obese twin donors [77]. These findings, however, have been brought into question as more recent meta-analyses have found inconsistencies across studies [43] or found the effect sizes to be moderate [42].

Some clinical studies have found evidence that CSD leads to increased adiposity in children [78]. This association of CSD with childhood obesity has been reported in numerous studies, with ORs ranging from 1.34 to 5.3 [7,78,79]. A meta-analysis based on 28 studies also found that CSD, when compared with VD, is associated with a 34% increased risk of obesity in children between the ages of 2 and 18 years [80]. Yet, as with immune diseases, there is conflicting evidence on the relation between microbiome and obesity. In fact, a Danish group observed a higher infant BMI at the age of 6 months (increase of 0.37) but not in later childhood [81]. Accordingly, Flemming and Masukume both agreed on an absence of association between CSD and childhood obesity [82,83]. These inconsistencies may arise from unaccounted confounders in some studies, such as maternal BMI. For instance, the meta-analysis by Kuhle et al. found that, in 11 studies that adjusted their models for maternal prepregnancy weight, the OR was reduced from 1.57 to 1.29 after adjusting for maternal prepregnancy weight [80]. Preterm infants or high fetal weight (macrosomia, >4 kg) are also strong confounders [79] that are sometimes neglected [84]. Overall, the relation between CSD, disrupted microbiome, and increased infant obesity appears inconclusive, and further work will be needed to answer this question.

Can we modulate the infant microbiome after C-section?

Vaginal microbial transfer

Given the impact of CSD in the neonatal microbiome, and the immune and health outcomes associated with it, efforts to restore microbial imbalance are a promising approach to lower the risk of such diseases. We have previously reported the results from a proof-of-principle study that transferred vaginal bacteria from healthy mothers to CSD infants [85]. The microbiome of a cohort of 18 infants (7 VD, 11 scheduled CSD) and their mothers was studied for the first month of life. Four of the CSD infants were exposed to maternal vaginal fluids immediately after birth using a gauze that had been previously introduced for an hour in the maternal vagina. Anal, oral, and skin samples were collected from all subjects from birth to 4 weeks of age. The microbiome of the CSD infants exposed to vaginal fluids resembled that of VD infants, especially during the first week of life. Specifically, there was an enrichment in *Lactobacillus* in skin and anal samples of the exposed CSD infants compared to unexposed CSD babies. This was followed by a bloom of *Bacteroides* from the second week, which again was not observed in CSD newborns who were not exposed to vaginal fluids. Although limited by a very small sample size and the lack of health outcomes, this result has paved the way for three ongoing clinical studies (clinicaltrials.gov: NCT03567707, NCT03298334, NCT03809390) that are testing whether vaginal exposure on CSD infants can significantly reduce the risk for allergies and obesity, and impact immune regulation and programming. As with any other intervention, it is important to note that certain risks are associated with this practice, including the potential risk of transferring pathogens to the newborn [86]. However, this risk can be mitigated by using adequate exclusion criteria that limit the inclusion of mothers to only those showing no evidence of infection before C-section [87].

Breastfeeding

Breastfeeding provides a natural approach to modify microbial communities in infants after CSD as it has been shown to support the growth of *Bifidobacterium longum* and inhibit other taxa (lower proportion of Firmicutes and Proteobacteria) in 4-week old Korean infants [88]. Moreover, a recent study has reported that, in breastfed infants, one fourth of the gut bacteria during the first month of life were predicted to be acquired from maternal breast milk, and that gut microbial changes are associated with the proportion of daily breast milk intake in a dose-dependent manner [89]. Of note, this study only considered breast milk or areolar skin as potential sources, and so the origin of most of the gut microbiota composition in this cohort remains unexplained. There was also considerable variability in the proportion of bacteria predicted to originate from breast milk, suggesting that other factors could also play an important role in shaping the gut microbiome during the first month of life.

Exclusive consumption of breast milk has also been reported to be involved in the restoration of a VD-like microbiome. For instance, Liu et al. show that the unweighted UniFrac distance between the VD and CSD exclusively breastfed infant groups was significantly smaller than between the VD and CSD mixed-fed infant groups [90]. In addition, Hill et al. acknowledge the importance of breastfeeding duration, as a comparison of infants breastfed for less than 4 months versus those breastfed for longer revealed a significant difference for CSD but not VD [20].

The beneficial effect of human milk is thought to be due to multiple interacting factors. First, short-chain fatty acids (SCFAs) in the newborn are produced through the bacterial fermentation of oligosaccharides that are present in breast milk (human milk oligosaccharides). While SCFAs are not directly digested by the host, they serve as an energy source for beneficial colonic bacteria such as *Bifidobacterium* spp. [91,92]. Lactoferrin (LF), which appears in highest concentration in the colostrum, can also favor the proliferation of *Bifidobacterium* and limits the growth of several strains of *Escherichia coli* by promoting exclusion of iron from bacteria [93]. LF also has an immunomodulatory effect as it can activate DC, which in turn induce Th1 helper cell population that promotes resistance against neonatal infection [94].

Probiotics

Several studies have investigated the use of probiotics in early life as an alternative approach to modify gut microbiome composition and, potentially, to reduce the immune and metabolic risks associated with cesarean delivery. Kuitunen et al. showed that consumption of a probiotic mixture (two lactobacilli, bifidobacteria, and propionibacteria) in VD and CSD infants conferred protection (IgE sensitization) only to cesarean-delivered children [95]. Similarly, the administration of formula containing *Lactobacillus reuteri* DSM17938 did not affect the microbiota of vaginally delivered infants but promoted restoration of microbial homeostasis in CSD infants [96]. However, other studies have not confirmed a positive effect on asthma associated with early administration of *L. reuteri* (OR of 1.16), *Lactobacillus rhamnosus* HN001 (OR of 0.95), or *Lactobacillus paracasei* F19 (relative risk of 1.05) [61,97,98]. The use of *Saccharomyces boulardii*, a yeast-based probiotic used to treat gastrointestinal disorders associated with abnormal bacterial growth [99], in CSD infants has also shown no significant difference with placebo [100].

It therefore appears that there is no substantial evidence of a positive impact of the use of probiotics to modify gut microbiome and reduce the risk of metabolic/immune disorders following CSD. We speculate that rationally designed combinations of bacteria, as opposed to single probiotics, could provide a more effective approach to modulate microbiome changes in CSD infants and improve health outcomes [101,102].

Concluding remarks

In this chapter, we explored the relationship between delivery mode, early infant microbiomes, and health outcomes (see Fig. 5.1). Different studies have reported transient differences in the microbiota of babies born by cesarean delivery compared to VD infants, with delay in colonization and succession by *Bacteroides*, *Bifidobacterium*, and *Lactobacillus* and a decrease in richness [22] and diversity [21]. These differences are transient, and as infants develop, they undergo a succession of microbial changes and gradual maturation over the first months of life. As a result, the differences between VD and CSD infants in microbiome composition become less apparent after 6 months, approximately at the time of introduction of solid foods. However, numerous studies have shown an association between early microbiome dysbiosis and an increase incidence of asthma, allergies, autoimmune, and metabolic diseases later in life. Therefore, understanding how delivery mode shapes the initial bacterial inoculum of the newborn remains of great importance for disease prevention.

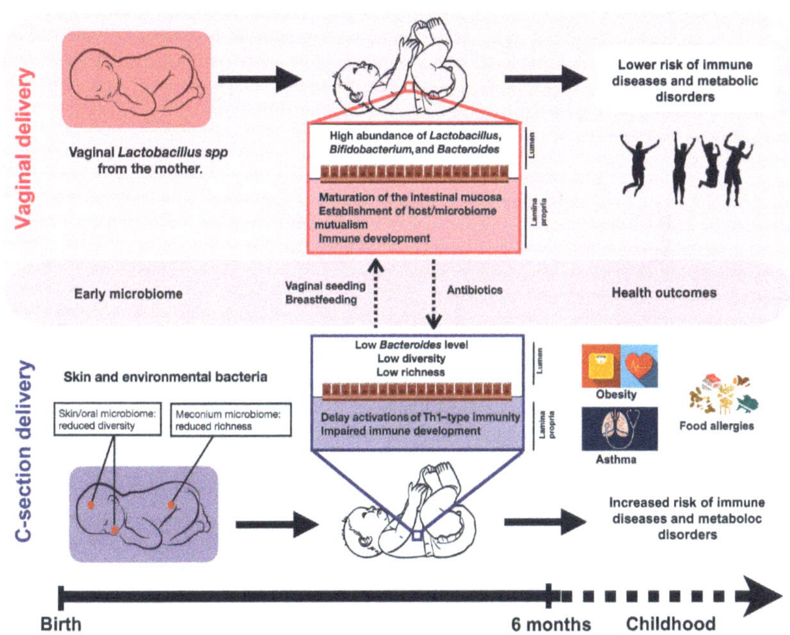

FIGURE 5.1 **Delivery mode and dynamics of the microbiome in early life.** Life stages from birth until childhood and the effect of delivery mode in microbiome and health outcomes for vaginal (*top*) and cesarean birthing (*bottom*).

We note that some of the signatures associated with CSD, such as low abundance of *Bacteroides* [15,22,28,38,39], have also been observed in VD infants [30]. What drives this variability? We hypothesize that maternal vaginal microbiome variability could partially be responsible for this heterogeneity. Vaginal bacterial communities differ significantly not only across ethnicities, but also between different women of the same ethnic group [41], suggesting that the observed differences in the microbiome of VD infants could be a consequence of different bacterial inocula.

Despite existing evidence linking CSD with deleterious outcomes, it is important to consider the high level of conflicting results. As we have mentioned, there are multiple and significant confounding factors present in most studies, and it is therefore highly unlikely that any single factor alone determines early-life microbial composition. We therefore strongly recommend the use of multivariate analysis and integrative models that consider antibiotic administration, labor onset, breastfeeding status, or maternal factors among others.

References

[1] Molina G, Weiser TG, Lipsitz SR, Esquivel MM, Uribe-Leitz T, Azad T, et al. Relationship between cesarean delivery rate and maternal and neonatal mortality. JAMA 2015;314:2263–70.

[2] WHO, H. (2015). WHO | WHO statement on caesarean section rates.

[3] Betrán AP, Ye J, Moller A-B, Zhang J, Gülmezoglu AM, Torloni MR. The Increasing Trend in Caesarean Section Rates: Global, Regional and National Estimates: 1990-2014. PLOS ONE 2016;11:e0148343.

[4] Liu S, Liston RM, Joseph KS, Heaman M, Sauve R, Kramer MS. Maternal mortality and severe morbidity associated with low-risk planned cesarean delivery versus planned vaginal delivery at term. CMAJ 2007;176:455–60.

[5] O'Shea TM, Klebanoff MA, Signore C. Delivery After Previous Cesarean: Long-Term Outcomes in the Child. Semin. Perinatol. 2010;34:281–92.

[6] Lagercrantz H. The good stress of being born. Acta Paediatr 2016;105:1413–16.

[7] Huh SY, Rifas-Shiman SL, Zera CA, Edwards JWR, Oken E, Weiss ST, et al. Delivery by caesarean section and risk of obesity in preschool age children: a prospective cohort study. Arch Dis Child 2012;97:610–16.

[8] Sevelsted A, Stokholm J, Bønnelykke K, Bisgaard H. Cesarean section and chronic immune disorders. Pediatrics 2015;135:e92–8.

[9] Bager P, Wohlfahrt J, Westergaard T. Caesarean delivery and risk of atopy and allergic disesase: meta-analyses. Clin Exp Allergy 2008;38:634–42.

[10] Lozupone CA, Stombaugh J, Gonzalez A, Ackermann G, Wendel D, Vázquez-Baeza Y, et al. Meta-analyses of studies of the human microbiota. Genome Res. 2013;23:1704–14.

[11] Yatsunenko T, Rey FE, Manary MJ, Trehan I, Dominguez-Bello MG, Contreras M, et al. Human gut microbiome viewed across age and geography. Nature 2012;486:222–7.

[12] Azad MB, Konya T, Maughan H, Guttman DS, Field CJ, Chari RS, et al. Gut microbiota of healthy Canadian infants: profiles by mode of delivery and infant diet at 4 months. CMAJ 2013;185:385–94.

[13] Bäckhed F, Roswall J, Peng Y, Feng Q, Jia H, Kovatcheva-Datchary P, et al. Dynamics and Stabilization of the Human Gut Microbiome during the First Year of Life. Cell Host Microbe 2015;17:690–703.

[14] Biasucci G, Benenati B, Morelli L, Bessi E, Boehm G. Cesarean delivery may affect the early biodiversity of intestinal bacteria. J. Nutr 2008;138:1796S–800S.
[15] Bokulich NA, Chung J, Battaglia T, Henderson N, Jay M, Li H, et al. Antibiotics, birth mode, and diet shape microbiome maturation during early life. Sci Transl Med 2016;8:343ra82.
[16] Chu DM, Ma J, Prince AL, Antony KM, Seferovic MD, Aagaard KM. Maturation of the infant microbiome community structure and function across multiple body sites and in relation to mode of delivery. Nat. Med. 2017;23:314–26.
[17] Dominguez-Bello MG, Costello EK, Contreras M, Magris M, Hidalgo G, Fierer N, et al. Delivery mode shapes the acquisition and structure of the initial microbiota across multiple body habitats in newborns. Proc Natl Acad Sci 2010;107:11971–5.
[18] Goedert JJ, Hua X, Yu G, Shi J. Diversity and Composition of the Adult Fecal Microbiome Associated with History of Cesarean Birth or Appendectomy: Analysis of the American Gut Project. EBioMedicine 2014;1:167–72.
[19] Grönlund MM. Fecal microflora in healthy infants born by different methods of delivery: permanent changes in intestinal flora after cesarean delivery. J. Pediatr. Gastroenterol. Nutr. 1999;28:19–25.
[20] Hill CJ, Lynch DB, Murphy K, Ulaszewska M, Jeffery IB, O'Shea CA, et al. Evolution of gut microbiota composition from birth to 24 weeks in the INFANTMET Cohort. Microbiome 2017;5:4.
[21] Jakobsson HE, Abrahamsson TR, Jenmalm MC, Harris K, Quince C, Jernberg C, et al. Decreased gut microbiota diversity, delayed Bacteroidetes colonisation and reduced Th1 responses in infants delivered by Caesarean section. Gut 2014;63:559–66.
[22] Martin R, Makino H, Yavuz AC, Ben-Amor K, Roelofs M, Ishikawa E, et al. Early-life events, including mode of delivery and type of feeding, siblings and gender, shape the developing gut microbiota. PLoS One 2016;11:e0158498.
[23] McCann A, Ryan FJ, Stockdale SR, Dalmasso M, Blake T, Ryan CA, et al. Viromes of one year old infants reveal the impact of birth mode on microbiome diversity. PeerJ 2018;6:e4694.
[24] Mueller NT, Shin H, Pizoni A, Werlang IC, Matte U, Goldani MZ, et al. Delivery mode and the transition of pioneering gut-microbiota structure, composition and predicted metabolic function. Genes 2017;8:364.
[25] Mueller NT, Shin H, Pizoni A, Werlang IC, Matte U, Goldani MZ, et al. Birth mode-dependent association between pre-pregnancy maternal weight status and the neonatal intestinal microbiome. Sci Rep 2016;6:23133.
[26] Nagpal R, Tsuji H, Takahashi T, Kawashima K, Nagata S, Nomoto K, et al. Sensitive quantitative analysis of the meconium bacterial microbiota in healthy term infants born vaginally or by cesarean section. Front Microbiol 2016;7.
[27] Salminen S, Gibson GR, McCartney AL, Isolauri E. Influence of mode of delivery on gut microbiota composition in seven year old children. Gut 2004;53:1388–9.
[28] Wampach L, Heintz-Buschart A, Hogan A, Muller EEL, Narayanasamy S, Laczny CC, et al. Colonization and succession within the human gut microbiome by archaea, bacteria, and microeukaryotes during the first year of life. Front Microbiol 2017;8.
[29] Wampach L, Heintz-Buschart A, Fritz JV, Ramiro-Garcia J, Habier J, Herold M, et al. Birth mode is associated with earliest strain-conferred gut microbiome functions and immunostimulatory potential. Nat. Commun. 2018;9.
[30] Yassour M, Vatanen T, Siljander H, Hämäläinen A-M, Härkönen T, Ryhänen SJ, et al. Natural history of the infant gut microbiome and impact of antibiotic treatment on bacterial strain diversity and stability. Sci Transl Med 2016;8:343ra81.

[31] Chu DM, Ma J, Prince AL, Antony KM, Seferovic MD, Aagaard KM. Maturation of the infant microbiome community structure and function across multiple body sites and in relation to mode of delivery. Nat Med 2017;23:314–26.

[32] Ardissone AN, de la Cruz DM, Davis-Richardson AG, Rechcigl KT, Li N, Drew JC, et al. Meconium microbiome analysis identifies bacteria correlated with premature birth. PLoS One 2014;9:e90784.

[33] Gosalbes MJ, Llop S, Vallès Y, Moya A, Ballester F, Francino MP. Meconium microbiota types dominated by lactic acid or enteric bacteria are differentially associated with maternal eczema and respiratory problems in infants. Clin Exp Allergy 2013;43:198–211.

[34] Jiménez E, Marín ML, Martín R, Odriozola JM, Olivares M, Xaus J, et al. Is meconium from healthy newborns actually sterile? Res Microbiol 2008;159:187–93.

[35] Salter SJ, Cox MJ, Turek EM, Calus ST, Cookson WO, Moffatt MF, et al. Reagent and laboratory contamination can critically impact sequence-based microbiome analyses. BMC Biol 2014;12.

[36] Eisenhofer R, Minich JJ, Marotz C, Cooper A, Knight R, Weyrich LS. Contamination in Low Microbial Biomass Microbiome Studies: Issues and Recommendations. Trends Microbiol. 2019;27:105–17.

[37] Ballenghien M, Faivre N, Galtier N. Patterns of cross-contamination in a multispecies population genomic project: detection, quantification, impact, and solutions. BMC Biol. 2017;15.

[38] Grönlund MM. Fecal microflora in healthy infants born by different methods of delivery: permanent changes in intestinal flora after cesarean delivery. J Pediatr Gastroenterol Nutr 1999;28:19–25.

[39] Vatanen T, Kostic AD, d'Hennezel E, Siljander H, Franzosa EA, Yassour M, et al. Variation in microbiome LPS immunogenicity contributes to autoimmunity in humans. Cell 2016;165:842–53.

[40] Bäckhed F, Roswall J, Peng Y, Feng Q, Jia H, Kovatcheva-Datchary P, et al. Dynamics and stabilization of the human gut microbiome during the first year of life. Cell Host Microbe 2015;17:690–703.

[41] Ravel J, Gajer P, Abdo Z, Schneider GM, Koenig SSK, McCulle SL, et al. Vaginal microbiome of reproductive-age women. Proc Natl Acad Sci 2011;108:4680–7.

[42] Lim ES, Zhou Y, Zhao G, Bauer IK, Droit L, Ndao IM, et al. Early life dynamics of the human gut virome and bacterial microbiome in infants. Nat Med 2015;21:1228–34.

[43] Sze MA, Schloss PD. Looking for a signal in the noise: revisiting obesity and the microbiome. MBio 2016;7 e01018-16.

[44] Walters WA, Xu Z, Knight R. Meta-analyses of human gut microbes associated with obesity and IBD. FEBS Lett 2014;588:4223–33.

[45] Fallani M, Amarri S, Uusijarvi A, Adam R, Khanna S, Aguilera M, et al. Determinants of the human infant intestinal microbiota after the introduction of first complementary foods in infant samples from five European centres. Microbiology 2011;157:1385–92.

[46] Koenig JE, Spor A, Scalfone N, Fricker AD, Stombaugh J, Knight R, et al. Succession of microbial consortia in the developing infant gut microbiome. Proc. Natl. Acad. Sci. 2011;108:4578–85.

[47] Stranges E, Wier LM, Elixhauser A. Complicating conditions of vaginal deliveries and cesarean sections, 2009: Statistical Brief #131. In: Healthcare Cost and Utilization Project (HCUP) Statistical Briefs [Internet]. Rockville (MD): Agency for Healthcare Research and Quality (US); 2006 Feb. 2012 May.

[48] Makino H, Kushiro A, Ishikawa E, Kubota H, Gawad A, Sakai T, et al. Mother-to-Infant Transmission of Intestinal Bifidobacterial Strains Has an Impact on the Early Development of Vaginally Delivered Infant's Microbiota. PLOS ONE 2013;8:e78331.

[49] Aloisio I, Mazzola G, Corvaglia LT, Tonti G, Faldella G, Biavati B, et al. Influence of intrapartum antibiotic prophylaxis against group B *Streptococcus* on the early newborn gut composition and evaluation of the anti-*Streptococcus* activity of *Bifidobacterium* strains. Appl Microbiol Biotechnol 2014;98:6051–60.

[50] Fallani M, Young D, Scott J, Norin E, Amarri S, Adam R, et al. Intestinal microbiota of 6-week-old infants across Europe: geographic influence beyond delivery mode, breast-feeding, and antibiotics. J Pediatr Gastroenterol Nutr 2010;51:77–84.

[51] Mazzola G, Murphy K, Ross RP, Gioia DD, Biavati B, Corvaglia LT, et al. Early gut microbiota perturbations following intrapartum antibiotic prophylaxis to prevent Group B Streptococcal disease. PLoS One 2016;11:e0157527.

[52] Azad MB, Konya T, Persaud RR, Guttman DS, Chari RS, Field CJ, et al. Impact of maternal intrapartum antibiotics, method of birth and breastfeeding on gut microbiota during the first year of life: a prospective cohort study. BJOG Int J Obstet Gynaecol 2016;123:983–93.

[53] Nogacka A, Salazar N, Suárez M, Milani C, Arboleya S, Solís G, et al. Impact of intrapartum antimicrobial prophylaxis upon the intestinal microbiota and the prevalence of antibiotic resistance genes in vaginally delivered full-term neonates. Microbiome 2017;5:93.

[54] Tun HM, Bridgman SL, Chari R, Field CJ, Guttman DS, Becker AB, Mandhane PJ, Turvey SE, Subbarao P, Sears MR, et al. Roles of Birth Mode and Infant Gut Microbiota in Intergenerational Transmission of Overweight and Obesity From Mother to Offspring. JAMA Pediatr 2018;172:368–77.

[55] Torres J, Hu J, Seki A, Eisele C, Nair N, Huang R, et al. Infants born to mothers with IBD present with altered gut microbiome that transfers abnormalities of the adaptive immune system to germ-free mice. Gut 2020;69(1):42–51. Available from: https://doi.org/10.1136/gutjnl-2018-317855. Epub 2019 Apr 29.

[56] Romagnani S. Regulation of the T cell response. Clin. Exp. Allergy 2006;36:1357–66.

[57] Hansen CHF, Nielsen DS, Kverka M, Zakostelska Z, Klimesova K, Hudcovic T, et al. Patterns of Early Gut Colonization Shape Future Immune Responses of the Host. PLOS ONE 2012;7:e34043.

[58] Olszak T, An D, Zeissig S, Vera MP, Richter J, Franke A, et al. Microbial Exposure During Early Life Has Persistent Effects on Natural Killer T Cell Function. Science 2012;336:489–93.

[59] Cherrier M, Eberl G. The development of LTi cells. Curr. Opin. Immunol. 2012;24:178–83.

[60] Maynard CL, Elson CO, Hatton RD, Weaver CT. Reciprocal interactions of the intestinal microbiota and immune system. Nature 2012;489:231–41.

[61] Round JL, Mazmanian SK. Inducible Foxp3+ regulatory T-cell development by a commensal bacterium of the intestinal microbiota. Proc. Natl. Acad. Sci. 2010;107:12204–9.

[62] Cox LM, Yamanishi S, Sohn J, Alekseyenko AV, Leung JM, Cho I, et al. Altering the intestinal microbiota during a critical developmental window has lasting metabolic consequences. Cell 2014;158:705–21.

[63] Abrahamsson TR, Jakobsson HE, Andersson AF, Björkstén B, Engstrand L, Jenmalm MC. Low diversity of the gut microbiota in infants with atopic eczema. J. Allergy Clin. Immunol. 2012;129:434–40 e2.

[64] Martínez I, Maldonado-Gomez MX, Gomes-Neto JC, Kittana H, Ding H, Schmaltz R, et al. Experimental evaluation of the importance of colonization history in early-life gut microbiota assembly. ELife 2018;7:e36521.

[65] Sudo N, Sawamura S, Tanaka K, Aiba Y, Kubo C, Koga Y. The requirement of intestinal bacterial flora for the development of an IgE production system fully susceptible to oral tolerance induction. J. Immunol. 1997;159:1739–45.

[66] Håkansson S, Källén K. Caesarean section increases the risk of hospital care in childhood for asthma and gastroenteritis. Clin. Exp. Allergy 2003;33:757–64.
[67] Eggesbø M, Botten G, Stigum H, Nafstad P, Magnus P. Is delivery by cesarean section a risk factor for food allergy? J. Allergy Clin. Immunol. 2003;112:420–6.
[68] Bager P, Simonsen J, Nielsen NM, Frisch M. Cesarean section and offspring's risk of inflammatory bowel disease: a national cohort study. Inflamm Bowel Dis 2012;18:857–62.
[69] Bruce A, Black M, Bhattacharya S. Mode of delivery and risk of inflammatory bowel disease in the offspring: systematic review and meta-analysis of observational studies. Inflamm Bowel Dis 2014;20:1217–26.
[70] Decker E, Engelmann G, Findeisen A, Gerner P, Laaß M, Ney D, et al. Cesarean delivery is associated with celiac disease but not inflammatory bowel disease in children. Pediatrics 2010;125:e1433–40.
[71] Dydensborg Sander S, Hansen AV, Størdal K, Andersen A-MN, Murray JA, Husby S. Mode of delivery is not associated with celiac disease. Clin Epidemiol 2018;10:323–32.
[72] Cardwell CR, Stene LC, Joner G, Cinek O, Svensson J, Goldacre MJ, et al. Caesarean section is associated with an increased risk of childhood-onset type 1 diabetes mellitus: a meta-analysis of observational studies. Diabetologia 2008;51:726–35.
[73] Renz-Polster H, David MR, Buist AS, Vollmer WM, O'Connor EA, Frazier EA, et al. Caesarean section delivery and the risk of allergic disorders in childhood. Clin. Exp. Allergy 2005;35:1466–72.
[74] Gitau R, Menson E, Pickles V, Fisk NM, Glover V, MacLachlan N. Umbilical cortisol levels as an indicator of the fetal stress response to assisted vaginal delivery. Eur. J. Obstet. Gynecol. Reprod. Biol. 2001;98:14–17.
[75] Ley RE, Bäckhed F, Turnbaugh P, Lozupone CA, Knight RD, Gordon JI. Obesity alters gut microbial ecology. Proc. Natl. Acad. Sci. 2005;102:11070–5.
[76] Turnbaugh PJ, Ley RE, Mahowald MA, Magrini V, Mardis ER, Gordon JI. An obesity-associated gut microbiome with increased capacity for energy harvest. Nature 2006;444:1027.
[77] Ridaura VK, Faith JJ, Rey FE, Cheng J, Duncan AE, Kau AL, et al. Gut microbiota from twins discordant for obesity modulate metabolism in mice. Science 2013;341:1241214.
[78] Blustein J, Attina T, Liu M, Ryan AM, Cox LM, Blaser MJ, et al. Association of caesarean delivery with child adiposity from age 6 weeks to 15 years. Int J Obes 2013;37:900–6.
[79] Zhou L, He G, Zhang J, Xie R, Walker M, Wen SW. Risk factors of obesity in preschool children in an urban area in China. Eur J Pediatr 2011;170:1401–6.
[80] Kuhle S, Tong OS, Woolcott CG. Association between caesarean section and childhood obesity: a systematic review and meta-analysis. Obes Rev 2015;16:295–303.
[81] Vinding RK, Sejersen TS, Chawes BL, Bønnelykke K, Buhl T, Bisgaard H, et al. Cesarean delivery and body mass index at 6 months and into childhood. Pediatrics 2017;139:e20164066.
[82] Flemming K, Woolcott CG, Allen AC, Veugelers PJ, Kuhle S. The association between caesarean section and childhood obesity revisited: a cohort study. Arch Dis Child 2013;98:526–32.
[83] Masukume G, McCarthy FP, Baker PN, Kenny LC, Morton SM, Murray DM, et al. Association between caesarean section delivery and obesity in childhood: a longitudinal cohort study in Ireland. BMJ Open 2019;9:e025051.
[84] Bar-Meir M, Friedlander Y, Calderon-Margalit R, Hochner H. Mode of delivery and offspring adiposity in late adolescence: The modifying role of maternal pre-pregnancy body size. PLoS One 2019;14:e0209581.
[85] Dominguez-Bello MG, De Jesus-Laboy KM, Shen N, Cox LM, Amir A, Gonzalez A, et al. Partial restoration of the microbiota of cesarean-born infants via vaginal microbial transfer. Nat. Med. 2016;22:250–3.

[86] Cunnington, A., Sim, K., Deierl, A., Kroll, J.S., Brannigan, E., and Darby, J. (2016). "Vaginal seeding" of infants born by Caesarean section. How should health professionals engage with this increasingly popular but unproven practice?.

[87] Clemente JC, Dominguez-Bello MG. Safety of vaginal microbial transfer in infants delivered by caesarean, and expected health outcomes. BMJ 2016;352:i1707.

[88] Lee SA, Lim JY, Kim B-S, Cho SJ, Kim NY, Kim OB, et al. Comparison of the gut microbiota profile in breast-fed and formula-fed Korean infants using pyrosequencing. Nutr. Res. Pract. 2015;9:242–8.

[89] Pannaraj PS, Li F, Cerini C, Bender JM, Yang S, Rollie A, et al. Association Between Breast Milk Bacterial Communities and Establishment and Development of the Infant Gut Microbiome. JAMA Pediatr 2017;171:647–54.

[90] Liu Y, Qin S, Song Y, Feng Y, Lv N, Xue Y, et al. The perturbation of infant gut microbiota caused by cesarean delivery is partially restored by exclusive breastfeeding. Front Microbiol 2019;10.

[91] Engfer MB, Stahl B, Finke B, Sawatzki G, Daniel H. Human milk oligosaccharides are resistant to enzymatic hydrolysis in the upper gastrointestinal tract. Am. J. Clin. Nutr 2000;71:1589–96.

[92] Yu Z-T, Chen C, Kling DE, Liu B, McCoy JM, Merighi M, et al. The principal fucosylated oligosaccharides of human milk exhibit prebiotic properties on cultured infant microbiota. Glycobiology 2013;23:169–77.

[93] Sherman MP. Lactoferrin and Necrotizing Enterocolitis. Clin. Perinatol. 2013;40:79–91.

[94] Spadaro M, Caorsi C, Ceruti P, Varadhachary A, Forni G, Pericle F, Giovarelli M. Lactoferrin, a major defense protein of innate immunity, is a novel maturation factor for human dendritic cells. FASEB J. 2008;22:2747–57.

[95] Kuitunen M, Kukkonen K, Juntunen-Backman K, Korpela R, Poussa T, Tuure T, et al. Probiotics prevent IgE-associated allergy until age 5 years in cesarean-delivered children but not in the total cohort. J. Allergy Clin. Immunol. 2009;123:335–41.

[96] Garcia Rodenas CL, Lepage M, Ngom-Bru C, Fotiou A, Papagaroufalis K, Berger B. Effect of Formula Containing Lactobacillus reuteri DSM 17938 on Fecal Microbiota of Infants Born by Cesarean-Section. J. Pediatr. Gastroenterol. Nutr. 2016;63:681–7.

[97] West CE, Hammarström M-L, Hernell O. Probiotics in primary prevention of allergic disease – follow-up at 8–9 years of age. Allergy 2013;68:1015–20.

[98] Wickens K, Stanley TV, Mitchell EA, Barthow C, Fitzharris P, Purdie G, et al. Early supplementation with Lactobacillus rhamnosus HN001 reduces eczema prevalence to 6 years: does it also reduce atopic sensitization? Clin. Exp. Allergy 2013;43:1048–57.

[99] Costalos C, Skouteri V, Gounaris A, Sevastiadou S, Triandafilidou A, Ekonomidou C, et al. Enteral feeding of premature infants with Saccharomyces boulardii. Early Hum. Dev. 2003;74:89–96.

[100] Zeber-Lubecka N, Kulecka M, Ambrozkiewicz F, Paziewska A, Lechowicz M, Konopka E, et al. Effect of Saccharomyces boulardii and Mode of Delivery on the Early Development of the Gut Microbial Community in Preterm Infants. PLOS ONE 2016;11:e0150306.

[101] Korpela K, Salonen A, Vepsäläinen O, Suomalainen M, Kolmeder C, Varjosalo M, et al. Probiotic supplementation restores normal microbiota composition and function in antibiotic-treated and in caesarean-born infants. Microbiome 2018;6.

[102] Atarashi K, Tanoue T, Oshima K, Suda W, Nagano Y, Nishikawa H, et al. T_{reg} induction by a rationally selected mixture of *Clostridia* strains from the human microbiota. Nature 2013;500:232–6.

CHAPTER

6

Early-life antibiotic exposure, the gut microbiome, and disease in later life

Samuli Rautava[1,2]

[1]Department of Pediatrics, University of Helsinki, Helsinki, Finland, [2]New Children's Hospital, Helsinki University Hospital, Helsinki, Finland

The discovery of antibiotics is without doubt one of the most important advances in medicine in the last century. Antibiotics have revolutionized the treatment of infectious diseases and made possible immense advances in invasive surgery, organ transplantation, and in the treatment of severe trauma and malignancies. In perinatal and neonatal medicine, prophylactic and therapeutic use of antimicrobial medication has contributed to the astonishing decrease in maternal and neonatal mortality, which has taken place in the last century. As a result, our view of infectious diseases and particularly bacterial infections has changed dramatically in the era of antibiotics. Interestingly, however, our understanding of microbes has also markedly shifted during the last decades. Specific bacteria naturally continue to play an important role as pathogens causing significant and sometimes life-threatening disease, but we have come to realize that human beings exist in a world dominated by microbes. The majority of the bacteria in our environment and on and in our bodies are innocuous, and some of them appear to be essential for our well-being. Understanding the significance of the indigenous microbiome to normal immune and metabolic development as well as the maintenance of healthy homeostasis has forced us to view the effects of antibiotics from a new angle. The gut colonization process during the first months of life is prone to

DOI: https://doi.org/10.1016/B978-0-12-818097-6.00006-7

external disturbances caused by cesarean section delivery, formula feeding, and antibiotics [1]. Aberrant gut colonization patterns in early life have been associated with increased risk of chronic noncommunicable diseases including atopic eczema and asthma [2–6] and overweight and obesity [7–9] later in childhood. In addition, early dysbiosis has been suggested to play a role in the pathogenesis of other noncommunicable diseases including autism spectrum disorders [10] and inflammatory bowel disease (IBD) [11]. Our knowledge of the establishment of the intestinal microbiome in early life is still incomplete, and we are only beginning to grasp the impact of antibiotic perturbations on the microbiome and, ultimately, on health and disease. The purpose of this chapter is to critically review the current body of knowledge regarding how antibiotic exposure during pregnancy, delivery, and early infancy may affect gut colonization and the risk of disease.

Antibiotic exposure during pregnancy

It is not clear how and to what extent bacteria might influence the developing fetus. It is evident, however, that intrauterine and other maternal infections may have devastating consequences, leading to fetal demise or preterm delivery. In this context, treatment with antibiotics may be lifesaving and, for instance, maternal antibiotic therapy delays but does not prevent preterm delivery in the case of preterm premature rupture of membranes [12]. Outside the context of pregnancy complications, there are data to suggest that microbes may modulate fetal physiology. Children of mothers living in a farming environment during pregnancy and thus presumably exposed to more bacteria than urban women reportedly have a lower risk of asthma [13]. The protective effect of this indirect fetal microbial exposure has been replicated in experimental animals by using just one specific bacterium isolated from cow sheds [14].

Whether more direct fetal bacterial exposure occurs during a healthy pregnancy is currently debated and discussed more thoroughly elsewhere in this volume. Several studies indicate that the placenta or amniotic fluid may contain bacteria or bacterial DNA [15–17], albeit not all studies have found reliable traces of bacteria in the intrauterine compartment [18,19]. Fetal contact with microbial components has recently been reported from both rodents [20] and ruminants [21]. It is therefore conceivable that the fetus receives immunomodulatory bacterial stimuli and that changes in the immediate microbial environment during fetal life caused by antibiotics may interfere with the normal maturation process and results in increased risk of disease.

Epidemiological studies suggest that maternal antibiotic use during pregnancy is associated with increased risk of asthma in the child. The observed correlations must be interpreted with caution. Antibiotic use during pregnancy is likely to reflect maternal infections, which may in turn either lead to fetal infection or, perhaps more often, maternal immune responses potentially affecting fetal immune development. It is also possible that antibiotic use is an indicator of respiratory morbidity in the mother, and thus hereditary factors may account for the increase in asthma risk in the offspring. Furthermore, the infectious disease leading to antibiotic use during pregnancy may increase the risk of preterm delivery, which is a well-established risk factor for childhood asthma and wheezing. Ideally, all these potentially confounding factors should be taken into consideration in epidemiological studies assessing the association between antibiotic use during pregnancy and the risk of developing asthma.

The risk of asthma hospitalization (hazard ratio 1.17, 95% confidence interval (CI) 1.00–1.36) and the use of inhaled corticosteroids (hazard ratio 1.18, 95% CI 1.10–1.27) were found to be significantly increased in children whose mothers had used antibiotics during pregnancy in the Danish National Birth cohort of more than 30,000 subjects [22]. In a smaller cohort of 411 children born to mothers with asthma, the authors found the risk of childhood asthma exacerbation to be almost twofold if the mothers had used antibiotics during the third trimester of pregnancy [22]. According to a recent systematic review and meta-analysis of eight studies, antibiotic use during the first, second, or third trimester of pregnancy are all associated with increased childhood asthma risk after adjusting for potential confounding factors including maternal age at delivery, marital status at pregnancy, race, educational level, parity, smoking during pregnancy, maternal asthma, and maternal drug allergy [23]. The associations were no longer statistically significant after additional adjustment for maternal infection during pregnancy, but the rationale for this adjustment can be questioned since antibiotics are in most cases used to treat infection, and thus a strong correlation between the two is to be expected.

The matter was further dissected in a register study of more than 900,000 children born in Denmark [24]. Maternal antibiotic use was found to have a dose-related association with childhood asthma after adjusting for potentially confounding factors including maternal asthma and gestational age. Interestingly, however, a similar association with asthma in the offspring was detected for maternal antibiotic use occurring before or after the pregnancy. These findings were corroborated in a population-based epidemiological study of more than 200,000 mother–infant pairs, which indicated that maternal antibiotic use before, during, and after pregnancy was associated with childhood asthma [25]. These data may be interpreted to suggest that maternal antibiotic use merely reflects proneness to infection, which may in turn explain the predisposition to immune-mediated disease in the child.

The association between maternal antibiotic use during pregnancy and childhood disease is not confined to asthma but similar data exist with regard to infectious disease [26] and body mass index (BMI) [27]. Again based on Danish population-based data, Miller et al. reported that the risk of hospitalization for infectious disease was significantly higher (hazard ratio 1.18, 95% CI 1.17–1.19) in children whose mothers had used antibiotics during pregnancy [26]. Interestingly, a similar risk increase was also seen in children born to mothers who had used antibiotics before pregnancy. On the other hand, particularly vaginally delivered children appeared to be at increased risk. While maternal proneness to infection may again at least partially explain the reported association, the latter observation suggests that the microbial inoculum received from the mother may also play a role. This notion may be interpreted to be corroborated by an observation according to which antibiotic exposure particularly late in pregnancy is associated with increased risk of otitis media and need for tympanostomy tubes [28].

Maternal antibiotic use particularly in early pregnancy was reported to correlate with child BMI at the age of 2 years in a cohort study of 527 subjects [27]. Unfortunately, no fecal samples were analyzed, and the mechanisms underlying the observed effect remain unknown. In general, no reliable data on the effect of prenatal antibiotic exposure on human gut colonization patterns are to our knowledge available. However, the impact of antibiotic use in pregnancy on the gut microbiome and risk of obesity in the offspring has been investigated in experimental animal models. Extrapolating these data to humans is not straightforward since the antibiotic exposure is often quite different from the occasional courses of antibiotics used in the clinical setting. Nonetheless, the experimental design aids in dissecting the effect of antibiotic exposure from that of proneness to or the immunologic consequences of infection. In a study by Yoshimoto et al. [29], female mice were administered a combination of low-dose vancomycin, penicillin V, and chlortetracycline from early life and throughout pregnancy. The abundance of Firmicutes and particularly members of the genus *Clostridium* was increased in the offspring gut microbiome. Furthermore, the pups exposed to antibiotics in prenatal life exhibited increased adiposity as compared to nonexposed pups, and the fat percentage in mice was detected to correlate with fecal clostridia. While the study hardly recapitulates the circumstances in real life, it provides mechanistic insight into the possible interactions between maternal antibiotic use during pregnancy, early-life gut microbiome, and development of overweight.

Intrapartum antibiotics

An estimated 33%–39% of women are administered antibiotics during delivery [30,31]. The reasons for intrapartum antibiotic use range from treatment of intrauterine or other infections to prophylactic use aiming to prevent infections in the mother or the child. The incidence of early-onset neonatal sepsis caused by group B *Streptococcus* (GBS) has decreased dramatically after risk-based or screening-based practices of antibiotic prophylaxis have been introduced [32]. Prophylactic antibiotics are also often routinely administered during cesarean section delivery. According to a Cochrane collaboration systematic review and meta-analysis [33], this practice significantly reduces the risk of uterine and wound infections in the mother, but it is of note that the authors conclude that there are no data on the effects on the child.

The impact of intrapartum antibiotic exposure on the developing gut microbiome in the child has been investigated in several studies but long-term data are still lacking. Studies assessing these associations must be devised carefully to consider potential confounding factors. The mode of delivery, the particular antibiotic used, and whether intrapartum antibiotics have been administered in the context of infection or prophylaxis should all ideally be taken into consideration. Aloisio et al. investigated the gut microbiome of 52 vaginally born term infants, 26 of whom had been exposed to ampicillin prophylaxis because of maternal GBS colonization [34]. At 6–7 days of age, the neonates exposed to maternal ampicillin had significantly fewer bifidobacteria in their stools by qualitative PCR as compared to nonexposed neonates. The same research group has later confirmed this finding in two studies. In a cohort of 84 term infants, the differences in bifidobacteria detected using qPCR at 7 days of age were no longer seen at 30 days of age [35]. In 2016, the fecal microbiota of 20 vaginally delivered neonates was analyzed by 16S rRNA gene sequencing at 6–7 days of age, and the abundance of the phylum Actinobacteria and the genus *Bifidobacterium* was lower in infants exposed to intrapartum antibiotics [36]. According to a prospective cohort study of 198 healthy term infants from Canada, the dysbiosis associated with intrapartum antibiotic prophylaxis is still detectable at the age of 3 months both in children born by cesarean and vaginal delivery [37]. The antibiotic-associated microbiome changes were no longer detectable at 12 months of age in vaginally delivered infants but remained significant in those born by cesarean section delivery. Most likely due to practices in clinical care, the study did not include any infants born by cesarean section and not exposed to antibiotics, which renders the data somewhat difficult to interpret. Taken together, these data indicate that intrapartum antibiotics cause a transient delay in early colonization with bifidobacteria. The clinical significance of this phenomenon remains to be determined, but it is important to note that reduced fecal bifidobacteria in the first weeks of life have been associated with the risk of atopic disease and asthma [2,4] and as well as overweight [8] later in childhood.

Few studies have directly investigated the long-term effects of intrapartum antibiotic use on infant and child health. The first direct demonstration of association between intrapartum antibiotic exposure and risk of disease comes from a case-control study of 77 infants in which antibiotic use during delivery was observed to be associated with an increased risk of infantile colic [38]. No reliable data to our knowledge exist regarding noncommunicable diseases such as asthma, autism spectrum disorders, or IBD, which all have been suggested to be linked to early-life dysbiosis.

An association between intrapartum antibiotic use and the abundance of fecal microbial antibiotic resistance genes in the infant has recently been reported in two independent studies. In a study of 40 term, vaginally delivered infants, Nogacka et al. detected a decreased abundance of Actinobacteria and Bacteroidetes and an increased abundance of Proteobacteria and Firmicutes in the early gut microbiome of infants exposed to intrapartum antibiotics [39]. Moreover, the results suggest a higher occurrence of beta-lactamase genes in the fecal samples obtained at the age of 1 month. We have recently reported that infants acquire microbial antibiotic resistance genes and mobile genetic elements from their mothers and that intrapartum antibiotic use is associated with higher abundances of specific antibiotic resistance genes in infant fecal samples obtained at the age of 1 month [40].

Antibiotic use in the neonatal period and early infancy

Antibiotic use is very common in children. It has been estimated that in the United States, on average children have received nearly three courses of antibiotics by the age of 2 years, and by the age of 10 years the number of courses increases to 10 [41]. Antibiotics have been an immense success story in reducing the mortality and morbidity associated with acute infectious diseases in neonates and infants, but it is likely that not all antibiotic use that occurs is necessary. The obvious negative consequences of antibiotic use in early life include cost to the family and the society and increased risk for the emergence and propagation of antibiotic resistance. The impact of early antibiotic administration on the developing gut microbiome and the potentially associated risk of noncommunicable disease has only recently become an active field of study.

It is well-established that antibiotics have a devastating impact on intestinal microbial ecology [42,43]. This may manifest as antibiotic-associated diarrhea, which, in particular when caused by *Clostridioides difficile*, causes significant morbidity and cost. In adults, the gut microbiome perturbations caused by antibiotics have been thought to resolve relatively rapidly. The intestinal microbiome reportedly begins to recover almost directly after the antibiotic insult, but for instance after repeated courses of ciprofloxacin, the recovery is incomplete, and even though after 10 months the gut microbiome has stabilized, it is distinct from the original state [43]. Even more long-term gut microbiome disturbances have been detected in children. Korpela et al. analyzed the gut microbiome in 142 children aged 2–7 years and discovered that the relative abundance of Actinobacteria was reduced and that of Proteobacteria and Bacteroidetes was increased 6 months after antibiotic use as compared to nonexposed children [44]. Macrolide antibiotics were observed to have a more devastating impact on the gut microbiota than penicillins, and even though some of the changes resolved 12 months after antibiotic exposure, differences were still detectable after 24 months [44].

Observational human studies are by design confounded by the fact that antibiotics are administered to treat infection. It is difficult to dissect between the impact of the antibiotic treatment, the infection, and the ensuing inflammatory reaction on the gut microbiome. To bypass this problem, studies have been conducted with healthy volunteers who have been administered antibiotics in the absence of infection. Before the advent of microbiome sequencing techniques, Jernberg et al. demonstrated that a 7-day course of clindamycin results in altered gut microbiome detectable at least 2 years after exposure in healthy individuals [45]. In a more recent study using sophisticated molecular microbiome analysis techniques [46], seven healthy adults were administered ciprofloxacin and metronidazole for 7 days. While drastic changes in the gut microbiome were detected after antibiotic exposure, the dysbiosis appeared to mostly have resolved by 21 days after the antibiotic course based on serial stool samples and lower intestinal microbiome specimens obtained through colonoscopy.

Neonatal antibiotic exposure and the developing gut microbiome

Antibiotic use is common in the neonatal period. Early onset sepsis is a devastating and life-threatening disease, which afflicts approximately 1 in 1000 newborns. The symptoms and signs of neonatal sepsis are often nonspecific. Since rapid commencement of antimicrobial therapy is paramount for reducing the mortality and morbidity caused by neonatal bacterial infection, empirical antibiotic therapy is currently recommended to all neonates with symptoms suggesting infection, and also asymptomatic newborns deemed to be at high risk based on risk factors such as maternal chorioamnionitis [47]. As a consequence, more than 5% of all neonates reportedly receive antibiotics [31]. It is important to note that a large number of neonates receiving empirical antibiotics do not have bacterial infection and the treatment should be rapidly discontinued after sepsis has been ruled out by negative blood cultures and reassuring results from other laboratory tests. Differentiating between neonates with bacterial infection and those who have received empirical antibiotics and in whom sepsis was subsequently ruled out would allow dissecting between the effects of antibiotics and the infection and inflammation on the gut microbiome and later health.

Several observational studies indicate that neonatal antibiotic exposure has a significant impact on the gut microbiome during the first weeks of life [48–50]. Most of the studies include a relatively small number of subjects, and the interpretation of the results is often complicated by confounding factors including prematurity and the related morbidities and differences in mode of birth and duration of breastfeeding. In a study of 18 term newborns, treatment with a combination of ampicillin and gentamicin was reported to be associated with an increased abundance of Proteobacteria and a decreased abundance of Actinobacteria and *Bifidobacterium* species at 4 weeks of age [48]. During the 8-week follow-up, the abundance of Actinobacteria and bifidobacteria recovered but the Proteobacteria remained more abundant in antibiotic-exposed infants as compared to nonexposed subjects. The long-term consequences of neonatal antibiotic exposure to gut colonization or later health have not been extensively studied. According to a recent report from the Netherlands, neonatal antibiotic exposure is associated with reduced growth during the first year of life [51]. Unfortunately, data on the composition of the gut microbiome were not reported in the study.

Early antibiotic exposure and the risk of disease

Several epidemiological studies have investigated the association between early-life antibiotic use and the subsequent development of noncommunicable diseases. The sometimes conflicting results of these studies must be interpreted with caution since it is again not at all clear whether the observed associations result from the antibiotic use, the physiological responses to infection, or from confounding factors. In some instances, the antibiotic use preceding asthma or IBD may actually reflect the disease process. All of this notwithstanding, the epidemiological studies constitute a valuable piece in the puzzle attempting to decipher the long-term impact of early antibiotic exposure.

Asthma

Childhood asthma has been linked to altered gut microbiome and particularly lower bacterial diversity in early infancy [3,5,6,52]. Reduced relative abundance of specific indigenous gut bacteria including bifidobacteria during the first weeks of life has been reported to be associated with the development of atopic sensitization and asthma at the age of 4 years [4]. Given the impact of early antibiotic use on the developing gut microbiome, it has been hypothesized that antibiotic exposure during the vulnerable period of infancy might increase the risk of developing asthma.

Already 20 years ago, Wickens et al. reported that antibiotic use was associated with childhood asthma in a study population of 456 children in New Zeeland [53]. The odds ratio for developing asthma after adjusting for potential confounding factors, including gender, family history of asthma, and parental smoking, was 4.05 (95% CI 1.55–10.59) if the child had been treated with antibiotics during the first year of life. Furthermore, the risk of developing asthma increased with the number of antibiotic courses during the first year of life. It is possible that antibiotic use is merely an indicator of frequent infections during the first year of life, which may play a role in the pathogenesis of asthma [54]. The association may also be explained of the fact that asthma and the preceding bronchial hyperactivity are often exacerbated by infections of the respiratory tract, which may result in more severe symptoms of common early-life infections and, consequently, more frequent prescription of antibiotics. Indeed, in a study of more than 4400 children, a significant association between antibiotic use during the first year of life and asthma between the ages of 1–2 years was detected [55]. The association was markedly less prominent, if infections of the lower respiratory tract were taken into consideration as a confounding factor and no link between early antibiotic use and asthma at the age of 5 years was evident. This led the authors to conclude that reverse causation was the most likely explanation for their data. Other studies investigating the association between early antibiotic use and asthma risk have produced conflicting results, and reverse causation remains a substantial problem when interpreting the results. In a systematic review and meta-analysis of 18 studies conducted in 2011, antibiotic use in infancy increased the risk of childhood asthma (OR 1.27, 95% CI 1.12–1.43) but the association was considerably weaker when the possibility of reverse causation and confounding by indication were taken into account [56]. More recently, the risk of asthma was found to be increased in infants and children treated with antibiotics during the first 3 years of life in a meta-analysis of 2 population-based cohort studies comprising of more than 9000 children [57].

It is likely that asthmatic children and those developing asthma are more likely to manifest with wheezing and more severe infections of the respiratory tract and hence be treated with antibiotics. Whether some of the risk increase may be attributed to true causation and mediated by altered composition of the gut microbiome remains unclear. Experimental animal studies have provided important insight into the causal links between early antibiotic exposure, perturbations of gut colonization, and the risk of asthma. Neonatal exposure to vancomycin resulted in marked shifts in the composition of the gut microbiome and increased susceptibility to experimental allergic asthma in a study conducted using a mouse model [58]. Interestingly, adult mice were not affected in a similar fashion, and neonatal treatment with streptomycin had a less pronounced impact. Whether the altered gut microbiota alone is sufficient to cause increased risk of asthma has to our knowledge not been systematically studied.

Overweight and obesity

The epidemic of overweight and obesity is undoubtedly caused by changes in the modern diet and a more sedentary lifestyle. Interestingly, however, there are compelling data to suggest that altered gut microbiome may mediate some of the effects of the high-energy diet by increased energy harvest and effects on host physiology [59]. There are also reports indicating that overweight children have manifested with altered gut microbiome already during early infancy [7–9]. In experimental studies, perinatal and early-life low-dose antibiotic exposure has been shown to not only disrupt the gut microbiome but also result in increased weight and adiposity particularly in male mice [60]. Furthermore, fecal transplant of the gut microbiome of antibiotic-exposed animals leads to weight gain in unexposed mice. This may be regarded as compelling evidence for a causal relationship between early-life antibiotic exposure, altered gut microbiome, and weight gain. It is also of note that the obese phenotype persists in the experimental animals after the antibiotic exposure has been discontinued and the gut microbiome has normalized. Taken together, these data raise concern for a possible connection between early antibiotic use and the risk of overweight and obesity in later life. Interestingly, however, higher dose [61]or broad-spectrum [62] antibiotic exposure results in impaired growth in a mouse model.

The association between early-life antibiotic use and risk of overweight and obesity has been investigated in a number of high-quality epidemiological studies. A modest but statistically significant association has relatively consistently been found between antibiotic use and either BMI, overweight, or obesity later in childhood. Recently, two independent systematic reviews and meta-analyses have been published. Rasmussen et al. systematically reviewed data from 13 studies with more than 500,000 subjects [63]. In a meta-analysis of eight studies, they found that antibiotic exposure in infancy was associated with increased risk of childhood overweight and obesity (OR 1.11, 95% CI 1.02–1.20). The effect was stronger in the case of repeated antibiotic use or exposure during the first 6 months of life. Miller et al., on the other hand, found a modestly increased risk of childhood overweight or obesity in subjects exposed to antibiotics (OR 1.05, 95% CI 1.00–1.11) in their meta-analysis of 12 studies [64]. The authors concluded that boys were more strongly affected than girls, and the association was strongest for repeated or broad-spectrum antibiotics. After the publication of these meta-analyses, Block et al. reported that, in a population of more than 360,000 children, 58% of the subjects had received at least one course of antibiotics during their first 24 months of life [65]. At 5 years of age, this antibiotic exposure was associated with a higher mean BMI Z score (0.04, 95% CI 0.03–0.05) and increased risk of overweight or obesity (OR 1.05, 95% CI 1.03–1.07). These epidemiological associations do not provide insight into the underlying mechanisms. A link to the gut microbiome was provided by a prospective study of 162 vaginally born, term infants in whom fecal microbiome profiles at the age of 3 months were predictive of later overweight [66]. In particular, lower relative abundance of fecal bifidobacteria and higher abundance of streptococci were associated with the development of overweight, and the links were strongest in subjects exposed to antibiotics.

Other noncommunicable diseases

As our understanding of the links between the gut microbiome in infancy and early childhood and the risk of chronic disease in later life has expanded, the links between early gut colonization patterns, antibiotic exposure, and disease have been explored. For instance, inappropriate immune responsiveness against indigenous intestinal bacteria is known to play a role in the pathogenesis of IBD [11]. Differences in the composition of the gut microbiome have been reported between individuals with and without IBD, and the perturbations appear to be detectable at the time of diagnosis in children [67]. Interestingly, children with IBD have reportedly used more antibiotics during their first year of life as compared to those who remained healthy in a study of 36 children with IBD and 360 matched controls [68]. Subsequent population-based studies conducted in Denmark [69] and Finland [70] indicate that antibiotic use is associated with increased risk of Crohn's disease but not ulcerative colitis. These epidemiological studies provide no mechanistic explanation for the observed association, and confounding by indication and reverse causation may explain the results. Intriguingly, however, transfer of antibiotic-perturbed intestinal microbiome to pregnant mice that resulted in altered gut colonization in the offspring without direct antibiotic exposure has been reported to lead to increased susceptibility to experimental colitis [71]. This sophisticated experimental study provides a convincing causal explanation for the association between antibiotic exposure, dysbiosis, and the development of IBD, which needs to be confirmed in prospective human studies.

Recently, the so-called gut–brain axis has become the focus of intensive research efforts. Disturbances in the gut microbiome have been proposed to play a role in the pathogenesis of neurological and psychiatric disorders ranging from depression and Parkinson's disease to autism spectrum disorders. Establishing a causal relationship with the early gut microbiome and its perturbations and diseases with onset years or even decades later is difficult because of confounding, reverse causation, and recall bias. Children with autism often experience gastrointestinal symptoms, and differences in the gut microbiome have been reported between affected and nonaffected children [10]. It is therefore not unfounded to hypothesize that early antibiotic use might be linked to autism. Still, no convincing evidence currently exists for a link between early antibiotic use and autistic spectrum disorder in children [72,73] despite experimental studies indicating that early-life treatment with antibiotics leads to behavioral changes in mice [74].

Conclusions

Antibiotic intervention is often vitally important and lifesaving in the perinatal and neonatal periods both for the mother and her child. Our understanding of the impact of antibiotic exposure during pregnancy, delivery, and early life on the developing gut microbiome and its role in health and disease is still rudimentary at best. Nonetheless, there are convincing data to indicate that intrapartum and neonatal antibiotic use is associated with significant perturbations of early gut colonization (Table 6.1). It is of particular note that, according to several studies, intrapartum [34–36] and neonatal [48] antibiotic exposure may result in reduced abundance of fecal bifidobacteria during the first weeks and months of life. Aberrant early gut colonization and particularly delayed

TABLE 6.1 Effect of antibiotic exposure during pregnancy, delivery, the neonatal period, and early infancy on the gut microbiome and disease risk in later life. Clinical observational studies indicate possible associations but cannot prove causality.

Antibiotic exposure	Gut microbiome	Epidemiological data	Experimental data
Pregnancy	Not established	↑ Asthma [22–25]	↑ Adiposity [29]
		↑ Infections [26,28]	
		↑ Obesity [27]	
Intrapartum	↓Actinobacteria [39]	↑ Infantile colic [38]	↑ Adiposity [59][a]
	↓ Bacteroidetes [39]		↑ Weight gain [59][a]
	↓ Bifidobacteria [34–36]		
	↑ Proteobacteria [39]		
	↑ Firmicutes [39]		
	↑ ARGs [39,40]		
Neonatal	↓Actinobacteria [48]	↓ Early growth [51]	↑ Asthma [57]
	↓ Bifidobacteria [48,49]		↑ Adiposity [59] [a]
	↑ Proteobacteria [48]		↑ Weight gain [59][a]
Early infancy	↓ Bifidobacteria [65]	↑ Asthma [52,55]	↑ IBD [70]
		↑ Obesity [62–65]	↑ Anxiety [73]
		↑ IBD [67–69]	

[a]*The experimental study design included both perinatal and neonatal antibiotic exposure.*
↑ denotes increase, ↓ denotes decrease; *ARG*, antibiotic resistance gene; *IBD*, inflammatory bowel disease.

colonization with bifidobacteria have in some previous studies been associated with increased risk of asthma [4] and obesity [8,9] later in childhood. In epidemiological studies, antibiotic use in early infancy has consistently been linked with subsequent asthma [55] and obesity [61] (Table 6.1). While these observations must be interpreted with caution and confounding factors and reverse causation are likely to explain some of the associations, it is possible that a causal chain from early antibiotic exposure resulting in altered gut colonization contributes to the development of disease. This notion has to some extent been corroborated in experimental animal models (Table 6.1).

Future studies should strive to investigate the link between early antibiotic use, the gut microbiome, and disease risk in adequately powered prospective cohorts. Ideally, follow-up should begin before or during early pregnancy, and data on possible confounding factors such as hereditary propensity to disease, breastfeeding, mode of birth, and number of siblings should be taken into consideration. It is likely that the impact of antibiotics on both the gut microbiome and disease risk is age-dependent. Intrapartum and neonatal antibiotic exposure are of particular interest since, despite being very common, its clinical long-term effects remain largely unexplored. Moreover, antibiotics are often administered during delivery and the early neonatal period to uninfected individuals, which provides an opportunity to investigate the effects of antibiotics without confounding by infection or inflammation. Finally, experimental studies are needed to demonstrate causality. Studies using human fecal transplant from antibiotic-exposed and non-exposed infants to germ-free animals would provide important new information on the contribution of antibiotic-induced gut microbiome perturbations to host physiology. Only by carefully dissecting the potentially detrimental impact of early antibiotic exposure, we can devise evidence-based guidelines for the safe use of these vitally important medical interventions.

References

[1] Rautava S, Luoto R, Salminen S, Isolauri E. Microbial contact during pregnancy, intestinal colonization and human disease. Nat Rev Gastroenterol Hepatol 2012;9 (10):565–76. Available from: https://doi.org/10.1038/nrgastro.2012.144.

[2] Kalliomäki M, Kirjavainen P, Eerola E, Kero P, Salminen S, Isolauri E. Distinct patterns of neonatal gut microflora in infants in whom atopy was and was not developing. J Allergy Clin Immunol 2001;107(1):129–34. Available from: https://doi.org/10.1067/mai.2001.111237.

[3] Abrahamsson TR, Jakobsson HE, Andersson AF, Björkstén B, Engstrand L, Jenmalm MC. Low gut microbiota diversity in early infancy precedes asthma at school age. Clin Exp Allergy 2014;44(6):559–66. Available from: https://doi.org/10.1111/cea.12253.

[4] Fujimura KE, Sitarik AR, Havstad S, et al. Neonatal gut microbiota associates with childhood multisensitized atopy and T cell differentiation. Nat Med 2016;22 (10):1187–91. Available from: https://doi.org/10.1038/nm.4176.

[5] Stiemsma LT, Arrieta M-C, Dimitriu PA, et al. Shifts in Lachnospira and *Clostridium* sp. in the 3-month stool microbiome are associated with preschool age asthma. Clin Sci 2016;130(23):2199–207. Available from: https://doi.org/10.1042/CS20160349.

[6] Stokholm J, Blaser MJ, Thorsen J, et al. Maturation of the gut microbiome and risk of asthma in childhood. Nat Commun 2018;9(1):141. Available from: https://doi.org/10.1038/s41467-017-02573-2.

[7] Vael C, Verhulst SL, Nelen V, Goossens H, Desager KN. Intestinal microflora and body mass index during the first three years of life: an observational study. Gut Pathog 2011;3(1):8. Available from: https://doi.org/10.1186/1757-4749-3-8.

[8] Kalliomäki M, COllado MC, Salminen S, Isolauri E. Early differences in fecal microbiota composition in children may predict overweight. Am J Clin Nutr 2008;87(3):534–8. Available from: https://doi.org/10.1093/ajcn/87.3.534.

[9] Dogra S, Sakwinska O, Soh SE, et al. Dynamics of infant gut microbiota are influenced by delivery mode and gestational duration and are associated with subsequent adiposity. MBio 2015;6(1). Available from: https://doi.org/10.1128/mBio.02419-14 e02419-14.

[10] Hughes HK, Rose D, Ashwood P. The gut microbiota and dysbiosis in autism spectrum disorders. Curr Neurol Neurosci Rep 2018;18(11):81. Available from: https://doi.org/10.1007/s11910-018-0887-6.

[11] Ananthakrishnan AN, Bernstein CN, Iliopoulos D, et al. Environmental triggers in IBD: a review of progress and evidence. Nat Rev Gastroenterol Hepatol 2018;15 (1):39–49. Available from: https://doi.org/10.1038/nrgastro.2017.136.

[12] Kenyon S, Boulvain M, Neilson JP. Antibiotics for preterm rupture of membranes. Cochrane Database Syst Rev 2013;(12). Available from: https://doi.org/10.1002/14651858.CD001058.pub3 CD001058.

[13] Douwes J, Cheng S, Travier N, et al. Farm exposure in utero may protect against asthma, hay fever and eczema. Eur Respir J 2008;32(3):603–11. Available from: https://doi.org/10.1183/09031936.00033707.

[14] Conrad ML, Ferstl R, Teich R, et al. Maternal TLR signaling is required for prenatal asthma protection by the nonpathogenic microbe *Acinetobacter lwoffii* F78. J Exp Med 2009;206(13):2869–77. Available from: https://doi.org/10.1084/jem.20090845.

[15] DiGiulio DB, Romero R, Amogan HP, et al. Microbial prevalence, diversity and abundance in amniotic fluid during preterm labor: a molecular and culture-based investigation. PLoS One 2008;3(8):e3056. Available from: https://doi.org/10.1371/journal.pone.0003056.

[16] Aagaard K, Ma J, Antony KM, Ganu R, Petrosino J, Versalovic J. The placenta harbors a unique microbiome. Sci Transl Med 2014;6(237):237ra65. Available from: https://doi.org/10.1126/scitranslmed.3008599.

[17] Collado MC, Rautava S, Aakko J, Isolauri E, Salminen S. Human gut colonisation may be initiated in utero by distinct microbial communities in the placenta and amniotic fluid. Sci Rep 2016;623129. Available from: https://doi.org/10.1038/srep23129.

[18] Rehbinder EM, Lødrup Carlsen KC, Staff AC, et al. Is amniotic fluid of women with uncomplicated term pregnancies free of bacteria? Am J Obstet Gynecol 2018;219(3): e1–289. Available from: https://doi.org/10.1016/j.ajog.2018.05.028.

[19] Leiby JS, McCormick K, Sherrill-Mix S, et al. Lack of detection of a human placenta microbiome in samples from preterm and term deliveries. Microbiome 2018;6(1):196. Available from: https://doi.org/10.1186/s40168-018-0575-4.

[20] Martinez KA, Romano-Keeler J, Zackular JP, et al. Bacterial DNA is present in the fetal intestine and overlaps with that in the placenta in mice. PLoS One 2018;13(5): e0197439. Available from: https://doi.org/10.1371/journal.pone.0197439.

[21] Alipour MJ, Jalanka J, Pessa-Morikawa T, et al. The composition of the perinatal intestinal microbiota in cattle. Sci Rep 2018;8(1):10437. Available from: https://doi.org/10.1038/s41598-018-28733-y.

[22] Stensballe LG, Simonsen J, Jensen SM, Bønnelykke K, Bisgaard H. Use of antibiotics during pregnancy increases the risk of asthma in early childhood. J Pediatr 2013;162 (4):832–838.e3. Available from: https://doi.org/10.1016/j.jpeds.2012.09.049.

[23] Bai L, Zhao D, Cheng Q, et al. Trimester-specific association between antibiotics exposure during pregnancy and childhood asthma or wheeze: the role of confounding. Ann Epidemiol 2018;30:1–8. Available from: https://doi.org/10.1016/j.annepidem.2018.10.003.

[24] Stokholm J, Sevelsted A, Bønnelykke K, Bisgaard H. Maternal propensity for infections and risk of childhood asthma: a registry-based cohort study. Lancet Respir Med 2014;2 (8):631–7. Available from: https://doi.org/10.1016/S2213-2600(14)70152-3.

[25] Loewen K, Monchka B, Mahmud SM, Jong G, Azad MB. Prenatal antibiotic exposure and childhood asthma: a population-based study. Eur Respir J 2018;52(1). Available from: https://doi.org/10.1183/13993003.02070-2017.

[26] Miller JE, Wu C, Pedersen LH, de Klerk N, Olsen J, Burgner DP. Maternal antibiotic exposure during pregnancy and hospitalization with infection in offspring: a population-based cohort study. Int J Epidemiol 2018;47(2):561–71. Available from: https://doi.org/10.1093/ije/dyx272.

[27] Cassidy-Bushrow AE, Burmeister C, Havstad S, et al. Prenatal antimicrobial use and early-childhood body mass index. Int J Obes 2017;42:1. Available from: https://doi.org/10.1038/ijo.2017.205.
[28] Pedersen TM, Stokholm J, Thorsen J, Mora-Jensen ARC, Bisgaard H. Antibiotics in pregnancy increase children's risk of otitis media and ventilation tubes. J Pediatr 2017;183:153–158.e1. Available from: https://doi.org/10.1016/j.jpeds.2016.12.046.
[29] Yoshimoto A, Uebanso T, Nakahashi M, Shimohata T, Mawatari K, Takahashi A. Effect of prenatal administration of low dose antibiotics on gut microbiota and body fat composition of newborn mice. J Clin Biochem Nutr 2018;62(2):155–60. Available from: https://doi.org/10.3164/jcbn.17-53.
[30] Stokholm J, Schjørring S, Pedersen L, et al. Prevalence and predictors of antibiotic administration during pregnancy and birth. PLoS One 2013;8(12):e82932. Available from: https://doi.org/10.1371/journal.pone.0082932.
[31] Persaud RR, Azad MB, Chari RS, Sears MR, Becker AB, Kozyrskyj AL. Perinatal antibiotic exposure of neonates in Canada and associated risk factors: a population-based study. J Matern Fetal Neonatal Med 2015;28(10):1476–7058. Available from: https://doi.org/10.3109/14767058.2014.947578.
[32] Russell NJ, Seale AC, O'Sullivan C, et al. Risk of early-onset neonatal Group B Streptococcal disease with maternal colonization worldwide: systematic review and meta-analyses. Clin Infect Dis 2017;65(Suppl. 2):S152–9. Available from: https://doi.org/10.1093/cid/cix655.
[33] Smaill FM, Grivell RM. Antibiotic prophylaxis versus no prophylaxis for preventing infection after cesarean section. Cochrane Database Syst Rev 2014;(10):CD007482. Available from: https://doi.org/10.1002/14651858.CD007482.pub3.
[34] Aloisio I, Mazzola G, Corvaglia LT, et al. Influence of intrapartum antibiotic prophylaxis against group B *Streptococcus* on the early newborn gut composition and evaluation of the anti-*Streptococcus* activity of *Bifidobacterium* strains. Appl Microbiol Biotechnol 2014;98(13):6051–60. Available from: https://doi.org/10.1007/s00253-014-5712-9.
[35] Corvaglia L, Tonti G, Martini S, et al. Influence of intrapartum antibiotic prophylaxis for group B *Streptococcus* on gut microbiota in the first month of life. J Pediatr Gastroenterol Nutr 2016;62(2):304–8. Available from: https://doi.org/10.1097/MPG.0000000000000928.
[36] Aloisio I, Quagliariello A, De Fanti S, et al. Evaluation of the effects of intrapartum antibiotic prophylaxis on newborn intestinal microbiota using a sequencing approach targeted to multi hypervariable 16S rDNA regions. Appl Microbiol Biotechnol 2016;100(12):5537–46. Available from: https://doi.org/10.1007/s00253-016-7410-2.
[37] Azad MB, Konya T, Persaud RR, et al. Impact of maternal intrapartum antibiotics, method of birth and breastfeeding on gut microbiota during the first year of life: a prospective cohort study. BJOG 2016;123(6):983–93. Available from: https://doi.org/10.1111/1471-0528.13601.
[38] Leppälehto E, Pärtty A, Kalliomäki M, Löyttyniemi E, Isolauri E, Rautava S. Maternal intrapartum antibiotic administration and infantile colic: is there a connection? Neonatology 2018;114(3):226–9. Available from: https://doi.org/10.1159/000489991.
[39] Nogacka A, Salazar N, Suárez M, et al. Impact of intrapartum antimicrobial prophylaxis upon the intestinal microbiota and the prevalence of antibiotic resistance genes in vaginally delivered full-term neonates. Microbiome 2017;5(1):1–10. Available from: https://doi.org/10.1186/s40168-017-0313-3.
[40] Pärnänen K, Karkman A, Hultman J, et al. Maternal gut and breast milk microbiota affect infant gut antibiotic resistome and mobile genetic elements. Nat Commun 2018;9(1):1–11. Available from: https://doi.org/10.1038/s41467-018-06393-w.

[41] Cox LM, Blaser MJ. Antibiotics in early life and obesity. Nat Rev Endocrinol 2015;11 (3):182–90. Available from: https://doi.org/10.1038/nrendo.2014.210.

[42] Dethlefsen L, Huse S, Sogin ML, Relman DA. The pervasive effects of an antibiotic on the human gut microbiota, as revealed by deep 16s rRNA sequencing. PLoS Biol 2008;6(11):e280. Available from: https://doi.org/10.1371/journal.pbio.0060280.

[43] Dethlefsen L, Relman DA. Incomplete recovery and individualized responses of the human distal gut microbiota to repeated antibiotic perturbation. Proc Natl Acad Sci U S A 2011;108(Suppl 1):4554–61. Available from: https://doi.org/10.1073/pnas.1000087107.

[44] Korpela K, Salonen A, Virta LJ, et al. Intestinal microbiome is related to lifetime antibiotic use in Finnish pre-school children. Nat Commun 2016;7:10410. Available from: https://doi.org/10.1038/ncomms10410.

[45] Jernberg C, Löfmark S, Edlund C, Jansson JK. Long-term ecological impacts of antibiotic administration on the human intestinal microbiota. ISME J 2007;1(1):56–66. Available from: https://doi.org/10.1038/ismej.2007.3.

[46] Suez J, Zmora N, Zilberman-Schapira G, et al. Post-antibiotic gut mucosal microbiome reconstitution is impaired by probiotics and improved by autologous FMT. Cell 2018;174(6):1406–1423.e16. Available from: https://doi.org/10.1016/j.cell.2018.08.047.

[47] Polin RA. Management of neonates with suspected or proven early-onset bacterial sepsis. Pediatrics 2012;129(5):1006–15. Available from: https://doi.org/10.1542/peds.2012-0541.

[48] Fouhy F, Guinane CM, Hussey S, et al. High-throughput sequencing reveals the incomplete, short-term recovery of infant gut microbiota following parenteral antibiotic treatment with ampicillin and gentamicin. Antimicrob Agents Chemother 2012;56(11):5811–20. Available from: https://doi.org/10.1128/AAC.00789-12.

[49] Arboleya S, Sánchez B, Milani C, et al. Intestinal microbiota development in preterm neonates and effect of perinatal antibiotics. J Pediatr 2015;166(3):538–44. Available from: https://doi.org/10.1016/j.jpeds.2014.09.041.

[50] Greenwood C., Morrow AL, Lagomarcino AJ, et al. Early empiric antibiotic use in preterm infants is associated with lower bacterial diversity and higher relative abundance of enterobacter. J Pediatr. 2014;165(1):23–29. Available from: 10.1016/j.jpeds.2014.01.010.

[51] Kamphorst K, Oosterloo BC, Vlieger AM, et al. Antibiotic treatment in the first week of life impacts the growth trajectory during the first year of life in term infants. J Pediatr Gastroenterol Nutr. 2019;69 (1): 131–136. Available from: 10.1097/MPG.0000000000002360.

[52] Arrieta MC, Stiemsma LT, Dimiriu PA, et al. Early infancy microbial and metabolic alterations affect risk of childhood asthma. Sci Transl Med 2015;7(307):307ra152. Available from: https://doi.org/10.1126/scitranslmed.aab2271.

[53] Wickens K, Pearce N, Crane J, Beasley R. Antibiotic use in early childhood and the development of asthma. Clin Exp Allergy 1999;29(6):766–71. Available from: https://doi.org/10.1046/j.1365-2222.1999.00536.x.

[54] Mattes J, Karmaus W. The use of antibiotics in the first year of life and development of asthma: which comes first? - Editorial. Clin Exp Allergy 1999;29(6):729–32. Available from: https://doi.org/10.1046/j.1365-2222.1999.00540.x.

[55] Celedón JC, Fuhlbrigge A, Rifas-Shiman S, Weiss ST, Finkelstein JA. Antibiotic use in the first year of life and asthma in early childhood. Clin Exp Allergy 2004;34 (7):1011–16. Available from: https://doi.org/10.1111/j.1365-2222.2004.01994.x.

[56] Penders J, Kummeling I, Thijs C. Infant antibiotic use and wheeze and asthma risk: a systematic review and meta-analysis. Eur Respir J 2011;38(2):295–302. Available from: https://doi.org/10.1183/09031936.00105010.

[57] Ahmadizar F, Vijverberg SJH, Arets HGM, et al. Early life antibiotic use and the risk of asthma and asthma exacerbations in children. Pediatr Allergy Immunol 2017;28 (5):430–7. Available from: https://doi.org/10.1111/pai.12725.

[58] Russell SL, Gold MJ, Hartmann M, et al. Early life antibiotic-driven changes in microbiota enhance susceptibility to allergic asthma. EMBO Rep 2012;13(5):440–7. Available from: https://doi.org/10.1038/embor.2012.32.

[59] Tremaroli V, Bäckhed B. Functional interactions between the gut microbiota and host metabolism. Nature 2012;489(7415):242–9. Available from: https://doi.org/10.1038/nature11552.

[60] Cox LM, Yamanishi S, Sohn J, et al. Altering the intestinal microbiota during a critical developmental window has lasting metabolic consequences. Cell 2014;158(4):705–21. Available from: https://doi.org/10.1016/j.cell.2014.05.052.

[61] Ge X, Ding C, Zhao W, et al. Antibiotics-induced depletion of mice microbiota induces changes in host serotonin biosynthesis and intestinal motility. J Transl Med 2017;15(1):13. Available from: https://doi.org/10.1186/s12967-016-1105-1104.

[62] Bourgeois T, Delezoide AL, Zhao W, et al. Safety study of Ciprofloxacin in newborn mice. Regul Toxicol Pharmacol 2016;74:161–9. Available from: https://doi.org/10.1016/j.yrtph.2015.11.002.

[63] Rasmussen SH, Shrestha S, Bjerregaard LG, et al. Antibiotic exposure in early life and childhood overweight and obesity: a systematic review and meta-analysis. Diabetes Obes Metab 2018;20(6):1508–14. Available from: https://doi.org/10.1111/dom.13230.

[64] Miller SA, Wu RKS, Oremus M. The association between antibiotic use in infancy and childhood overweight or obesity: a systematic review and meta-analysis. Obes Rev 2018;19(11):1463–75. Available from: https://doi.org/10.1111/obr.12717.

[65] Block JP, Bailey LC, Gillman MW, et al. Early antibiotic exposure and weight outcomes in young children. Pediatrics 2018;142(6):e20180290. Available from: https://doi.org/10.1542/peds.2018-0290.

[66] Korpela K, Zijlmans MAC, Kuitunen M, et al. Childhood BMI in relation to microbiota in infancy and lifetime antibiotic use. Microbiome 2017;5(1):26. Available from: https://doi.org/10.1186/s40168-017-0245-y.

[67] Gevers D, Kugathasan S, Denson LA, et al. The treatment-naive microbiome in new-onset Crohn's disease. Cell Host Microbe 2014;15(3):382–92. Available from: https://doi.org/10.1016/j.chom.2014.02.005.

[68] Shaw SY, Blanchard JF, Bernstein CN. Association between the use of antibiotics in the first year of life and pediatric inflammatory bowel disease. Am J Gastroenterol 2010;105(12):2687–92. Available from: https://doi.org/10.1038/ajg.2010.398.

[69] Hviid A, Svanström H, Frisch M. Antibiotic use and inflammatory bowel diseases in childhood. Gut 2011;60(1):49–54. Available from: https://doi.org/10.1136/gut.2010.219683.

[70] Virta L, Auvinen A, Helenius H, Huovinen P, Kolho KL. Association of repeated exposure to antibiotics with the development of pediatric crohn's disease - A nationwide, register-based Finnish case-control study. Am J Epidemiol 2012;175(8):775–84. Available from: https://doi.org/10.1093/aje/kwr400.

[71] Schulfer AF, Battaglia T, Alvarez Y, et al. Intergenerational transfer of antibiotic-perturbed microbiota enhances colitis in susceptible mice. Nat Microbiol 2018;3 (2):234–42. Available from: https://doi.org/10.1038/s41564-017-0075-5.

[72] Atladottir HO, Henriksen TB, Schendel DE, Parner ET. Autism after infection, febrile episodes, and antibiotic use during pregnancy: an exploratory study. Pediatrics 2012;130(6):1447–54. Available from: https://doi.org/10.1542/peds.2012-1107.

[73] Axelsson PB, Clausen TD, Petersen AH, et al. Relation between infant microbiota and autism? – Results from a national cohort sibling-design study. Epidemiology 2019;30(1):52–60. Available from: https://doi.org/10.1097/EDE.0000000000000928.

[74] Leclercq S, Mian FM, Stanisz AM, et al. Low-dose penicillin in early life induces long-term changes in murine gut microbiota, brain cytokines and behavior. Nat Commun 2017;8:15062. Available from: https://doi.org/10.1038/ncomms15062.

CHAPTER

7

The intestinal microbiome of preterm infants

Lybil Mendoza, Max Van Belkum and Josef Neu

Department of Pediatrics, College of Medicine, University of Florida, Gainesville, FL, United States

Structure and function of the developing intestine

The gastrointestinal (GI) tract, although considered to be primarily a digestive-absorptive organ, plays a much broader role. As one of the largest surfaces of the body interacting with the external environment, the intestinal immune system must precisely discriminate between friend and foe. The gut's ability to maintain structural integrity and display tolerance to commensal bacteria determines the degree of inflammatory responses and immunity throughout one's entire lifetime and beyond. In addition, the developing GI tract provides neural, endocrine, and exocrine functions that play crucial roles for health throughout one's lifetime.

Here we will briefly discuss some of the developmental aspects of the GI tract that are most germane to host–microbe interactions. The human primitive gut is established by 4 weeks of gestation. It is divided into three embryological structures: foregut, midgut, and hindgut [1]. These differentiate into the proximal, middle, and distal intestine. These regions are diverse both histologically and functionally. In the stomach, the mucosal surface produces acid and enzymes via parietal, chief, and neuroendocrine cells. More distally, the intestinal epithelium acts as a barrier to antigens and microbes. This is largely accomplished by the inter-enterocyte tight junctions. Breakdown of tight junctions is associated with the development of several diseases including celiac disease, type 1 autoimmune diabetes, and inflammatory bowel disease [2]. Enterocytes constitute most of the surface epithelial cells of the small

The Human Microbiome in Early Life
DOI: https://doi.org/10.1016/B978-0-12-818097-6.00007-9

intestine and are responsible for absorbing nutrients. In the small intestinal villus, there is a villus to crypt gradient, with cells undergoing mitotic birth in the crypt regions, migrating to the villus tip, with subsequent extrusion into the lumen. One cell type, the Paneth cells remain in the deep crypt region, where they secrete protective peptides that serve a protective role for the stem cells found adjacent to the Paneth cells in the crypts. Turnover rate and sloughing of cells vary by location, individual factors, and age. In the rat, for example, the cell turnover from crypt to villus is considerably faster in the adult than in the preweaned animal [3].

There is an aboral differentiation of function where the ileal mucosa is highly differentiated for immune functions, and the colon plays a major bioreactor (or metabolomic) role. The fuel requirements for the different parts of the intestinal mucosa differ in that the small intestine relies largely on glutamate [4], and the large intestine is largely dependent on short-chain fatty acids (SCFAs) such as butyrate [5].

Prenatal microbial effects on the developing intestine

The fetal GI tract develops rapidly. Its lumenal surface is exposed to large quantities of antigens present in swallowed amniotic fluid, the volume of which may reach up to 150 mL/kg/day [6]. There are a large number of microbe-derived metabolites as well as inflammatory mediators produced in the mother's GI tract that have a capability to affect the fetus. In addition, microbes may actually hematogenously translocate from the mother's GI tract to the fetus and harbor in the fetal GI tract [7].

The maternal GI tract microbes may influence fetal physiology and well-being. For example, maternal diet, drugs, and other environmental exposures may shift her intestinal microbes and their metabolic capacity. Studies using transfaunation of microbes from early versus late stages of pregnancy into germ-free mice have shown fascinating effects [8]. If the microbes are provided from mothers who are in the first trimester of pregnancy, no effects are seen in the mice in terms of obesity or insulin resistance. If germ-free mice are transfaunated with microbes from third trimester pregnant mother's stools, the mice develop obesity and insulin resistance [8].

Maternal intestinal microbial colonization also affects development of innate immunity in the newborn. With transient colonization of previously germ-free pregnant female mice, it was shown that the certain types of innate lymphoid cells were increased in the offspring. The mechanism of this effect is unclear but appear to be partially dependent on maternal antibodies that retain microbial molecules and transmit them to the offspring during pregnancy and in milk [9].

Intestinal functions in the infant born preterm

Significant transitions occur within the neonatal GI tract during pregnancy. The last trimester represents a critical developmental window for the fetal intestine that is shortened by preterm birth. If the infant is born preterm, intestinal motility, villus and crypt anatomy, digestive-absorptive function, innate immunity, and various other aspects of intestinal maturity are not yet fully developed to the same degree as would be present had the infants been born at term. The preterm intestine, although swallowing large quantities of amniotic fluid daily, does not have the functional capability for interaction with food and the microbial environment as seen in the infant born at term.

Early microbial exposures

Several studies suggest that the fetal GI tract is not sterile, and exposure of the intestinal mucosa to microbes occurs before parturition. This is based on studies showing microbial DNA present in meconium [10,11]. After birth, gut colonization is thought to be influenced from the birth canal in addition to microbes from mother's milk, skin, and other environmental exposures [12]. This colonization of the intestine by commensal bacteria is a vital part of development of the mucosal barrier and modulates risk of developing systemic inflammation-related pathologies. In the preterm infant, abnormal colonization of the intestinal tract (dysbiosis), immature motility, an underdeveloped intestinal surface, and an untrained immune system together may set the stage for destructive processes. These include microbial translocation of pathogenic microbes and stimulation of inappropriate inflammatory responses. These, in turn can affect the neonate through systemic inflammation-related catabolism, and breakdown of various other developmental processes that can lead to acute disease as well as more chronic long-term health problems.

Preterm infants have lower fecal relative abundance of protective bacterial genera such as *Bifidobacterium*, *Lactobacillus* and increased proportions of potentially pathogenic bacteria such as *Clostridium difficile* and bacteria of the γ-proteobacterial class, namely *Pseudomonas*, *Klebsiella*, and *Escherichia coli*. It is suggested that this "dysbiosis" results in outcomes such as necrotizing enterocolitis (NEC), late-onset sepsis, and developmental delay [13].

Microbe–host interactions

The immature GI tract of premature infants is less capable to defend against pathogenic microorganisms. Epithelial mucosal barriers, motility patterns and gastric acid, proteolytic enzymes, secretory IgA, and antimicrobial peptides production capabilities are not yet fully developed. This contributes to an increased risk of dysbiosis, bacterial overgrowth [14], bacterial translocation, and a local as well as systemic inflammatory responses that could lead to disease, such as NEC [15].

Commensal bacteria develop a close relationship with the GI epithelium and gut-associated lymphoid tissue. Interaction between bacterial metabolic products acts to maintain immune reactivity and tolerance, which partially depend on the type of microbe–host response. A balance between immunoreactivity and tolerance mechanisms allows for coexistence with GI microbes without developing inappropriate inflammatory responses that might eliminate the commensal microbes. Toll-like receptors (TLRs) are involved with cell signaling in response to bacterial cell wall components such as lipopolysaccharides (LPS). Preterm infants have higher activity of TLR4, an innate pattern recognition receptor for LPS that signals pro-inflammation, which correlates with reduced ability to recover from mucosal injury and barrier dysfunction [16]. Dysregulation of the GI microbiome has been associated with a variety of diseases [17]. Dysregulated inflammatory responses may affect the entire body and are thus called systemic inflammatory responses. With this in mind, the intestine has been termed "The Motor that Drives Systemic Inflammation and Critical Illness" [18]. In this paradigm, cytokines and chemokines incite a pro-inflammatory response, which promotes intestinal permeability with further local GI and systemic immune activation, which in turn creates a cyclic feedback process. This is one of the major proposed mechanism for the increased risk of NEC in preterm infants [19].

Environmental factors that affect the microbiome

Factors that affect the intestinal microbiome of infants include mode of delivery, diet, gestational age, and antibiotic exposure [20]. Because of their prolonged hospital courses, preterm infants tend to have higher exposure to bacteria that are indigenous to the hospital environment. While many factors are capable of inciting changes in the microbiome and causing dysbiosis, the following select factors will be discussed further in this review: the effects of antibiotic exposure, use of parenteral nutrition, and feeding patterns.

The neonatal intensive care unit environment

Antibiotics

Routine antibiotic use in preterm infants is firmly ingrained in neonatology despite risks associated with antibiotics and the extremely low incidence of early-onset sepsis (EOS) which antibiotics are used to prevent. There are several questions that are highly germane to this area:

Does routine antibiotic use immediately after birth prevent EOS? Antibiotics can be lifesaving when used appropriately, but they can cause harm when used inappropriately. They are by far the most common drugs prescribed to infants in the neonatal intensive care unit (NICU), largely because of the concern for EOS [21–23]. EOS has a very low incidence and is poorly defined as a disease [24]. Presenting clinical symptoms is not specific, and the biomarker positive predictive values and sensitivities are poor [25]. Despite recent initiatives to limit antibiotic use, a majority of infants born less than 33 weeks' gestational age who are likely uninfected are exposed to treatment. There are currently no adequately powered randomized controlled trials evaluating safety and efficacy of postpartum prophylaxis against EOS.

Can antibiotics cause harm? In the preterm very low birthweight (VLBW) infant, the rationale to routinely give antibiotics shortly after birth is speculative rather than evidence based. Speculation is driven by concerns over the immature immune system, the possibility that preterm delivery may have been caused by infection, immaturity-related respiratory distress (which is not readily distinguished from infectious pneumonia), and that antibiotic use is safe. These assumptions are not based on solid evidence and may be incorrect. The average length of treatment of this "standard of care" practice is between 5 and 7 days [26,27]. Yet, observational studies show an association between duration of early antibiotic use with an increased chance of developing NEC [26,28,29], a disease with high mortality and morbidity [19,29]. We [30–33] and others [34,35] have shown that NEC and late-onset sepsis are preceded by an intestinal dysbiosis, which may partially be due to antibiotic usage [33]. Others have shown an increase in the composite of adverse outcomes with antibiotic treatment [36–39]. However, a more recent large observational analysis of data from the NIH Neonatal Research Network suggests that prolonged early antibiotic use is not significantly associated with NEC or death, contrary to previous results. This same retrospective study shows a strong relationship of early antibiotics with late-onset fungal sepsis [40]. None of these studies were done in a randomized prospective manner and are thus prone to error due to confounding factors, such as greater initial degree of illness in the treated infants. In summary, the potential harm of antibiotics in preterm infants is concerning [41–45]. However, much of this information is based on retrospective, observational studies, in which confounding covariates challenge accurate statistical evaluation, and thus requires a prospective blinded randomized study that obviates these confounders.

What intermediate multiomic mechanisms may relate antibiotic administration to outcomes? It has been shown that microbial taxonomy is significantly affected by the early use of antibiotics, and that this effect often persists after discontinuation of the antibiotic. It is thus very concerning that infants who received 5–7 days of antibiotics in their first week postpartum display signs of intestinal dysbiosis, including increased relative abundance of Enterobacteriaceae and low fecal bacterial alpha diversity in the following 2 weeks of life [46]. In fact, only 2 days of antibiotic use after birth has been associated with an intestinal dysbiosis well after these antibiotics have been discontinued [42,47]. These studies show long-term elevations of Proteobacteria and decreased Firmicutes in treated infants, which are microbial taxonomic changes that we have found to be associated with the development of NEC [33]. These infants also experienced more sepsis and death than those not exposed [46]. However, these associations shed little light on mechanisms. Multiomic approaches that integrate the microbiome, metabolome, inflammatory mediators, disease, and environmental perturbations are needed to provide a more robust analysis and visualization of the interaction of these processes [48]. This allows a picture of causality more informative than that seen with single-omic associations.

H-2 blocker therapy

Another interesting relationship exists between the development of NEC and the use of H2 blockers in premature infants [49]. The practice of H2 blocker use a few years ago was widespread in many NICUs because of the concept that many of the apneas and bradycardias seen often in preterm neonates were caused by acid reflux. Studies found increased incidences of sepsis and NEC in babies receiving H2 blockers in addition to the fact that most episodes of apnea and bradycardia in preterm neonates were secondary to central nervous system immaturity rather than reflux, calling into question the rationale of why H2 blockers are still given prophylactically in these instances [50]. Furthermore, H2 blockers were found to affect the intestinal microbiome [49], with a shift toward Proteobacteria predominance over the Firmicutes, similar to what is seen before the development of NEC in these preterm infants [33].

Nutrition and feedings

Human mother's milk

The composition of human milk is uniquely suited for human infants. It contains numerous factors that stimulate growth, provides immunoglobulins, immune cells, lactose and oligosaccharides, enzymes, stem cells, and numerous other factors that likely play a role in health of the baby [51]. Unfortunately, there are often factors that mitigate against a mother providing milk for her own baby and if a baby is born preterm, some of the infant's capabilities to directly consume milk from the mother are not feasible because of suck-swallow and related immaturities. This necessitates tube feeding, and thus the milk needs to be frozen for later use. In addition, when the mother is not able to provide her own milk for her infant, either formula or donor milk is provided. Provision of donor milk does provide for human milk protein, carbohydrates, and lipids, but frequently the protein concentration is lower than if the baby would be obtaining the milk from its' mother. Most donor milk in the United States and parts of Europe undergoes pasteurization or similar process, which kills most of the live components including enzymes, cells, and microbes [52].

The presence of microbes in human milk was at one time thought to be problematic, but it is becoming increasingly clear that the microbial composition of human milk may play an important role in establishment of the infant intestinal microbiota and play a role in protection against the establishment of pathogenic microbes via promoting the growth of certain taxa of beneficial bacteria within the infant GI tract [53]. Metabolic products of certain beneficial microbes in milk may also benefit the infant via colonic fermentation of residual lactose (after digestion and absorption in the small intestine) and oligosaccharides into SCFAs, which are utilized as a major fuel by the colonocyte [54].

Although there is evidence to support the widely held belief that provision of human milk to preterm infants is associated with improved outcomes including less NEC, retinopathy of prematurity, and sepsis, and improved neurodevelopmental outcomes [55], we should consider some caveats about donor milk being inferior to maternal milk in terms of growth and possibly even neurodevelopment [56]. Whether microbial composition, the pasteurization process, or other factors such as protein content play a major role in these findings remains speculative.

Human milk oligosaccharides (HMOs) are the second most predominant carbohydrate in human milk after lactose, and they have a chemical composition and properties that are unique for humans [57]. They are resistant to GI hydrolysis and digestion by pancreatic and brush-border enzymes and are therefore not absorbed in significant amounts. Instead, HMOs serve as carbon sources for mutualistic microbes and thus shape the developing microbiome and mucosal immune system by selecting for beneficial gut bacteria and preventing pathogens from binding to the intestinal epithelial cells (IECs) [58]. They promote the growth of Bifidobacteria, which have a high capability of production of the SCFA acetate and folate metabolites, the latter of which may act as a potent epigenetic regulator. In contrast to human milk-fed infants, many infant formula products do not contain lactose as a major carbohydrate, but rather contain sucrose, maltose, and glucose or other carbohydrates that are not as readily hydrolyzed to SCFAs by the intestinal microbiota.

The breast milk microbiome consists of over 700 different species of microbes and varies widely between regions and individuals [59,60]. If possible, exclusive consumption of the mother's own milk (MOM) via direct breastfeeding and skin-to-skin (STS) [61] contact for the first 6 months to 1 year postpartum is the ideal way to establish symbiosis in the developing preterm gut, which is key to establishing developmental continuity. Milk-derived mutualistic bacterial genera *Lactobacillus* and *Bifidobacterium* and their metabolites are major components of this process. When breastfeeding is not possible, other nutritional options (each with more drawbacks than the one before it) include donor milk, cow's milk formula, soy formula, and parenteral nutrition. Infants given exclusively MOM develop a potentially more immunoregulatory and protective intestinal community composition than infants fed any of the other ways. Routine pasteurization of donor milk, in an attempt to prevent infection, sepsis, and other responses to pathogenic microbes, results in loss of some beneficial biologically active components including the aforementioned indigenous milk microbes that may actually be unique for each mother's own infant. Recent studies using both culture methods and DNA-based sequencing of human milk microbes show that donor milk, which is largely devoid of live microbes, may be faunated with baby's own mother's milk microbes under very specific conditions [62]. This is an area of intense investigation.

Lack of enteral feeding: total parenteral nutrition

Due to GI immaturity, VLBW infants often require at least partial parenteral nutrition. Because these infants are critically ill after birth and often have poor tolerance to enteral feedings, delaying enteral feeds has been a common practice. Lack of enteral feeding, however, results in a dearth of critical fuel that should be provided at the mucosal surface for not only microbiota development but also intestinal growth, stimulation of motility, and secretion of trophic hormones. As a result, mucosal atrophy with a resultant increase in bacterial translocation, systemic inflammatory responses, and sepsis becomes more likely [63].

Studies in animals have shown a reversal of total parenteral nutrition (TPN)-associated bacterial translocation and inflammation with enteral feeds [64]. Such studies further suggest that TPN may alter bacterial colonization of the intestinal tract and contribute to complications. A recent study was carried out with 47 human infants, 25 of whom received TPN, and 22 infants did not (control) [65]. By week 4, fecal bacterial alpha diversity was significantly lower in the TPN group compared to controls. The TPN group had significantly lower Bacteroidetes and higher Verrucomicrobia abundance compared to controls, and these differences became more pronounced over time. Thus the loss of biodiversity and alterations in the pattern of intestinal colonization in infants on TPN may lead to a dysbiosis, which in turn may predispose infants to adverse NICU outcomes.

Premature infants often require TPN for long periods. Prolonged lack of enteral nutrition and exposure to certain components of TPN, puts them at risk of developing parenteral nutrition (PN)-associated cholestasis (PNAC). A recent study compared fecal microbiomes of premature infants who developed PNAC compared to those who did not [66]. PNAC versus control samples demonstrated differences in relative abundance at the phylum level. On a more specific level, the genus *Bacillus* was more prevalent in samples from the PNAC group, whereas the genus *Escherichia–Shigella* was more prevalent in control. With feeding advances, weaning of PN, and resolution of PNAC, most differences in the microbiome resolved. In this study, it was concluded that premature neonates who develop PNAC have a different microbiome compared to those who do not, suggesting a role for the microbiome in the pathogenesis of this disease.

Skin microbiome

As with the microbes colonizing the neonatal intestine, infant skin microbiota serves as a microbial reservoir that influences immune development and may also contribute to or protect from neonatal infections. A recent study characterized the skin microbiota of preterm and full-term infants who were at birth hospitalization [67]. Preterm infant skin microbiota had lower alpha diversity and higher abundances of *Staphylococcus* and *Escherichia* than that of full-term infants. Differences in epidermal microbial community structure were observed at each of five body sites (though there was considerable exchange of microbes between sites) and were influenced by postnatal and gestational ages [67]. The skin of preterm infants, thus, harbors microbes that are commonly observed on hospitalized infants.

As mentioned previously, exchange of microbes from one skin site to another, as well as from skin sites to oral and fecal microbiomes, occurs frequently on the infant. It is also possible to transfer protective microbes from the maternal skin to the infant by the physical act of breastfeeding and by STS [61] contact between mother and child. STS care is associated with shorter infant hospitalization and an increased rate of maturation of the preterm infant's oral bacterial community [68]. STS care of preterm infants is also associated with improved GI function, tolerance to feeding, and lower rates of diarrhea and pneumonia [68–71]. All of this implies a link between being exposed to (and potentially consuming) maternal skin microbes and beneficial immune responses in the preterm infant.

Necrotizing enterocolitis

NEC is a debilitating intestinal disease affecting primarily very and extremely premature infants. Despite being a subject of intense research for decades, its etiology remains poorly understood. We do not know how to effectively treat this disease, and it continues to cause great mortality and morbidity. The specific cause of what is commonly referred to as NEC disease has been an enigma largely because what we are calling NEC represents more than one disease with multifactorial pathogenic mechanisms. There are several disease entities that are termed NEC which do not really represent the classic form of the disease. When looking at the more classic forms of this disease, it is clear that the greater the degree of prematurity of the infant, the longer it takes to develop NEC. Thus the time from birth to the onset of NEC is inversely proportional with length of gestation [72]. There are several key factors associated with the development of NEC including intestinal immaturity, enteral feeds, the intestinal microbiome, inflammation, local ischemia, and reperfusion injury, development of the microvasculature, and the potential for a robust inflammatory response. The time of greatest NEC occurrence is around 30 weeks postmenstrual age (PMA) [33,73–75].

The concept of NEC associated with inappropriate colonization of the premature intestine was first introduced by Claude and Walker [76], who suggested a cycle of microbe-induced neutrophil activation leading to inflammatory mediator release, vasoconstriction due to a response to inflammatory mediators, and disruption of the intestinal barrier. Although the concept of dysbiosis being a major factor in NEC pathogenesis is not new [76], the identification of specific causative microbes or metabolites remains elusive [77].

Although several mechanisms are thought to be involved in the proinflammatory immunological cascade in response to commensal bacteria that leads to NEC, TLR4 overactivation by LPS is thought to be a major player. Mutations in inflammation-related genes may predispose the premature infant to developing NEC. Loss of function mutations in SIGIRR, a gene encoding a negative regulator of TLR signaling, predisposes infants to NEC. In addition, allelic variants NFKB1 and NFKBIA that encode subunit variants of the inflammatory transcription factor nuclear factor-kappa B-light-chain-enhancer of activated B cells have differential effects on the development of NEC [78]. However, there are many other pathogenic pathways that appear to be in play, and these remain to be elucidated. New "omic" based technologies may help us better understand these complex mechanisms.

Studies utilizing 16S rDNA for microbiome evaluation suggested that relative increases in the phylum Proteobacteria when compared to other major microbial phyla such as Firmicutes, Bacteroidetes, and Negativicutes are associated with the development of NEC [33]. This intriguing relationship suggests that these may become the targets of microbial intervention against NEC.

A recent study measuring bacterial compositions and metabolomic profiles of fecal samples from relatively healthy and diseased preterm infants found that preterm infants' microbiomes and metabolomes were personalized for each infant rather than finding specific pathogenic microbes in all diseased infants [77]. One general conclusion they made, though, was that the preterm infant gut microbiome is enriched with microbes that commonly dominate in the presence of antibiotic selection pressure [77]. In addition to the aforementioned antibiotics, retrospective database reviews and secondary analyses of large multicenter studies show that the use of antacids, both H2 blockers and proton pump inhibitors, increase the risk for the development of NEC.

Metabolome in necrotizing enterocolitis

SCFAs' butyrate, propionate, and acetate are bacterial metabolites that exert beneficial downstream immunomodulatory, neural, and endocrine effects via binding and activating G-protein coupled receptors 41, 43, and 109A on the surface of IECs [79]. Acetate and propionate (but not butyrate) are more abundant in fecal samples from MOM-fed preterm infants than of samples from formula-fed preterm infants [80]. Of these, especially butyrate maintains intestinal tight junction integrity and reduces permeability, stopping potentially antigenic molecules and live microbes from finding their way into the bloodstream and thus preventing systemic inflammation in both breastfed and formula-fed infants, though this metabolite is most likely diminished in neonates relying on intravenous nutrition. SCFA concentrations per kilogram of mouse weight is over 100-fold lower in the cecum of germ-free mice compared to mice with conventional microbiota, demonstrating the role of gut bacteria in producing these metabolites [81]. Butyrate is also the major fuel for colonocytes and is a major stimulus for tight junction formation and integrity [69]. Butyrate and propionate, but not acetate, have been shown to stimulate histone acetylation of *forkhead box P3* (FOXP3) locus on naïve $CD4^+$ T-cells, which increases FOXP3 expression and promotes differentiation of the naïve cells toward antiinflammatory Tregs [82]. *Faecalibacterium prausnitzii* and *Eubacterium rectale/Roseburia* species (Firmicutes) produce butyrate, which regulates gene expression by histone modifications [83].

Bronchopulmonary dysplasia

Bronchopulmonary dysplasia (BPD) has a clinical operational definition based on oxygen receipt at a specific postnatal time point, usually 36 weeks' PMA, but this criterion has limited correlation with underlying disease phenotype [84].

The role of the airway microbiome has generated considerable interest [85]. Preterm and full-term lung and gut microbiomes codevelop systemically in a regulated but mechanistically enigmatic fashion [86]. 16S gene sequence-based microbiome analysis in preterm and term intubated babies [87] showed that airways of infants with BPD had increased abundance of Proteobacteria and decreased *Lactobacillus*. Decreased lung *Lactobacillus* abundance starts at birth in babies who subsequently developed BPD. In addition to its known beneficial effects in the gut, this genus has been known to have strong antiinflammatory properties and regulates lung development. Despite these studies demonstrating associations between dysbiosis and BPD, causality has not been established. Further mechanistic studies are necessary to better understand this phenomenon, and then there will be a need for further clinical studies.

The microbiome and the developing brain

Interaction between microbes and the brain has been thought to be rare except in cases where infectious agents penetrate the blood–brain barrier and cause disease such as rabies or infectious meningitis. However, more recently, it is becoming clear that there are associations between microbial communities in the GI tract and the brain. Several lines of evidence support this concept including: (1) regulation of tryptophan production in the serotonin pathway and gut–brain axis has profound effects on neurological and behavioral disorders [88], (2) intestinal microbes affect immune tolerance, therefore have an influence on neurologically implicated autoimmune diseases, for example, multiple sclerosis [89], (3) studies in animals show influence of intestinal microbes on anxiety-related behavior [90], and (4) in autism, schizophrenia, and other neural disorders, there are reports of specific bacterial species in the intestine that are altered [91].

There is speculation and accumulating evidence that the brain–gut–microbe axis is especially important during the early stages of life. Sudo et al.

were one of the first groups to investigate the role that the microbiome played in the hypothalamic–pituitary–adrenal system. Germ-free-reared mice exposed to restraint stress had exaggerated stress–anxiety behavior with increased corticosterone and adrenocorticotropin levels compared with specific-pathogen-free (SPF) mice [92]. This effect was partially normalized by colonizing the germ-free mice with fecal matter of SPF animals.

Several years later, additional studies in germ-free mice again demonstrated less anxious behavior than mice colonized with natural microbes [93]. When the brains of these animals were examined by histologic section and immunostaining, it was found that the striatum had higher turnover of neurochemicals including serotonin that are associated with anxious behavior. Introducing adult germ-free mice to nonsterile environments failed to normalize the behavior. However, the offspring of these conventionalized mice showed return to normal behavior, suggesting the presence of an early "critical window" during which microbes could affect a difference.

It is thought that microbial host interaction-derived metabolites may play an important role in gut–brain–microbe interactions. Serotonin is one of the most likely metabolic candidates. It should be noted that some antidepressant drugs work by promoting serotonin at the junctions between neurons. There are certain cells lining the colon termed enterochromaffin cells that produce large quantities of serotonin. Certain microbes promote serotonin production from these cells [94,95]. These cells produce 60% of peripheral serotonin in mice and more than 90% in humans. Subsequent studies linked microbial dysbiosis to autism spectrum disorders [96]. Another microbial metabolic product, butyrate, has been shown to play a role in gut–brain interactions. A recent study showed that gavage of butyrate in mice resulted in enhanced sleep [97].

This information suggests that the gut–brain–axis may play an important role in early life in terms of long-term development and begs the question of how various environmental perturbations that affect the intestinal microbiome might affect the subsequent neurodevelopment of the infant. This is underlined by a recent observation that human preterm infants exposed to antibiotics showed electroencephalography abnormalities at corrected term gestational age [98]. Although this was an observational trial and is subject to confounding factors, it raises an important question about the very common routine use of antibiotics in this population and the long-term effects.

These recently discovered relationships remain in their early stages of investigation and are primarily associations that require further in-depth studies to show causality and provide a basis for effective therapeutic measures. Fig. 7.1 summarizes what is currently known about the gut–brain axis.

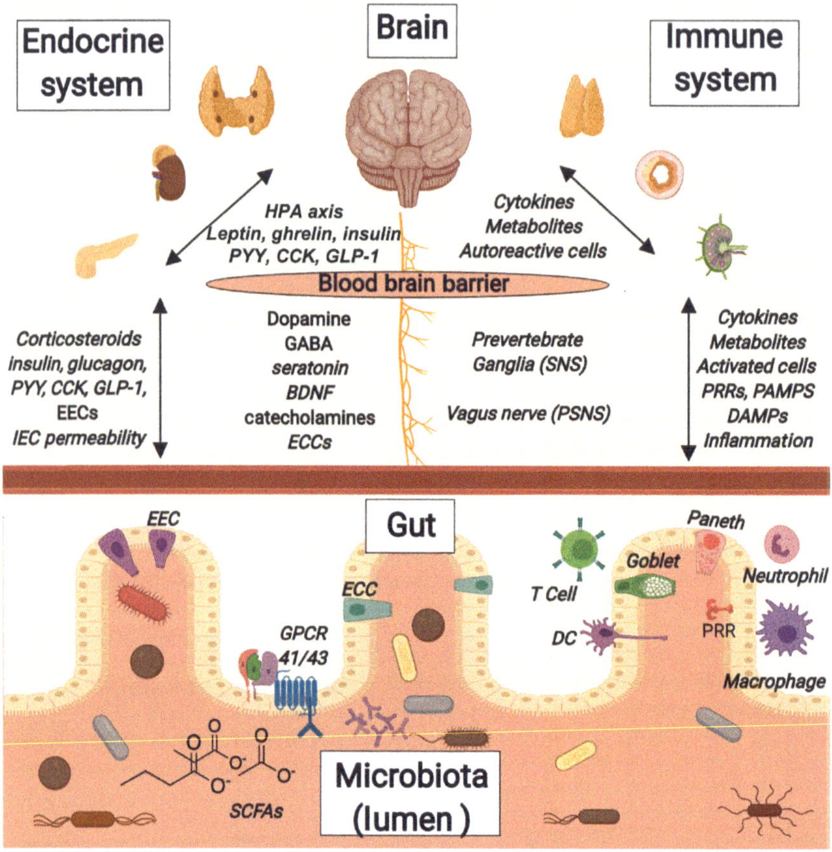

FIGURE 7.1 Microbiota gut–brain axis. *BDNF*, brain-derived neurotrophic factor; *CCK*, cholecystokinin; *DAMPS*, damage-associated molecular patterns; *DC*, dendritic cell; *ECCs*, enterochromaffin cells; *EECs*, enteroendrocrine cells; *GABA*, gamma-amino butyrate; *GLP-1*, glucagon-like peptide 1; *GPCR 41/43*, G-protein coupled receptor 41/43; *HPA*, hypothalamic pituitary adrenal axis; *PAMPS*, pathogen (or microbial)-associated molecular patterns; *PRRs*, pattern recognition receptors; *PSNS*, parasympathetic nervous system; *PYY*, peptide YY; *SCFAs*, short-chain fatty acids; SNS, sympathetic nervous system.

Summary

Commensal microbes inhabiting neonatal skin, lung, and intestine play critical roles in shaping endocrine, neurological, and local/systemic immune responses. Various factors are able to perturb the balance of commensal and "dysbiotic" microbes and thus disrupt homeostasis as well as short-term outcomes during infancy and long-term outcomes throughout the life span of the individual. Fig. 7.2 recapitulates what is currently known about these interactions.

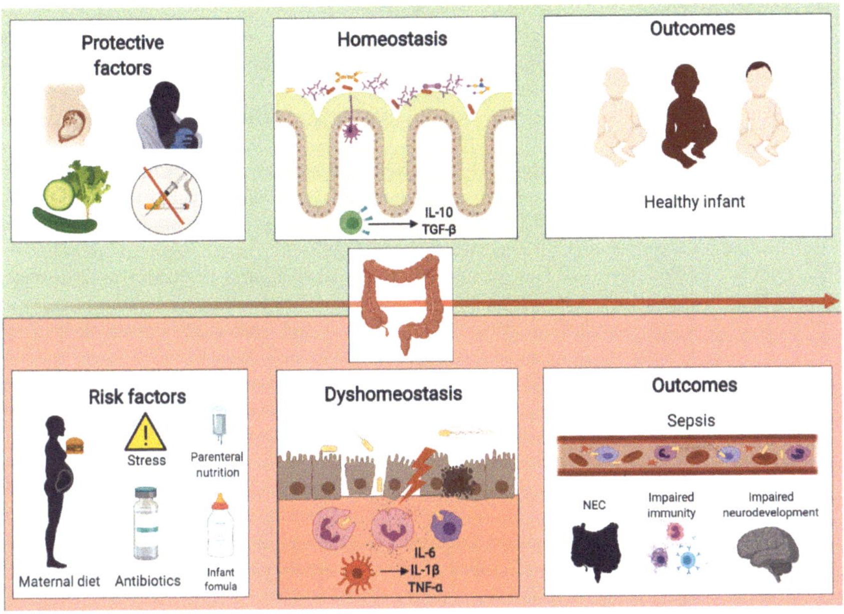

FIGURE 7.2 Factors contributing to microbial homeostasis and dyshomeostasis in the infant born preterm.

References

[1] Stainier DY. No organ left behind: tales of gut development and evolution. Science 2005;307(5717):1902–4.

[2] Vancamelbeke M, Vermeire S. The intestinal barrier: a fundamental role in health and disease. Expert Rev Gastroenterol Hepatol 2017;11(9):821–34.

[3] Koldovsky O, Sunshine P, Kretchmer N. Cellular migration of intestinal epithelia in suckling and weaned rats. Nature 1966;212(5068):1389–90.

[4] Burrin DG, Janeczko MJ, Stoll B. Emerging aspects of dietary glutamate metabolism in the developing gut. Asia Pac J Clin Nutr 2008;17(Suppl 1):368–71.

[5] Hamer HM, Jonkers D, Venema K, et al. Review article: the role of butyrate on colonic function. Aliment Pharmacol Ther 2008;27(2):104–19.

[6] Pritchard JA. Fetal swallowing and amniotic fluid volume. Obstet Gynecol 1966;28 (5):606–10.

[7] Tan Q, Xu H, Xu F, et al. Survival, distribution, and translocation of *Enterococcus faecalis* and implications for pregnant mice. FEMS Microbiol Lett 2013;349(1):32–9.

[8] Koren O, Goodrich JK, Cullender TC, et al. Host remodeling of the gut microbiome and metabolic changes during pregnancy. Cell 2012;150(3):470–80.

[9] Gomez de Aguero M, Ganal-Vonarburg SC, Fuhrer T, et al. The maternal microbiota drives early postnatal innate immune development. Science 2016;351(6279) 1296–302.

[10] Ardissone AN, Cruz DM, Davis-Richardson AG, et al. Meconium microbiome analysis identifies bacteria correlated with premature birth. PLoS One 2014;9.

[11] Jimenez E, Marin ML, Martin R, et al. Is meconium from healthy newborns actually sterile? Res Microbiol 2008;159(3):187–93.

[12] Tanaka M, Nakayama J. Development of the gut microbiota in infancy and its impact on health in later life. Allergol Int 2017;66(4):515–22.

[13] Ho TTB, Groer MW, Kane B, et al. Dichotomous development of the gut microbiome in preterm infants. Microbiome 2018;6(1):157.

[14] Claud EC. Neonatal necrotizing enterocolitis -inflammation and intestinal immaturity. Antiinflamm Antiallergy Agents Med Chem 2009;8(3):248–59.

[15] Martin CR, Walker WA. Intestinal immune defences and the inflammatory response in necrotising enterocolitis. Semin Fetal Neonatal Med 2006;11(5):369–77.

[16] Mihi B, Good M. Impact of Toll-like receptor 4 signaling in necrotizing enterocolitis: the state of the science. Clin Perinatol 2019;46(1):145–57.

[17] Lynch SV, Pedersen O. The human intestinal microbiome in health and disease. N Engl J Med 2016;375(24):2369–79.

[18] Clark JA, Coopersmith CM. Intestinal crosstalk: a new paradigm for understanding the gut as the "motor" of critical illness. Shock 2007;28(4):384–93.

[19] Neu J, Walker WA. Necrotizing enterocolitis. N Engl J Med 2011;364(3):255–64.

[20] Björkstén B. Effects of intestinal microflora and the environment on the development of asthma and allergy. Springer Semin Immunopathol 2004;25(3-4):257–70.

[21] Clark RH, Bloom BT, Spitzer AR, Gerstmann DR. Reported medication use in the neonatal intensive care unit: data from a large national data set. Pediatrics 2006;117 (6):1979–87.

[22] Klingenberg C, Kornelisse RF, Buonocore G, et al. Culture-negative early-onset neonatal sepsis - at the crossroad between efficient sepsis care and antimicrobial stewardship. Front Pediatr 2018;6:285.

[23] Hsieh EM, Hornik CP, Clark RH, et al. Medication use in the neonatal intensive care unit. Am J Perinatol 2014;31(9):811–21.

[24] Wynn JL, Polin RA. Progress in the management of neonatal sepsis: the importance of a consensus definition. Pediatr Res 2018;83(1-1):13–15.

[25] Ng PC, Ma TP, Lam HS. The use of laboratory biomarkers for surveillance, diagnosis and prediction of clinical outcomes in neonatal sepsis and necrotising enterocolitis. Arch Dis Child Fetal Neonatal Ed 2015;100(5):F448–52.

[26] Cotten CM, Taylor S, Stoll B, et al. Prolonged duration of initial empirical antibiotic treatment is associated with increased rates of necrotizing enterocolitis and death for extremely low birth weight infants. Pediatrics 2009;123(1):58–66.

[27] Stoll BJ, Gordon T, Korones SB, et al. Early-onset sepsis in very low birth weight neonates: a report from the National Institute of Child Health and Human Development Neonatal Research Network. J Pediatr 1996;129(1):72–80.

[28] Alexander VN, Northrup V, Bizzarro MJ. Antibiotic exposure in the newborn intensive care unit and the risk of necrotizing enterocolitis. J Pediatr 2011;159(3):392–7.

[29] Esmaeilizand R, Shah PS, Seshia M, et al. Antibiotic exposure and development of necrotizing enterocolitis in very preterm neonates. Paediatr Child Health 2018;23(4):e56–61.

[30] Torrazza RM, Ukhanova M, Wang X, et al. Intestinal microbial ecology and environmental factors affecting necrotizing enterocolitis. PLoS one 2013;8(12):e83304.

[31] Mai V, Torrazza RM, Ukhanova M, et al. Distortions in development of intestinal microbiota associated with late onset sepsis in preterm infants. PLoS one 2013;8(1): e52876.

[32] Mai V, Young CM, Ukhanova M, et al. Fecal microbiota in premature infants prior to necrotizing enterocolitis. PLoS One 2011;6(6):e20647.

[33] Pammi M, Cope J, Tarr PI, et al. Intestinal dysbiosis in preterm infants preceding necrotizing enterocolitis: a systematic review and meta-analysis. Microbiome 2017;5(1):31.

[34] Wang Y, Hoenig JD, Malin KJ, et al. 16S rRNA gene-based analysis of fecal microbiota from preterm infants with and without necrotizing enterocolitis. ISME J 2009;3 (8):944–54.

[35] Carl MA, Ndao IM, Springman AC, et al. Sepsis from the gut: the enteric habitat of bacteria that cause late-onset neonatal bloodstream infections. Clin Infect Dis 2014;58 (9):1211–18.
[36] Kuppala VS, Meinzen-Derr J, Morrow AL, Schibler KR. Prolonged initial empirical antibiotic treatment is associated with adverse outcomes in premature infants. J Pediatr 2011;159(5):720–5.
[37] Ting JY, Roberts A, Sherlock R, et al. Duration of initial empirical antibiotic therapy and outcomes in very low birth weight infants. Pediatrics 2019;143(3).
[38] Fajardo C, Alshaikh B, Harabor A. Prolonged use of antibiotics after birth is associated with increased morbidity in preterm infants with negative cultures. J Matern Fetal Neonatal Med 2018;1–7.
[39] Cantey JB, Pyle AK, Wozniak PS, et al. Early antibiotic exposure and adverse outcomes in preterm, very low birth weight infants. J Pediatr 2018;203:62–7.
[40] Greenberg RG, Chowdhury D, Hansen NI, et al. Prolonged duration of early antibiotic therapy in extremely premature infants. Pediatr Res 2019;85(7):994–1000.
[41] Dardas M, Gill SR, Grier A, et al. The impact of postnatal antibiotics on the preterm intestinal microbiome. Pediatric Res 2014;76(2):150–8.
[42] Arboleya S, Sanchez B, Milani C, et al. Intestinal microbiota development in preterm neonates and effect of perinatal antibiotics. J Pediatr 2015;166(3):538–44.
[43] Fricke WF. The more the merrier? Reduced fecal microbiota diversity in preterm infants treated with antibiotics. J Pediatr 2014;165(1):8–10.
[44] DiGiulio DB. Prematurity and perinatal antibiotics: a tale of two factors influencing development of the neonatal gut microbiota. J Pediatr 2015;166(3):515–17.
[45] Tzialla C, Borghesi A, Serra G, et al. Antimicrobial therapy in neonatal intensive care unit. Italian J Pediatr 2015;41(1):27.
[46] Greenwood C, Morrow AL, Lagomarcino AJ, et al. Early empiric antibiotic use in preterm infants is associated with lower bacterial diversity and higher relative abundance of Enterobacter. J Pediatr 2014;165(1):23–9.
[47] Fouhy F, Guinane CM, Hussey S, et al. High-throughput sequencing reveals the incomplete, short-term recovery of infant gut microbiota following parenteral antibiotic treatment with ampicillin and gentamicin. Antimicrob Agents Chemother 2012;56(11):5811–20.
[48] Neu J. Multiomics-based strategies for taming intestinal inflammation in the neonate. Curr Opin Clin Nutr Metab Care 2019;22(3):217–22.
[49] Gupta RW, Tran L, Norori J, et al. Histamine-2 receptor blockers alter the fecal microbiota in premature infants. J Pediatr Gastroenterol Nutr 2013;56(4):397–400.
[50] Terrin G, Passariello A, Canani RB, et al. Minimal enteral feeding reduces the risk of sepsis in feed-intolerant very low birth weight newborns. Acta Paediatr 2009;98(1):31–5.
[51] Neu J. Mother's own milk: how does it differ from donor milk for the baby. Breastfeed Med 2019;14(S1):S3–4.
[52] Merlino-Barr S, Groh-Wargo S. Donor breast milk for the preterm infant: your questions answered!. Neonatal Netw 2019;38(1):7–16.
[53] Gomez-Gallego C, Garcia-Mantrana I, Salminen S, Collado MC. The human milk microbiome and factors influencing its composition and activity. Semin Fetal Neonatal Med 2016;21(6):400–5.
[54] Gomez-Gallego C, Morales JM, Monleon D, et al. Human breast milk NMR metabolomic profile across specific geographical locations and its association with the milk microbiota. Nutrients 2018;10(10).
[55] Lucas A. Scientific evidence for breastfeeding. Nestle Nutr Workshop Ser 2019;90:1–12.
[56] Hard AL, Nilsson AK, Lund AM, et al. Review shows that donor milk does not promote the growth and development of preterm infants as well as maternal milk. Acta Paediatr 2019;108(6):998–1007.

[57] Oozeer R, van Limpt K, Ludwig T, et al. Intestinal microbiology in early life: specific prebiotics can have similar functionalities as human-milk oligosaccharides. Am J Clin Nutr 2013;98(2):561s–71s.
[58] Bering SB. Human milk oligosaccharides to prevent gut dysfunction and necrotizing enterocolitis in preterm neonates. Nutrients 2018;10(10).
[59] Cabrera-Rubio R, Collado MC, Laitinen K, et al. The human milk microbiome changes over lactation and is shaped by maternal weight and mode of delivery. Am J Clin Nutr 2012;96(3):544–51.
[60] Lackey KA, Williams JE, Meehan CL, et al. What's normal? Microbiomes in human milk and infant feces are related to each other but vary geographically: the INSPIRE study. Front Nutr 2019;6:45. Available from: https://doi.org/10.3389/fnut.2019.00045.
[61] Drell T, Lutsar I, Stsepetova J, et al. The development of gut microbiota in critically ill extremely low birth weight infants assessed with 16S rRNA gene based sequencing. Gut Microbes 2014;5(3):304–12.
[62] Cacho NT, Harrison NA, Parker LA, et al. Personalization of the microbiota of donor human milk with mother's own milk. Front Microbiol 2017;8:1470.
[63] Hermsen JL, Sano Y, Kudsk KA. Food fight! Parenteral nutrition, enteral stimulation and gut-derived mucosal immunity. Langenbecks Arch Surg 2009;394(1):17–30.
[64] Wildhaber BE, Yang H, Spencer AU, et al. Lack of enteral nutrition--effects on the intestinal immune system. J Surg Res 2005;123(1):8–16.
[65] Dahlgren AF, Pan A, Lam V, et al. Longitudinal changes in the gut microbiome of infants on total parenteral nutrition. Pediatr Res 2019;86(1):107–14.
[66] Mokha JS, Davidovics ZH, Maas K, et al. Fecal microbiomes in premature infants with and without parenteral nutrition-associated cholestasis. J Pediatr Gastroenterol Nutr 2019;69(2):224–30.
[67] Younge NE, Araujo-Perez F, Brandon D, Seed PC. Early-life skin microbiota in hospitalized preterm and full-term infants. Microbiome 2018;6(1):98.
[68] Hendricks-Muñoz KD, Xu J, Parikh HI, et al. Skin-to-skin care and the development of the preterm infant oral microbiome. Am J Perinatol 2015;32(13):1205–16.
[69] Lawn JE, Davidge R, Paul VK, et al. Born too soon: care for the preterm baby. Reprod Health 2013;10(Suppl 1):S5.
[70] Charpak N, Ruiz JG, Zupan J, et al. Kangaroo mother care: 25 years after. Acta Paediatr 2005;94(5):514–22.
[71] Flacking R, Ewald U, Wallin L. Positive effect of kangaroo mother care on long-term breastfeeding in very preterm infants. J Obstet Gynecol Neonatal Nurs 2011;40(2):190–7.
[72] Gonzalez-Rivera R, Culverhouse RC, Hamvas A, et al. The age of necrotizing enterocolitis onset: an application of Sartwell's incubation period model. J Perinatol 2011;31(8):519–23.
[73] Yee WH, Soraisham AS, Shah VS, et al. Incidence and timing of presentation of necrotizing enterocolitis in preterm infants. Pediatrics 2012;129(2):e298–304.
[74] Gordon PV, Clark R, Swanson JR, Spitzer A. Can a national dataset generate a nomogram for necrotizing enterocolitis onset? J Perinatol 2014;34(10):732–5.
[75] Eaton S, Rees CM, Hall NJ. Current research on the epidemiology, pathogenesis, and management of necrotizing enterocolitis. Neonatology 2017;111(4):423–30.
[76] Claud EC, Walker WA. Hypothesis: inappropriate colonization of the premature intestine can cause neonatal necrotizing enterocolitis. FASEB J 2001;15(8):1398–403.
[77] Wandro S, Osborne S, Enriquez C, et al. The microbiome and metabolome of preterm infant stool are personalized and not driven by health outcomes, including necrotizing enterocolitis and late-onset sepsis. mSphere 2018;3(3).
[78] Isani MA, Delaplain PT, Grishin A, Ford HR. Evolving understanding of neonatal necrotizing enterocolitis. Curr Opin Pediatr 2018;30(3):417–23.

[79] Li X, Watanabe K, Kimura I. Gut microbiota dysbiosis drives and implies novel therapeutic strategies for diabetes mellitus and related metabolic diseases. Front Immunol 2017;8:1882.
[80] Pourcyrous M, Nolan VG, Goodwin A, et al. Fecal short-chain fatty acids of very-low-birth-weight preterm infants fed expressed breast milk or formula. J Pediatr Gastroenterol Nutr 2014;59(6):725–31.
[81] Hoverstad T, Midtvedt T. Short-chain fatty acids in germfree mice and rats. J Nutr 1986;116(9):1772–6.
[82] Arpaia N, Campbell C, Fan X, et al. Metabolites produced by commensal bacteria promote peripheral regulatory T-cell generation. Nature 2013;504(7480):451–5.
[83] Rivière A, Selak M, Lantin D, et al. Bifidobacteria and butyrate-producing colon bacteria: importance and strategies for their stimulation in the human gut. Front Microbiol 2016;7:979.
[84] Jobe AH, Bancalari EH. Controversies about the definition of bronchopulmonary dysplasia at 50 years. Acta Paediatr 2017;106(5):692–3.
[85] Warner BB, Hamvas A. Lungs, microbes and the developing neonate. Neonatology 2015;107(4):337–43.
[86] Grier A, McDavid A, Wang B, et al. Neonatal gut and respiratory microbiota: coordinated development through time and space. Microbiome 2018;6(1):193.
[87] Lal CV, Bhandari V, Ambalavanan N. Genomics, microbiomics, proteomics, and metabolomics in bronchopulmonary dysplasia. Semin Perinatol 2018;42(7):425–31.
[88] O'Mahony SM, Clarke G, Borre YE, et al. Serotonin, tryptophan metabolism and the brain-gut-microbiome axis. Behav Brain Res 2015;277:32–48.
[89] Wasko NJ, Nichols F, Clark RB. Multiple sclerosis, the microbiome, TLR2, and the hygiene hypothesis. Autoimmun Rev 2020;19(1):102430.
[90] Lach G, Schellekens H, Dinan TG, Cryan JF. Anxiety, depression, and the microbiome: a role for gut peptides. Neurotherapeutics 2018;15(1):36–59.
[91] Mangiola F, Ianiro G, Franceschi F, et al. Gut microbiota in autism and mood disorders. World J Gastroenterol: WJG 2016;22(1):361–8.
[92] Sudo N, Chida Y, Aiba Y, et al. Postnatal microbial colonization programs the hypothalamic-pituitary-adrenal system for stress response in mice. J Physiol (Lond) 2004;558(1):263–75. Available from: https://doi.org/10.1113/jphysiol.2004.063388.
[93] Diaz Heijtz R, Wang S, Anuar F, et al. Normal gut microbiota modulates brain development and behavior. Proc Natl Acad Sci U S A 2011;108(7):3047–52.
[94] Yano JM, Yu K, Donaldson GP, et al. Indigenous bacteria from the gut microbiota regulate host serotonin biosynthesis. Cell 2015;161(2):264–76.
[95] Hsiao EY, McBride SW, Hsien S, et al. Microbiota modulate behavioral and physiological abnormalities associated with neurodevelopmental disorders. Cell 2013;155(7):1451–63.
[96] Vuong HE, Hsiao EY. Emerging roles for the gut microbiome in autism spectrum disorder. Biol Psychiatry 2017;81(5):411–23.
[97] Szentirmai E, Millican NS, Massie AR, Kapas L. Butyrate, a metabolite of intestinal bacteria, enhances sleep. Sci Rep 2019;9(1):7035.
[98] Firestein MR, Myers MM, Austin J, et al. Perinatal antibiotics alter preterm infant EEG and neurobehavior in the Family Nurture Intervention trial. Dev Psychobiol 2019;61(5):661–9.

SECTION III

Infancy – establishment of the microbiome and healthy development

CHAPTER

8

The compositional development of the microbiome in early life

Avital Cher[1] *and Moran Yassour*[1,2]

[1]Faculty of Medicine, The Hebrew University of Jerusalem, Jerusalem, Israel, [2]The Rachel and Selim Benin School of Computer Science and Engineering, The Hebrew University of Jerusalem, Jerusalem, Israel

Introduction

The infant gut microbial composition has been found to influence the lifelong health of the individual, far into adulthood. Many studies have been conducted in order to study what causes the different trends in bacterial composition throughout early development, as well as their implications. The observed similarities found by studying large cohorts have led to the suggestion of an average profile, depicting the appearance and disappearance of different bacterial groups, as well as their abundance. By studying the different genes encoded by these various bacterial groups, we can see shifts in different pathways that are expressed, such as carbon utilization pathways, which seem to be driven by the type of early-life diet of the infants. One concept that seems to be clear is that by tracking the different groups of bacteria that engraft themselves in the infant gut, we can see the development of ecological niches, or groups of bacteria that are specialized to perform tasks. This progression into ecological niches is what seems to ultimately determine which bacteria successfully make their home in the infant gut, while others get washed away. Nonetheless, it is clear that this process of selection is not entirely black and white, meaning keep the good bacteria and lose the bad. Instead, the infant gut composition can be affected by the mother's microbiome during birth, as well as

The Human Microbiome in Early Life
DOI: https://doi.org/10.1016/B978-0-12-818097-6.00008-0

throughout early life. However, here too, it seems that not all bacteria that can be obtained from the mother are regarded as worthy for the infant gut microbiome, and instead a type of selection process occurs that chooses which bacteria are most beneficial for the infant microbiome. The compositional development of the microbiome in early life is clearly a complex process, but by zooming into the specifics, we can use our knowledge to possibly help this process along when needed.

Bacterial composition: average profile of the infant gut

As discussed in earlier chapters, factors such as the mode of delivery and early use of antibiotics influence the microbial composition of infants. The canonical microbial signature of infants born by C-section is the lack of *Bacteroides* species in the first 6–18 months of life (Fig. 8.1), with ~20% of vaginally delivered children presenting the atypical, C-section-like pattern. Regardless of delivery mode, the following question remains open: What

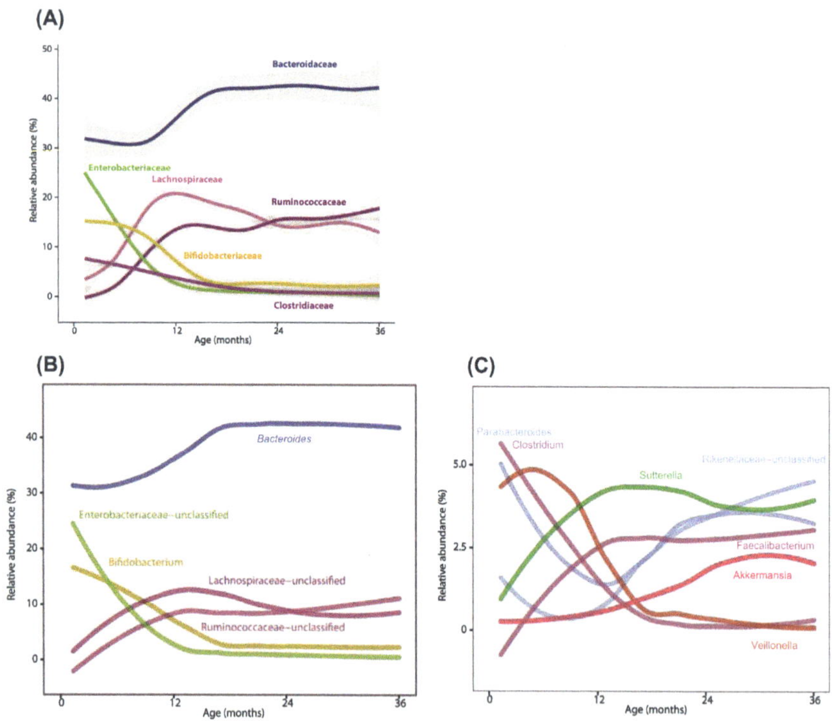

FIGURE 8.1 (A) Average profiles of highly abundant families. (B) Average profiles of highly abundant genera, (C) Average profiles of lowly abundant genera.

orchestrates the progression of the initial infant microbiome composition to the adult state in the first 2–3 years of life? When Yassour et al. tracked the gut microbial community dynamics of 39 (mostly vaginally delivered) Finnish children on a monthly basis at 2–36 months of age, they found that the succession of bacterial populations in the gut communities was overall similar among all vaginally delivered infants [1]. At the family level, nearly all infants (87%) had significant levels of Enterobacteriaceae (average relative abundance, 25%), Bifidobacteriaceae (15%), and Clostridiaceae (8%) at age 2 months, with each family steadily decreasing to an average relative abundance of 1%, 3%, and 2.5%, respectively, by the age of 18 months (Fig. 8.2). In contrast, they found an opposing trend with other species. Members of the Lachnospiraceae and Ruminococcaceae families were barely present at 2 months (4% and 0%, respectively), but they increased in the first 12 months before stabilizing at 12% to 20% of the microbial community. Besides these patterns, many more similarities were found in the

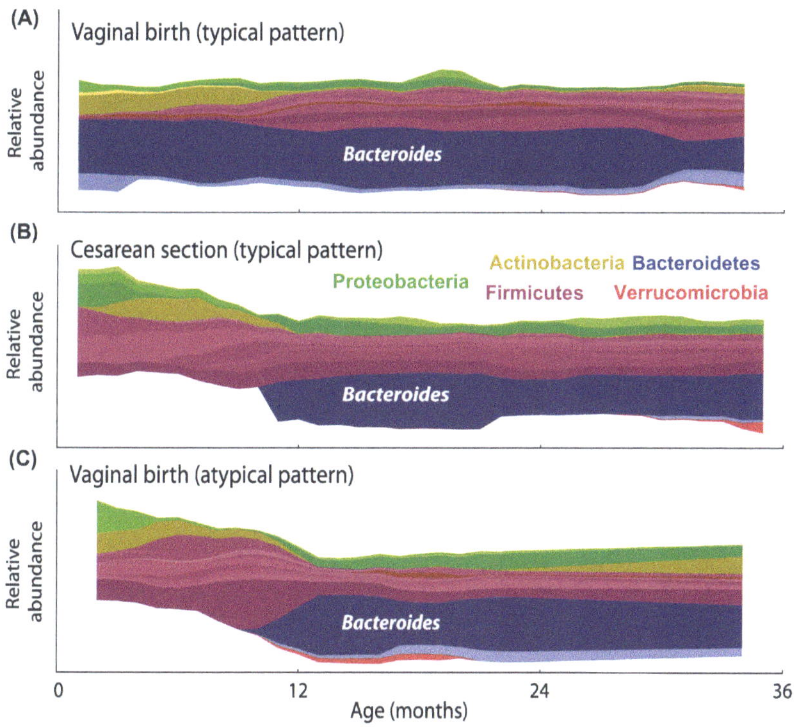

FIGURE 8.2 A to C Stream plots of individual microbial trajectories of three children over time (x axis), where each genus is colorcoded by its phylum. Shown are typical trajectories for a vaginally born child (A) (where species of the Bacteroides genus are present from an early age), a child born by cesarean section (B) (in which members of the Bacteroides genus are undetectable in the first 10 months), and an atypical trajectory for a vaginally born child (C) (which appears similar to the cesarean section signature).

succession of bacterial populations among the infants, including some which have been observed in previous studies [2,3]. As in many microbiome studies, geographic locations, lifestyle, and diet affect the gut microbial composition, and only recently several nonwestern cohorts have been established to study the establishment and development of the healthy infant gut microbiome, but still at a smaller scale [4–6].

While Yassour et al. analyzed samples beginning at 2 months of age, subsequent studies collected longitudinal sampling of infant gut, starting at birth. Two such studies suggested two opposing models for community establishment [7,8]. While Yassour et al. observed a very low-complexity microbial community soon after birth (within 24 hours), which slowly increased in diversity in the first 2 weeks of life, Ferretti et al. observed an opposite trend, where the initial birth samples were highly diverse, and decreased their richness overtime. These two distinct observations represent two opposing models, where either (1) we are slowly colonized with gut commensals, or (2) we are rapidly colonized followed by a period of selection of the most fit species.

Ferreti et al. also classified the early-colonizing species according to their ability to survive in an aerobic environment: (1) oxygen-utilizing organisms (also known as facultative anaerobes, which can grow with or without oxygen) and (2) organisms that cannot use oxygen as a terminal electron acceptor (namely, aerotolerant and obligate anaerobes), which are both unable to grow in oxygen; however, unlike obligate anaerobes, aerotolerant bacteria are not rapidly poisoned by atmospheric oxygen. They found that at day 1, there were more strict anaerobic species (on average 4.7 species per samples) than aerobic (0.4) or facultative aerobic species (4.4), although present at rather low abundances (16.2% in relative abundance). On the third day, the strict anaerobes had drastically decreased in number by 26%, and during this 3-day time period, the abundance of facultative anaerobes increased by +9.1%. One potential explanation for the shift of the oxygen-utilizing bacteria was proposed to be the changes in oxygen levels in the infant gut during the first few months of life. Although the infant gut is under anaerobic conditions in utero, following birth, the newborn initially has a high level of oxygen in the gastrointestinal tract, as the gut barrier is not fully intact yet. This favors the appearance of facultative anaerobes as some of the first colonizers. These early colonizers decrease the oxygen concentration of the gut through the consumption of oxygen for energy production. Therefore facultative anaerobes mediate the shift from aerobic to anaerobic conditions, which is associated with the adult microbiome state [9]. Similarly, in a study by Backhed et al., where the infants were also sampled during the first days of life, they found that a number of facultative anaerobes positively correlated with many obligate anaerobes in newborns, but they had reduced or negative correlations at 4 months [10]. This reflects upon the changing environment of the infant gut to a more anaerobic state. It also suggests the progression of the developing microbial composition into a developed niche, housing the strains of bacteria that most suit the conditions of the environment, as well as the needs of the host.

Rate of stability: intraspecies diversity—the world of strains...

While the infant gut microbiome continues to develop and eventually stabilize to the adult state, it is possible to visualize the development of ecological niches by analyzing the strain stability. This is because when the strains with higher fitness are engrafted, they may have a higher rate of persistence in the gut. Vatanen et al. [11] characterized dominant strains for the most abundant species in the metagenomic data by calling single-nucleotide polymorphisms (SNPs) on conserved and unique species-specific marker genes selected from their core genomes (genes shared across all strains within the species). In this way, the researchers generated a marker gene-based SNP haplotype of the dominant strain per species, referred to as the SNP haplotype. SNP haplotype comparison by sequence similarity could then be performed within the subject and between different subjects. They found that intrasubject comparisons showed more similar strains compared to intersubject comparisons, but there was a wide range of strain diversities among the investigated bacterial species. For example, *Haemophilus parainfluenzae* and *Faecalibacterium prausnitzii* were among the most diverse species, with strains having less than 95% sequence similarity in SNP haplotype comparisons, while all other species analyzed had an average intersubject similarity of 98.9%.

When characterizing the strain space of each species, Vatanen et al. [11] were surprised to find an unexpected high level of observed sequencing identity in *Bacteroides* species. It was then speculated that SNP haplotypes do not capture all means of microbial genetic diversification for these species, including lateral gene transfer (LGT) and niche adaptation. Therefore the genome diversity of *Bacteroides* was analyzed by investigating the bacteria–phage interactions in the gut by reconstructing metagenomic CRISPR arrays that harbor spacer sequences targeting phages to which bacteria have adapted a response. Interestingly, they found that the number of CRISPR spacers in the metagenomes reflected the increasing diversity and overall maturation of the microbiomes, and the extensity of species harboring CRISPR spacers suggested that *Bacteroides* species were exposed to an extensive phage repertoire in the children's guts. These observed bacteria–phage interactions in the gut also provide a possible mechanism for the increased genomic plasticity in *Bacteroides*.

Bacteroides species and *Escherichia coli* harbored the largest assembled pangenomes, but the magnitude of variability was much higher for gene content than for SNP haplotypes. The lack of correlation between SNP haplotype and gene content similarities may be due to the difference in the rates at which the two events occur. Genetic adaptation can occur rapidly due to promiscuous LGT and gene loss, whereas long-term SNP acquisition occurs more slowly. Expanding our understanding of the effects of these genetic adaptations and their effect on strain fitness or interaction with the host may allow us to more clearly understand why the composition of the microbial gut exhibits certain trends, and how these trends can influence the health and development of children.

Interspecies diversity

The development of ecological niches is further supported in Ferretti et al. through examining the strain heterogeneity for each species in the microbiome of infants and comparing to their mothers [8]. At day 1 postpartum, the early infant microbiome was characterized by a high species diversity, as well as a high strain diversity. This reflects the microbial environment of the infant gut before the development of ecological niches of specialized strains contributing to different metabolic pathways. At later time points, there was a decrease in the intraspecies polymorphic rates, up to 1 month, and reaching to levels that were comparable to the mothers. This was accompanied by a higher relative frequency of the dominant strain. Interestingly, the samples collected from the infants at 4 months of age suggested that the strain diversity is increasing and reached levels that were now significantly higher than that observed in the mothers. This suggests that although the data supports the occurrence of initial high bacterial richness at birth that slowly decreases with the development of ecological niches, there are other factors that can affect this progress, such as the increased exposure of the infant to other sources of microbial seeding from the environment.

Other factors that affect ecological niche development

While environmental exposure and early intimate reactions (e.g., with the mother) are factors determining the infant microbial acquisition [8], many other associations have been found that are also thought to have an influence on the infant gut composition. For example, Vatanen et al. found that height at age 3 and growth rate during the first 3 years of life were positively associated with microbial diversity. In addition, the mode of delivery, gestational age at birth, and treatment by antibiotics for the mother and/or child were found to be crucial factors. While previous studies [12,13] have shown that preterm neonates gestational age at birth was the major factor shaping the microbial composition,

Backhed et al. found that the mode of delivery and feeding patterns have the largest effects on that of term infants.

Transmission from the mother

The large effect that birth mode seems to have on the gut microbiome suggests that the mother serves as a source from which the bacteria that reside in the infant's gut originate. There is currently a debate as to when transmission from the mother begins, prepartum or beginning at delivery. Multiple studies have demonstrated that the maternal microbial reservoir is an important source for the acquisition of microbial species and strains in the infant [14–16]. Ferreti et al. characterized the transfer of microbes from mothers to infants in their first 4 months of life. They sampled the stool, as well as four additional body sites: the oral cavity, the skin, the vagina, and breast milk, of 25 healthy pregnant women who vaginally delivered. They found that the infant gut contained a large proportion of species in common with their mothers. For example, at day 1, about 50% of the microbial population in the infant gut belonged to species also present in at least one of the sampled maternal body sites. Previous studies have suggested that maternal transmission of bacteria in vaginally delivered children are mostly from the maternal rectal and vaginal microbiome [17–19]. In this study, while all children were vaginally delivered, their gut microbiomes did not resemble solely one of the maternal body sites that were sampled, but perhaps a mixture of these maternal sources. This high subject diversity and lack of uniformity suggest a nondeterministic relation between the initial exposure of the infant to its maternal sources and the resulting engraftment of microbes, as each maternal microbiome (vagina, skin, oral, and fecal) influenced the infants to a varying degree.

Despite the large influence the maternal microbiome sources have on the developing infant gut, most of the species that colonized the infant gut on day 1 were lost before the consecutive time point. Interestingly, most of these lost species were shared with at least one body site of the respective mother. Many of these shared microbes likely originated from maternal body sites other than the maternal stool, suggesting that microbes from these sources are not suited to reside in the infant gut. In contrast, species that are considered to be fecal related were found to persist from birth up to at least 4 months of age. Although it may seem likely that the bacterial strains that are most abundant in the mother's fecal microbiome would have the greatest effective engraftment to her infant, the shared species between the mother and infant were actually in lower relative abundance in mothers, compared with their infants, potentially due to the environmental differences of the adult and infant gut. Therefore it is the fitness of the microbes in relation to the infant, rather than the number of bacteria available for transfer from the host, that most heavily influences the engraftment success. Nonetheless, Yassour et al. found six species in both the mother and child that had 5% relative abundance or greater, in both the mother and child: three *Bacteroides* species (*Bacteroides uniformis, Bacteroides vulgatus,* and *Bacteroides dorei*), two *Bifidobacterium* species (*Bifidobacterium adolescentis* and *Bifidobacterium longum*), and *E. coli.* This supports the idea that there are some species that could be equally beneficial in both the infant and mother.

Interestingly, the number of strains found by Ferretti et al. to be shared between the mother and infant gradually decreased overtime, again suggesting the development of ecological niches specific to the infant microbiome overtime. However, in this case, the results exhibit a selective process of strains from the microbial source that the mother offers to her infant. The strains that lacked evidence as the mother being the source, but were present at more than one time point, were found to usually be replaced and had low retention, whereas most of the vertically transferred strains had no replacement. This suggests that the vertically transferred strains from the mother have a higher level of fitness compared to strains that originate from other environmental sources. Therefore it is clear that the infant benefits from the transfer from strains from its mother during delivery and through intimate reactions in early life.

Another study examining mother-to-child bacterial transmission [7] developed a novel strain-level approach to study maternal bacterial transmission patterns, thereby enabling them to determine if they can distinguish benefits for the infant gained from each transfer. In order to do this, they defined the "dominant strain" within a microbial species in a sample to be the single most common haplotype detectable by metagenomic sequencing, and all the nondominant strains were denoted as the collection of "secondary strains." At any given genomic position, the dominant strain has the highest read count, and the nucleotide in this position is called the "major nucleotide." The secondary strain contributed to the second most frequent nucleotide at each position, and this nucleotide is denoted as the "minor nucleotide." They examined patterns of single-nucleotide variants (SNV) in genomic regions that are unique to each species within and across families. In this way, they were able to identify a potential strain inheritance from mother to child, which was suggested by highly similar nucleotide variation patterns in two individuals. They found that in many cases, the dominant strain was inherited, but occasionally the secondary strain was inherited instead. In addition, they found that certain species showed dominant strain transmission from mother to child most often (*B. vulgatus and B. adolescentis*), while for *B. uniformis*, there was a variance between the families, of either dominant strain, secondary, or no inheritance.

They hypothesized that there is a lack of fitness or advantage of the dominant strain, resulting in the selection of the secondary instead. In order to test this, they examined strain transmission of *B. uniformis*, as this species had different inheritance patterns across families, allowing for analysis to determine what may cause this variance.

They searched for genes in this species that were observed to be present in both the mother and child samples of families who shared a dominant strain and that were present only in the child samples of families where the mother's secondary strain was transmitted. In other words, they were searching for the genes that were missing in the mother's dominant strains and that perhaps prompted the colonization of the maternal secondary strain. In this way, they identified nine such genes that vary by mode of transmission, suggesting that they may confer an advantage to the infant, since they are inherited from the secondary strain when absent from the dominant strain (Fig. 8.3). Interestingly, these genes are located in a single genomic region, and this genomic region appears to be a starch utilization system, suggesting it has a potential selective advantage in breaking down the glycans from the mother's milk. This supports its necessity in the infant gut microbiome, for enhancing the infant's digestion efficiency, as well as contributing to the microbiome maturation. Therefore when the dominant strain in the mother lacks this gene, the secondary strain that confers a higher fitness will be better a colonizer and maintained in the infant's gut.

Yassour et al. observed that the infant and maternal microbiomes have striking differences on the species level, suggesting that it is not possible to conclude that the infant gut composition resembles that of their mothers. Despite this lack of similarity, by zooming into the species level, Yassour et al. was successful in identifying strain similarities in both mother and child samples of the same family. Although the transfer of the mother's strain to her infant has been demonstrated in other work, this study is unique in that it also tracked the transmission of the less dominant strain, or the secondary strain, further supporting the idea that the fitness of strains, determined by the genes that they encode, is what drives the development of gut ecological niches.

Metagenomics: potential functions of the infant gut

An additional approach to understand which microbes make their presence and at which time periods is to examine the functions of the bacteria that appear at different stages of development, which can be inferred by their encoded genes. We expect the potential function of the infant gut microbiome to be different than that of the adult microbiome as the available nutrients in the gut are different among these two

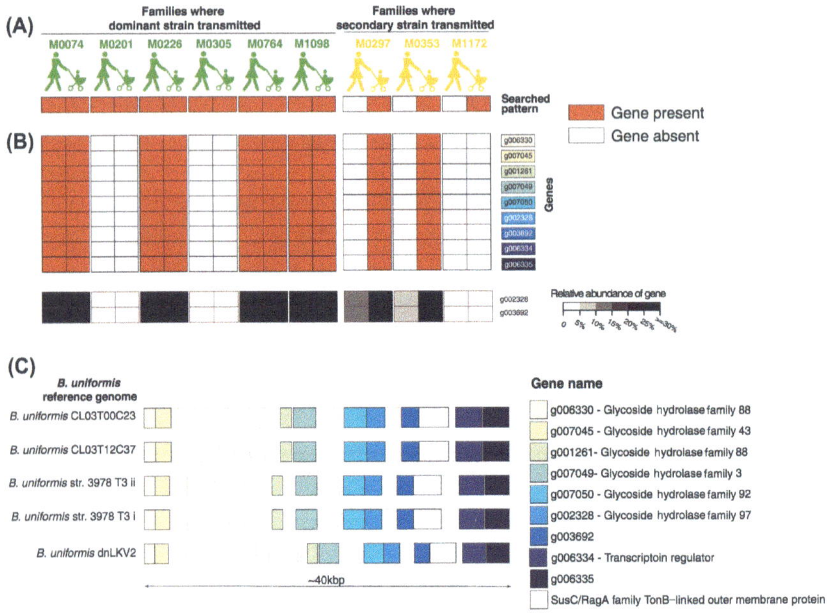

FIGURE 8.3 **B. uniformis Starch Utilization System.** (A) A schematic showing the gene presence/absence (red/white) patterns we inspected to identify functions that potentially drive the secondary strain inheritance. (B) Presence/absence profiles of nine B. uniformis genes that differ based on their inheritance patterns in a majority of mothers (red/white), together with qPCR results for two of the nine genes (grayscale). (C) The genomic location of the nine differential genes (yellow to blue) in five B. uniformis reference genomes, together with the SusC gene found in that location (stripes). White boxes are other genes.

groups. Overall, the infant nutrition is mostly breast- and formula-feeding, whereas the adult nutrition is much more diverse.

By examining the functions of the infant microbiome through mapping metagenomic reads to the KEGG orthology groups, Backhed et al. found different utilizations of energy sources across the developmental time period of the infant's gut microbiome. For example, they found that phosphotransferase system genes for carbohydrate uptake were enriched in the newborn microbiome, while the lactose-specific transporter was most abundant in 4-month-old infants, which would be expected with a diet dominated by milk. In addition, while the microbiomes of newborns and 4-month-old infants were enriched in genes required for degradation of sugars from breast milk, -glucoside-specific transporters were most abundant in the 4-month-old and 12-month-old infants. In addition, the 12-month microbiome was enriched in genes involved in the degradation of complex sugars and starch. These results suggest that the genes available for nutrient degradation and absorption specialize overtime, as the infant adapts to the source of nutrition provided to them.

However, beyond the functions of carbon uptake and utilization, other pathways of the bacteria can also enhance its likelihood of engrafting itself into the infant's gut. In a study analyzing an integrated data set from DIABIMMUNE Finnish cohort of 300 children, Vatanen et al. identified functional adaptation that can be interpreted from the succession of bacterial species cycling through the gut. For each examined function, they asked how many different species contribute to that function, namely have sequenced reads that map to the genes of a related pathway. A highly contributional-diverse function would be one that has many species encoding for it. Vatanen et al. found that some functions increase their contributional diversity overtime, whereas others are decreased, indicating that some functions are common and are found in many of the gut species, whereas others are being more and more specialized overtime and are found only in a small fraction of the gut microbiome. For example, they found that most gene ontology terms displayed increasing within-sample functional diversity with increasing age, including sporulation, glycolysis, and riboflavin biosynthesis. However, a few specific pathways displayed the opposite trend, including the aerobic electron transport chain, viral release from host cells, and the siderophore biosynthetic process. These two differing trends may suggest the positive and negative selections that occur during the process of ecological niche formation. For example, Vatanen et al. found that a single dominant species per community contributes to siderophore biosynthesis, whereas siderophore transport-related genes were more widely distributed across community members. This is suggested to be due to the fact that the ability to produce energy costly molecules is under negative selection. In this way, the minimum number of bacteria species encode the genes that are necessary to produce the total amount of protein required by the entire bacterial community. Therefore having one dominant species create and secrete the siderophore and many species then utilizing siderophore transport-related gene products results in the most cost-effective energy expenditure. However, these shifts in contributional diversities per pathway eventually stabilize overtime, as Vatanen et al. found that the between-sample contributional diversities (beta-diversity) decreased with increasing age. This demonstrates the shift toward a more stable adult microbial composition through processes of negative and positive selections, eventually leading to a cease in the competition and the formation of a synchronized bacterial community, with each member specialized to their designated tasks or functional pathways.

Vatanen et al. also discuss taxonomic diversity and found an increase in the taxonomic diversity with increasing age, which is a trend consistent with what has been observed of a maturing microbiome (such as Yassour et al. as described previously).

Citing the review [20], "One of the limitations of metagenomic approaches is that the microbiome data do not provide information on whether or not genes are expressed at any given time." Indeed, the perfect view into the microbiome function would be through the lens of meta-transcriptomic sequencing; however, at least to date, there are not that many studies performing sequencing of the RNA content of the gut microbiome, especially not in infants.

The abundance of *Bacteroides*

As mentioned previously, one function that allows researchers to observe the development of the microbiome overtime is the carbon source utilization. Many of the bacterial species that contribute to the newborn's carbon source utilization belong to the *Bacteroides* genus. Previous studies have reported that children born by cesarean section initially lack members of the *Bacteroides* genus in their gut microbiome. While the results by Yassour et al. supported this finding, they also found that a substantial portion of vaginally born children (7 out of 35) exhibited a "low-*Bacteroides*" signature, similar to that exhibited by children born by C-section. They found that in the "low-*Bacteroides*" children, *Bacteroides* species were detected for the first time only between 6 and 18 months of age, yet still without understanding the reason for its appearance. The low-*Bacteroides* signature was associated with a decrease in alpha-diversity, or richness of the microbiome, which was evident by 6 months of age. While previous studies have shown that the cesarean section delivery is associated with decreased microbial diversity, the results by Yassour et al. demonstrate that the decrease in microbial diversity is seen across all children that express the low-*Bacteroides* signature, regardless of their mode of birth. One question that remains unanswered is why 20% of vaginally delivered children have the "low-*Bacteroides*" microbial signature. With only seven children in this group, the researchers tried to determine the clinical variable that might explain this microbial signature, such as: gestational age, antibiotic treatment of the mother, duration of delivery, the mother's use of enema before delivery, or time spent at hospital. Although none of these variables statistically significantly correlated with the difference, it may be relevant to identify the reason for the low-*Bacteroides* signature, as it has been suggested that decreased community diversity can limit the education of the immune system, including its ability to recognize commensal bacteria [20]. The decreased microbial diversity seen in low-*Bacteroides* children continued to persist even at 36 months, which was when the *Bacteroides* had become highly abundant. This further emphasized the need to understand the implications of the associated decrease in microbial diversity, in order to increase our understanding of how to prevent cesarean section-associated health consequences, including asthma, obesity, and allergies.

Bacteroides abundance effect on *Bifidobacterium*

In the first few months of life, species from the *Bacteroides* and *Bifidobacterium* genera are among the most abundant in the infant gut. *Bifidobacterium* species have been extensively studied in relation to their ability to digest complex sugars from breast milk, called human milk oligosaccharides (HMOs). While these HMOs are the third most abundant component of breast milk, they are unable to be digested by the infant, and thereby serve as food for their developing gut microbiome. Interestingly, whereas previous studies reported a decreased abundance of the *Bifidobacterium* genus in the gut microbiomes of children born by cesarean section, Yassour et al. found that children from the low-*Bacteroides* group had a higher abundance of *Bifidobacterium* species in the first 6 months of life compared to all other individuals. Since *Bacteroides* are also believed to have the ability to break down HMOs, it is possible that the *Bifidobacterium* species occupy the available ecological niche left due to the lack of *Bacteroides*. Analysis of the metagenomic data allowed for the quantification of the relative contribution of each species to the breakdown of HMOs, and it was computationally identified that *Bifidobacterium* species were the main contributors to HMO breakdown in the low-*Bacteroides* group, whereas *Bacteroides* species were the dominant contributors for all other children. This suggests that regardless of the birth mode, a niche is formed so that the breakdown of HMOs is efficiently conducted in the infant's gut, where one bacterial species is the major contributor to HMO breakdown. The development of this niche is crucial to the growth of the infant as they themselves are unable to digest the HMOs that are provided from the mother's breast milk.

Antibiotic resistant genes

In order to further determine what may play a role in the selection of the strains that persist in the gut and thereby become a member of the developed ecological niche, Yassour et al. [7] searched for genes in the noncore genome, as these genes often contribute to important functional differences, such as antibiotic resistance, or as shown previously, carbon source utilization. They found antibiotic resistant (AR) genes to be more prevalent in the infant samples compared to mothers samples, as was found in previous work. Backhed et al. suggested that the presence of

AR genes in the newborn may be a consequence of the relative high abundance of Proteobacteria DNA, since their genomes contain high levels of AR genes. The most common AR genes were present in too many samples in order to accurately determine which are the result of transmission events versus random occurrence. Therefore to study transmission they focused on less common AR genes, for a total of 249 genes. Of these genes, 59 had at least one significant family transmission. Notably, these transmitted AR genes were not classified as a specific type of resistance and were not limited to a subset of families.

In a previous study by Yassour et al. [1], they investigated what is the impact of antibiotic treatment on the appearance of AR genes. They found that the abundance of the chromosomally encoded AR genes rose extremely rapidly during antibiotic treatment and then decreased swiftly after antibiotic withdrawal. On the other hand, AR genes encoded episomally increased their abundance upon antibiotic treatment, and their presence continued for much longer periods of time (Fig. 8.4). This was suggested to be due to the fact that episomally encoded elements can be shared among multiple species, while chromosomally encoded elements are nonmobile.

It is believed that repetitive antibiotic treatment in early childhood can lead to later hyper-immune response diseases, such as asthma, irritable bowel syndrome, and diabetes. Yassour et al. investigated the effects of multiple antibiotic treatments in the first 3 years of life, and they found that children who did not receive antibiotics had a richer microbial community compared to children who did receive repetitive treatment. However, this difference was deemed relatively modest, and it was only evident after the first year of life. The most striking difference between the two groups of children, those with and without antibiotic treatment, was actually the intraspecies diversity. Namely, the richness of strains within each species. They found that the microbiome of antibiotic-treated children had significantly more species dominated by a single strain compared to the no treatment group (Fig. 8.5).

In many cases, antibiotics are crucial to the infant's or child's survival, and therefore despite their negative effect on the microbial community structure, treatment by antibiotics is often deemed as the better option by health care providers and parents. Unfortunately, problems arise when antibiotics are used excessively, without weighing the costs and benefits, and without giving the microbiome a chance to recover from the harm that was inflicted from the previous treatment. This can be compared to how in ecology, repeated perturbations to an ecosystem are recognized as particularly harmful if their rate of occurrence exceeds the time necessary for recovery from the previous perturbation. Yassour et al. compared the microbial composition at all pairs of time points for each child by calculating the Jaccard index, where higher values

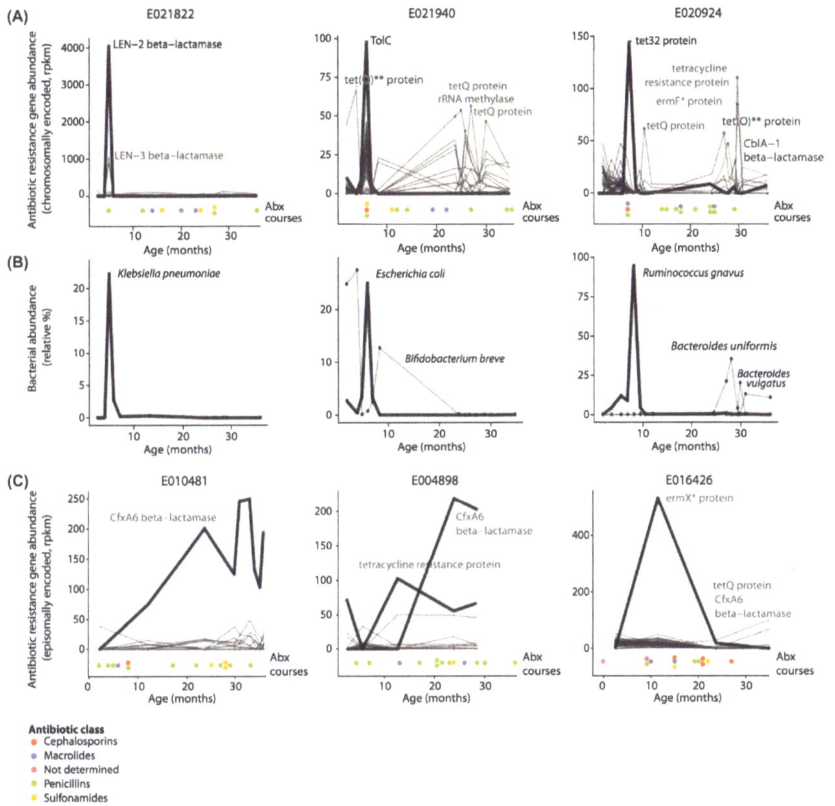

FIGURE 8.4 **Antibiotic resistance gene profiles.** (A) Abundance of antibiotic resistance gene products [mostly chromosomally encoded, reads per kilobase of transcript per million mapped reads (rpkm) values] in three children over time, together with the timing of individual antibiotic courses (colored dots; see fig. S10 for plots for all children). (B) Relative abundance of species that most correlated with the resistance gene profiles of (A). (C) Examples as in (A) of antibiotic resistance genes that are found on mobile elements and that are present in the gut for longer time periods. In (A) and (C), the number and order of antibiotic courses are shown, with each antibiotic class indicated by color.

indicate similar microbial composition. They found that while the Jaccard index of the samples from untreated children was consistently high, antibiotic treatment resulted in short-term decrease in stability around the time of antibiotic treatment. In addition, they found that children who received antibiotic treatment had significantly less stable overall microbial communities, which was defined as the median of the Jaccard indices for consecutive time points. The variation observed in the stability of antibiotic-treated children was much higher than the children without treatment. This may be due to unique factors of each individual, such as the previous microbial composition, host

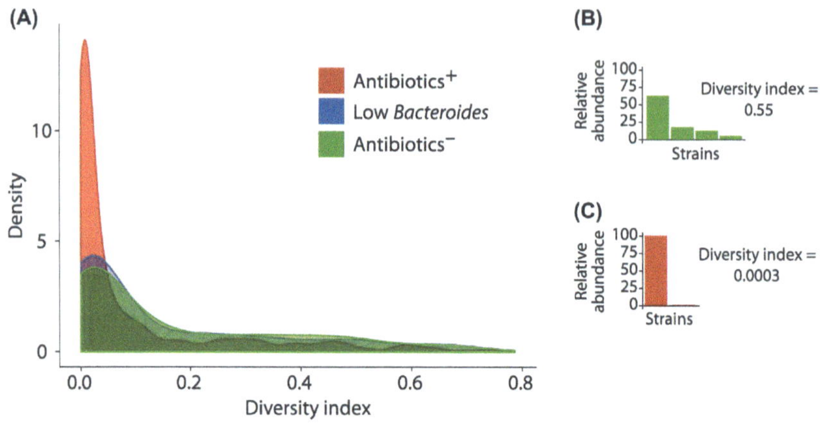

FIGURE 8.5 **Diversity and strain similarities of the infant gut microbiota.** (A) Diversity index of the strain distribution within each species shown for all metagenomic samples. Samples are colored according to three groups: children who received antibiotics (red), children with low Bacteroides(blue), and children who received no antibiotics (green). (B and C) Strain distributions of two selected samples (y axis is relative abundance of each strain) and their calculated diversity indices.

genetics, or the overall disease state. While this study focused on children born vaginally, a different study studied the effects of antibiotic treatment on a study group that contained many participants born by cesarean section [19]. They found that the mode of delivery had a stronger effect than repeated treatment of antibiotics, again suggesting that a compound of factors can either enhance or diminish the effect of antibiotics on the gut microbiome's stability.

Effect on maturity by breast milk versus formula, and transition to adult state

In addition to the mode of delivery affecting the bacterial composition, the source of feeding was also found to have an effect, specifically on the maturity of the microbiome. Backhed et al. trained the random forest model using relative abundances of all MetaOTUs in exclusively breastfed newborns who were vaginally delivered and ceased breastfeeding between 4 and 12 months, and 64 MetaOTUs were selected as markers for gut microbiome age. Using this method, they found that newborns who received formula in addition to breast milk received an older age estimate compared to the exclusively breastfed newborns. Interestingly, at 4 months, the differences between infants who received exclusive breastfeeding versus exclusive formula-feeding became clearer. The cessation of breastfeeding at 12 months showed a greater rate in maturation of the microbiome, as the infants who ceased breastfeeding had microbiota consisting of *Bacteroides, Bilophila, Roseburia, Clostridium,* and *Anaerostipes,* which is characteristic of an adult-like composition. These results suggest that the cessation of breastfeeding, rather than the introduction of solid food, is the major contributor to the development toward an adult microbiome composition. However, a different study [21] found that the introduction of solid foods prior to 6 months led to early maturation of the microbiome. In addition, while it was previously believed that the introduction of solid food drives the appearance of *Bacteroides* [22], Yassour et al. [1], found that many individuals in their studied cohort showed a significant *Bacteroides* presence at the earliest time points, before the introduction of solid food.

Metagenomic shotgun sequencing on fecal samples from full-term Swedish infants and their mothers by Backhed et al. revealed that at 12 months, the infants were more similar to their mothers than were newborn infants or 4-month old infants. In addition, the signature genera at 12 months included many types of microbes that are efficient degraders of dietary fibers and producers of short-chain fatty acids. These bacteria were only present at 12 months, and thereby suggest a shift around this age toward an adult-like intestinal environment, which is associated with an increased functional capacity for carbohydrates degradation. However, there were still differences between the gut microbiome between the 1-year-old children and the mothers, both compositionally and functionally, supporting the idea that microbial gut composition maturation continues past 1 year of age, even in full term infants.

Conclusion

In the first year of the infant's development, we can observe similar trends in bacterial abundance trajectories that reflect the development of ecological niches. The infant is exposed to new sources of bacteria from birth and throughout its early life, but the shared patterns sometimes seen with both types of delivery styles leave us questioning if there is room for intervention, perhaps through probiotics or selective formula. The ecological niches formed in the infant gut shift and then specialize as the nutritional source of the infant transitions. They eventually stabilize as the infant microbiome reaches an adult state, but throughout this process, they are impacted by the infant's surroundings, especially its mother. Many other details of the infant gut microbiome have not been discussed in this chapter, such as the virome and the effect of bacterial by-products. However, by utilizing the methods discussed in this chapter, as well as new methods that continue to be developed, we are getting closer to understanding the infant gut compositional development, and how to find evidence-based intervention to push the infant gut community to a more "healthy" state.

References

[1] Yassour M, Vatanen T, Siljander H, Hämäläinen A-M, Härkönen T, Ryhänen SJ, et al. Natural history of the infant gut microbiome and impact of antibiotic treatment on bacterial strain diversity and stability. Sci Transl Med 2016;8:343ra81. Available from: https://doi.org/10.1126/scitranslmed.aad0917.

[2] Subramanian S, Huq S, Yatsunenko T, Haque R, Mahfuz M, Alam MA, et al. Persistent gut microbiota immaturity in malnourished Bangladeshi children. Nature 2014;510:417–21. Available from: https://doi.org/10.1038/nature13421.

[3] Kostic AD, Gevers D, Siljander H, Vatanen T, Hyötyläinen T, Hämäläinen AM, et al. The dynamics of the human infant gut microbiome in development and in progression toward type 1 diabetes. Cell Host Microbe 2015;17:260–73. Available from: https://doi.org/10.1016/j.chom.2015.01.001.

[4] Lackey KA, Williams JE, Meehan CL, Zachek JA, Benda ED, Price WJ, et al. What's normal? Microbiomes in Human milk and infant feces are related to each other but vary geographically: the INSPIRE study. Front Neurol 2019;6. Available from: https://doi.org/10.3389/fnut.2019.00045.

[5] Attri S, Nagpal R, Goel G. High throughput sequence profiling of gut microbiome in Northern Indian infants during the first four months and its global comparison. Meta Gene 2018;17:184–91. Available from: https://doi.org/10.1016/j.mgene.2018.06.011.

[6] Ayeni FA, Biagi E, Rampelli S, Fiori J, Soverini M, Audu HJ, et al. Infant and adult gut microbiome and metabolome in rural Bassa and urban settlers from Nigeria. Cell Rep 2018;23:3056–67. Available from: https://doi.org/10.1016/j.celrep.2018.05.018.

[7] Yassour M, Jason E, Hogstrom LJ, Arthur TD, Tripathi S, Siljander H, et al. Strain-level analysis of mother-to-child bacterial transmission during the first few months of life. Cell Host Microbe 2018;24:146–154.e4. Available from: https://doi.org/10.1016/j.chom.2018.06.007.

[8] Ferretti P, Pasolli E, Tett A, Asnicar F, Gorfer V, Fedi S, et al. Mother-to-infant microbial transmission from different body sites shapes the developing infant gut microbiome. Cell Host Microbe 2018;24:133–145.e5. Available from: https://doi.org/10.1016/j.chom.2018.06.005.

[9] Chong CYL, Bloomfield FH, O'Sullivan JM. Factors affecting gastrointestinal microbiome development in neonates. Nutrients 2018;10:1–17. Available from: https://doi.org/10.3390/nu10030274.

[10] Bäckhed F, Roswall J, Peng Y, Feng Q, Jia H, Kovatcheva-Datchary P, et al. Dynamics and stabilization of the human gut microbiome during the first year of life. Cell Host & Microbe 2015;17(5):690–703.

[11] Vatanen T, Plichta DR, Somani J, Münch PC, Arthur TD, Hall AB, et al. Genomic variation and strain-specific functional adaptation in the human gut microbiome during early life. Nat Microbiol 2019;4:470–9. Available from: https://doi.org/10.1038/s41564-018-0321-5.

[12] La Rosa PS, Warner BB, Zhou Y, Weinstock GM, Sodergren E, Hall-Moore CM, et al. Patterned progression of bacterial populations in the premature infant gut. Proc Natl Acad Sci U S A 2014;111:12522–7. Available from: https://doi.org/10.1073/pnas.1409497111.

[13] Korpela K, Blakstad EW, Moltu SJ, Strømmen K, Nakstad B, Rønnestad AE, et al. Intestinal microbiota development and gestational age in preterm neonates. Sci Rep 2018;8:2453. Available from: https://doi.org/10.1038/s41598-018-20827-x.

[14] Korpela K, Costea P, Coelho LP, Kandels-Lewis S, Willemsen G, Boomsma DI, et al. Selective maternal seeding and environment shape the human gut microbiome. Genome Res 2018;28:561–8. Available from: https://doi.org/10.1101/gr.233940.117.

[15] Asnicar F, Manara S, Zolfo M, Truong DT, Scholz M, Armanini F, et al. Studying vertical microbiome transmission from mothers to infants by strain-level metagenomic profiling. mSystems 2017;2. Available from: https://doi.org/10.1128/MSYSTEMS.00164-16 e00164-16.

[16] Jost T, Lacroix C, Braegger CP, Rochat F, Chassard C. Vertical mother-neonate transfer of maternal gut bacteria via breastfeeding. Environ Microbiol 2014;16:2891–904. Available from: https://doi.org/10.1111/1462-2920.12238.

[17] Knight R, Dominguez-Bello MG, Fierer N, Magris M, Costello EK, Hidalgo G, et al. Delivery mode shapes the acquisition and structure of the initial microbiota across multiple body habitats in newborns. Proc Natl Acad Sci U S A 2010;107:11971–5. Available from: https://doi.org/10.1073/pnas.1002601107.

[18] Biasucci G, Benenati B, Morelli L, Bessi E, Boehm G. Cesarean delivery may affect the early biodiversity of intestinal bacteria. J Nutr 2008;138:1796S–800S. Available from: https://doi.org/10.1093/jn/138.9.1796S.

[19] Bokulich NA, Chung J, Battaglia T, Henderson N, Jay M, Li H, et al. Antibiotics, birth mode, and diet shape microbiome maturation during early life. Sci Transl Med 2016;8:343ra82. Available from: https://doi.org/10.1126/scitranslmed.aad7121.

[20] Milani C, Duranti S, Bottacini F, Casey E, Turroni F, Mahony J, et al. The first microbial colonizers of the human gut: composition, activities, and health implications of the infant gut microbiota. Microbiol Mol Biol Rev 2017;81(4). Available from: https://doi.org/10.1128/MMBR e00036-17.

[21] Pannaraj PS, Li F, Cerini C, Bender JM, Yang S, Rollie A, et al. Association between breast milk bacterial communities and establishment and development of the infant gut microbiome. JAMA Pediatr 2017;171:647–54. Available from: https://doi.org/10.1001/jamapediatrics.2017.0378.

[22] Koenig JE, Spor A, Scalfone N, Fricker AD, Stombaugh J, Knight R, et al. Succession of microbial consortia in the developing infant gut microbiome. Proc Natl Acad Sci U S A 2011;108(Suppl 1):4578–85. Available from: https://doi.org/10.1073/pnas.1000081107.

CHAPTER

9

Microbes, human milk, and prebiotics

Magdalena Kujawska[1], *Maria Carmen Collado*[2] *and Lindsay J. Hall*[1,3]

[1]Gut Microbes and Health, Quadram Institute Bioscience, Norwich, United Kingdom, [2]Institute of Agrochemistry and Food Science, Spanish National Research Council, Valencia, Spain, [3]Chair of Intestinal Microbiome, School of Life Sciences, Technical University of Munich, Freising, Germany

Introduction: diet and early-life development

Infancy is defined as the period from birth to approximately 2 years of age. This early-life stage is characterized by rapid growth, development, and maturation of organs and systems [1]. The quality and the quantity of nutrients consumed by infants during the first year of life can permanently affect, or "program," the developing infant. This concept of "nutritional programming" was first introduced in the early 1990s [2] to describe a general process, whereby exposure to specific stimuli or insults at a critically important period of development can exert long-lasting effects on the host.

Over the years, numerous studies have provided strong evidence linking nutrition in early life to health in adulthood, which has laid the foundation for health promotion and nutrition programs around the world. The 2009 report published by the British Medical Association Board of Science [3] recognized the importance of infant nutrition and its link to lifelong health. Compelling evidence for a role of early-life nutrition in modulating risk of such medical conditions as coronary heart disease, type 2 diabetes, osteoporosis, asthma, lung disease, and certain types of cancer was further backed by the 2011 report of the UK

The Human Microbiome in Early Life
DOI: https://doi.org/10.1016/B978-0-12-818097-6.00009-2

Scientific Advisory Committee on Nutrition [4], with recommendations to optimize the diet and body composition of young women, and to promote and support breastfeeding. In 2014, global targets and guidelines for the nutrition of mothers, infants, and young children were published by the World Health Organization (WHO), aiming at reducing the prevalence of low birthweight and increasing the rate of exclusive breastfeeding [1,5]. These were followed by further recommendation that breastfeeding should be initiated early and preferably within 1 hour of birth to reduce cases of neonatal mortality and morbidity [6].

More recently, further studies have provided additional evidence that infant wellness and development, resulting in lifelong health benefits, can be supported via balanced co-maturation of gut-associated microbial communities (i.e., the microbiome) and host. It is now well recognized that diet, along with factors such as genetics, mode of birth, and antibiotic usage, plays a significant role in shaping microbiomes across different body sites. Crucially, establishment of the early-life microbiome within the gastrointestinal tract occurs in parallel with a key period of host growth and immune development [7–9], and thus drives key aspects of host development [10]. Notably, recent experimental data have linked gut microbiome composition to the risk of developing several chronic diseases, including type 2 diabetes [11,12], cardiovascular disease [13], and cancer [14,15].

The infant gut microbiome is highly unstable, and only reaches composition similar to that of an adult by 2–3 years of age. This is consistent with a move away from a milk-alone diet to a more diverse diet based on solid foods [16]. Indeed, diet is proposed to be one of the main factors influencing early-life microbiome development, with the type of feeding (breast vs formula) and the time of the introduction of solid foods [16] crucial to maturation of these burgeoning microbial communities. In contrast to the infant microbiome, adults harbor a more "resilient" microbiome, which is relatively stable to external factors that might lead to large changes in structure; however, recent experimental data have shown that dietary changes can lead to temporary microbial shifts within hours even in adults [17].

Introduction: a role for prebiotics in optimizing early-life nutrition

The manipulation of the intestinal microbiome for health improvement is currently being investigated worldwide. One widely used approach is the inclusion of specific dietary components, that is, prebiotics, that have the potential to preferentially "feed" and act upon beneficial members of the gut microbiome. The concept of prebiotics was first introduced by Gibson and Roberfroid [18], who defined them as "non-digestible food ingredients that beneficially affect the host by selectively stimulating the growth and/or activity of one or a limited number of bacteria in the colon, thus improving host health." In 2010 the International Scientific Association for Prebiotics and Probiotics (ISAPP) updated this definition to incorporate the functional aspect of prebiotics and described them as "selectively fermented ingredient that results in specific changes in the composition and/or activity of the gastrointestinal microbiome, thus conferring benefit(s) upon host health" [19]. More recently, Bindels et al. [20] have suggested an even more comprehensive definition: "a non-digestible compound that, through its metabolization by microorganisms in the gut, modulates composition and/or activity of the gut microbiome, thus conferring a beneficial physiologic effect on the host." Over time, the definition of prebiotics has been updated not only to accommodate scientific considerations but also to reflect its importance for regulatory bodies, industry, and consumers around the world [21,24]. According to Bindels et al. [20], all well-known prebiotics are complex carbohydrates; however, other nutrients, for example, bioactive secondary plant metabolites such as polyphenols, may also have prebiotic properties. Prebiotics are stable compounds, and thus they can easily be added to different types of food, such as yogurt, ice cream, biscuits, breads, cereals, spreads, and drinks [23]. A compound needs to fulfill the following criteria to be classified as a prebiotic: (1) it needs to resist gastric acidity, hydrolysis by mammalian digestive enzymes, and gastrointestinal absorption, (2) it has to be fermented by intestinal bacteria, and (3) it needs to selectively stimulate the growth and/or activity of intestinal bacteria linked with health and well-being [24].

To date, a number of prebiotics with various origin and chemical properties have been named, reviewed, and classified based on a set of common criteria [25]. These will be discussed in more detail later, but briefly; properties of breast milk-derived human milk oligosaccharides (HMOs) as first prebiotics available to infants have been widely described [26–28]. Other well-established compounds with prebiotic properties include inulin, fructo-oligosaccharides (FOS), galacto-oligosaccharides (GOS), and lactulose whereas isomalto-oligosaccharides, xylo-oligosaccharides (XOS), and lactitol are categorized as emerging prebiotics [25,29–32]. Chicory root inulin-derived FOS, wheat bran-derived arabinoxylo-oligosaccharides (AXOS), and XOS have been shown to have applications in human and animal welfare [33,34], whereas mannitol, maltodextrin, raffinose, lactulose, and sorbitol have been identified to exert beneficial effects on bacterial members of human microbiome, including growth promotion and production of antimicrobial compounds [35–37]. In addition, the fermentability of resistant starches (RSs) and other dietary fibers, such as beta-glucan and guar gum, by intestinal bacteria to short-chain fatty acids (SCFAs) has suggested potential prebiotic application of these compounds in promoting human well-being [38–40].

Breast milk: gold standard early-life nutrition

Human milk is considered the gold standard for infant nutrition during the first months of life since it is perfectly adapted to neonatal nutritional requirements and supports infant growth and development [41,42]. These key traits are supported by numerous international organizations (e.g., WHO) who recommend exclusive breastfeeding for at least 6 months, followed by continued breastfeeding as complementary foods are introduced, for 1 year or longer [43,44]. Breastfeeding represents a key postnatal link between mother and infant; driving beneficial microbial, metabolic, and immunological programming [45–47]. Human milk contains a complex and "personalized" combination of biologically active components, including immunoglobulins, chemokines, growth factors, cytokines, bioactive lipids, oligosaccharides, microRNAs, hormones, immune cells, and microorganisms [45,48–50]. The concentration of the human milk components varies among individuals and depends on several factors, such as maternal genetics (genotype), health status, geographical location, gestational age, diet, and time of lactation stage [45,51–54]. Experimental data suggest that certain human milk components can be modulated through dietary intake either directly or indirectly, via a range of metabolic pathways [55,56]. Notably, concentrations of fat- and water-soluble vitamins, such as vitamin A, C, B-6, and B-12, and fatty acids (FAs), in breast milk have been reported to reflect the intake of these nutrients in the maternal diet [51,57,58]. In contrast, the macronutrient and macromineral content of breast milk, for example, levels of carbohydrate, protein, fat, calcium and iron, is generally considered less affected by maternal dietary intakes [51,59]. Effects of maternal nutrition in pregnancy and during breastfeeding on long-term health outcomes in breastfed infants have been studied in recent years [60,61]. For example, fish consumption during pregnancy and lactation was correlated with higher concentration of long-chain polyunsaturated fatty acids (LC-PUFAs) in breast milk, especially that of docosahexaenoic acid, which in turn was linked to infant intelligence quotient later in life [62,63]. Offspring of mothers with excessive gestational weight gain and a BMI $> 30\ kg/m^2$ have been reported to be more frequently obese at age 16 in comparison to offspring of mothers with normal weight gain, even after adjusting for sex and socioeconomic status [64]. Furthermore, maternal malnutrition in pregnancy has been suggested to lead to suboptimal infant immune development, with the likelihood of reduced immune responses to routine vaccinations in the first year of life [65].

Crucially, breastfeeding has been associated with decreased risk for numerous diseases, which has been linked to differences in immune system development [66,67]. Biologically active factors in human milk have been shown to directly impact immune responses [66,68]. Oligosaccharides in human milk have been reported to prevent attachment of respiratory pathogens, such as *Streptococcus pneumoniae*, to respiratory epithelium, and milk glycoproteins have been found to prevent intestinal pathogens, such as *Vibrio cholerae* and *Escherichia coli*, from colonizing [69]. HMOs have also been shown to directly interact with glycan-binding proteins expressed on the epithelial cells and cells of the innate immune system [70]. For example, 2′-fucosyllactose has been reported to bind human dendritic cell-specific intercellular adhesion molecule-3-grabbing nonintegrin, a C-type lectin receptor present on the surface of macrophages and dendritic cells [71]. High levels of two bioactive molecules in human milk have been associated with reduced risk of certain noncommunicable diseases. Higher levels of TGF-β1 or TGF-β2 in colostrum or mature milk have been associated with reduced risk of atopic outcomes in infants [72]. The authors suggested that TGF-β in human milk may be involved in maintaining homeostasis in the intestine, regulating inflammation, and promoting oral tolerance, all leading to the reduced risk of allergies. Similarly, reduced levels of soluble CD14 in human milk, a bacterial pattern recognition receptor for cell wall components such as lipopolysaccharide, have been linked to allergic outcomes in infants [73,74]. This molecule has been identified as one of the immune factors actively secreted into mother's milk and suggested to play an important role in protection against allergy manifestations [73–75]. Furthermore, it has been proposed that adipokines present in human milk may be involved in regulating infant energy intake and thus influence the long-term risk for obesity [76], while LC-PUFAs may affect blood pressure and insulin resistance later in life [77].

Breast milk components also facilitate colonization of early microbial pioneers, which further modulate immunity; enhancing the synthesis and secretion of polymeric IgA, the coating antibody that protects mucosal surfaces against bacterial invasion, and helps drive homeostatic immune cell programming including the balance between Th1 versus Th2 responses, which if skewed are associated with certain clinical outcomes, such as atopic disease and Crohn's disease (discussed in more detail later) [68].

Breast milk: a source of beneficial microbes

Historically, breast milk was considered a sterile fluid, and the presence of bacteria was associated with milk contamination, mammary gland infections, or even to poor hygiene. However, this dogma was challenged after the isolation of viable bacteria from breast milk samples obtained from healthy women [78,79]. Traditional culture methods have reported the presence of lactic acid bacteria such as *Lactobacillus*, *Lactococcus*, *Leuconostoc*, and also in some cases *Bifidobacterium*, as well as other bacteria including *Streptococcus*, *Enterococcus*, and *Staphylococcus*; [79–81]. The presence of lactic acid bacteria with potential beneficial or "probiotic" properties raised research interest in exploring the breast milk microbiome and the potential biological implications for maternal and infant health. Previous studies have also utilized culture-independent methodologies to profile the breast milk microbiome including specific PCR and qPCR to identify specific or total bacteria [82–84], gradient gel electrophoresis, for example, denaturing gradient gel electrophoresis and temperature gradient gel electrophoresis to determine bacterial diversity [78,84,85], and more recently next-generation sequencing (NGS) methodologies targeting the 16S ribosomal RNA gene [86–88] or via metagenomic approaches [89]. Previous NGS studies have indicated high microbial diversity and bacteria that are typical inhabitants of the oral cavity, such as *Veillonella* and *Prevotella*; skin bacteria such as *Staphylococcus* and *Propionibacterium*; lactic acid bacteria, including *Enterococcus*, *Streptococcus*, *Leuconostoc*, and *Weissella*; and Gram-negative bacteria, such as *Pseudomonas*, *Ralstonia*, and *Klebsiella*, with high inter- and intraindividual specific profiles [53,86–88,80,90–91]. In a systematic review covering studies between 1964–2015, a universal predominance of *Streptococcus* and *Staphylococcus* in breast milk samples was identified (both within the Firmicutes phylum), which was independent of methodologies used and location [92]. Moreover, it has been reported that >820 species have been identified in human milk, which indicates that the neonate is exposed to a significant diversity and concentration of bacteria during breastfeeding: "human milk constitutes one of the main sources of bacteria to the breastfed infant gut since an infant consuming approximately 800 mL/day of milk would ingest between 1×10^5 and 1×10^7 bacteria daily" [93,94]. Specific shared bacteria or bacterial DNA has been reported between breast milk, the maternal gut, and infant feces and oral samples [78,95–98], with specific strains belonging to *Bifidobacterium* and *Staphylococcus* spp. transmitted to infants [99–102]. These results support vertical transmission of breast milk bacteria to the infant gut, which likely enhance development of the early-life immune system, maintenance of tolerance to commensal microbial members, and intestinal host defense to pathogens. Currently, there is limited evidence on the potential effect of breast milk microorganisms in infant health, but it has been speculated that they may play several important functions including acting as key early-life colonizers and helping to shape the developing mucosal immune system [48,94,103]. More specifically, the breast milk microbiome may promote intestinal immune homeostasis, encouraging a shift from the predominant intrauterine T helper cell 2 (Th2) immune milieu to a Th1/Th2 balanced response and to trigger regulatory T cell differentiation [48,104]. Indeed, early microbial exposure provides essential antigenic stimuli that promote the normal maturation of the intestinal and systemic immune systems [105]. Some studies have reported the in vitro immunomodulatory activity of specific milk isolates. *Lactobacillus fermentum* CECT5716 and *Lactobacillus salivarius* CECT5713 were found to modulate immune responses by activating NK cells, T cells (including T cell subsets $CD4^+$, $CD8^+$, and regulatory T cells), and inducing the production of a broad array of cytokines and IgA. Notably, these effects were not reported in nonbreast milk *Lactobacillus* strains [106]. With respect to colonization resistance, in vitro studies have shown that several breast milk isolates are able to inhibit *Staphylococcus aureus* growth [79], with showing bacteriostatic and/or bactericidal effects on an enteropathogenic *Salmonella enterica* strain, both in vitro and in vivo [107]. Furthermore, breast milk-isolated *L. fermentum* CECT5716 was able to reduce the incidence of infant gastrointestinal and respiratory infections [108].

Interestingly, more recent studies have indicated that apart from bacteria, breast milk also contains other microorganisms including yeasts, fungi, viruses, and bacteriophages. Recent studies have reported the presence of a wide range of yeast and fungi, including *Saccharomyces, Malassezia*, and *Candida* spp., in breast milk samples from healthy women [109,110] with qPCR data showing that 89% of samples had detectable levels of fungal DNA, with an estimated load of $\sim 10^5$ cells/mL. It is well appreciated that breast milk is a vehicle for transmission of specific pathogenic viruses such as herpes, *Varicella zoster*, hepatitis, cytomegalovirus, zika, and so on. [111]. However, recent data has also reported specific transference of bifidobacterial phages from mother to neonates during breastfeeding [112], and another recent study showed the presence of bacteriophages from the family Myoviridae and a higher viral diversity in breast milk when compared to infant stool [113]

Factors influencing breast milk microorganisms

There are numerous studies exploring how specific factors including genetics, maternal health and nutrition, mode of delivery, breastfeeding, and geographic location influence the human gut microbiome; however, little is known about the impact of these factors on human breast milk microbial communities [45]. Crucially, potential changes in milk microbiome composition may have biological implications for infant microbial colonization and for the development and maturation of the infant immune system.

Previous studies have indicated that lactation stage shapes the microbiome present in colostrum, transitional, and mature milk [82,86,87]; however, other studies have reported no differences [114]. It has also been demonstrated that the HMO profiles are influenced by lactation stage [115,116], and milk glycosylation patterns change in relation to time [117] (HMOs are discussed in more detail in the following sections). Maternal obesity, allergy, celiac disease, and HIV status also shape milk microbiome composition and diversity [86,118,119]. Mothers who are obese and/or have atopic diseases were reported to have lower bacterial diversity and reduced levels of *Bifidobacterium* spp., when compared with healthy mothers [86,120,121]. HIV-positive African breast milk samples were found to contain higher bacterial diversity and *Lactobacillus* spp. but lower levels of *Staphylococcus* spp. than healthy women [118]. Breast milk from female celiac patients harbored lower *Bifidobacterium* spp. and *Bacteroides* spp. compared to the abundance observed in healthy mothers [119], and women suffering lactational mastitis had a different microbiome compared to milk samples from healthy women [91,122,123].

Although diet is known to significantly influence breast milk composition, there are limited studies on the impact on the milk microbiome. A recent study showed the impact of macronutrient and micronutrient intake on milk microbial communities [124]. A higher abundance of Firmicutes was associated with lower intake of total carbohydrates, disaccharides, and lactose; with higher protein consumption associated with a higher abundance of *Gemella*, and higher intakes of saturated and monounsaturated fatty acids were associated with lower abundances of *Corynebacterium* in milk samples [124]. Breast milk microbiome is also influenced by gestational age [82]. Lower counts of *Enterococcus* spp. and higher counts of *Bifidobacterium* spp. were detected in milk samples from term deliveries. Furthermore, mode of birth can also influence the milk microbiome; however, contradictory and opposite findings can be found in the literature. Some studies have shown a higher microbial diversity and relative abundance of *Bifidobacterium* and *Lactobacillus* spp. in breast milk samples from women after vaginal birth compared to C-section [86,88,125,126], and other studies have indicated that nonelective C-section milk samples showed a similar microbial profile to those observed in vaginal deliveries, whereas nonemergency C-section showed a microbial profile close to the skin and oral microbial communities [127]. These data suggest a role for the physiological labor process, stress, and/or hormonal signals that may modulate milk microbiome composition. However, other studies have reported no differences [114,125], so further research is required to determine the importance of mode of delivery in breast milk microbiome composition.

Intrapartum antibiotic exposure (and mode of delivery) has also been associated with overall microbial composition in breast milk samples at 1 month postpartum [126]. Moreover, Soto et al. [129] demonstrated a lower occurrence of lactobacilli and bifidobacteria in milk samples from mothers treated with antibiotics during pregnancy or lactation. More recently, it has been reported that milk microbiome profiles are different between countries that have contrasting environments, climate, diet, and lifestyle [53,126,131]. However, further studies are needed to ascertain the influence of the environment, and the other factors described earlier, on the milk microbiome in order to identify potential strategies to beneficially modulate breast milk microbial composition.

Breast milk "feeds" specific members of the infant gut microbiome

During the early-life developmental window, maternal breast milk acts as a natural prebiotic feeding certain bacterial strains and species of the gut microbiome. Specific breast milk components, for example, HMOs, are poorly digested by the infant but are favored growth substrates mainly for *Bifidobacterium*, but also a few *Bacteroides* and *Lactobacillus* strains that can enzymatically degrade these complex dietary constituents. The term HMOs collectively describes nondigestible carbohydrates (NDCs) with the degree of polymerization equal or above 3, which are the third most abundant component of breast milk after lactose and lipids [132]. The milk of a single mother can contain more than 200 different HMOs, each composed of a common lactose reduced end linked to *N*-acetyllactosamine (Galβ1−4GlcNAc) or lacto-*N*-biose I units (LNB) (Galβ1−3GlcNAc) [133], although there are several HMOs that comprise the major types in breast milk, for example, 2′FL and LNT (Fig. 9.1). These structures can additionally be sialylated or/and

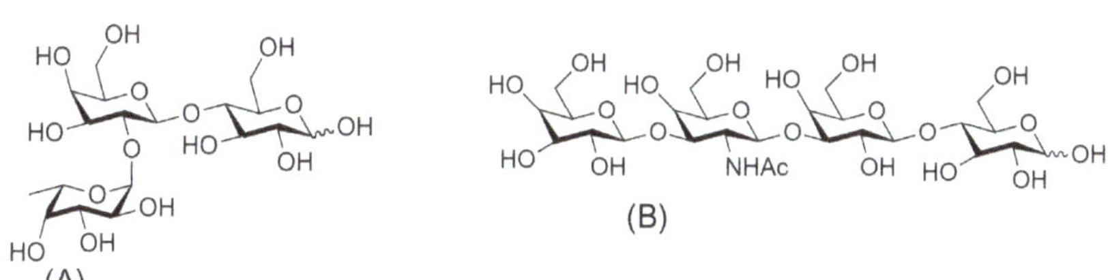

FIGURE 9.1 Examples of human milk oligosaccharides. Chemical structures of (A) 2′-fucosyllactose and (B) lacto-*N*-tetraose.

fucosylated. HMO fucosylation depends on the presence of functional enzymes, fucosyltransferases, in the mother's mammary glands, which is regulated by Lewis and secretor genes [133,134].

Dominance of *Bifidobacterium* in the gut microbiome during infancy and lower overall bacterial diversity are strongly linked to nutrient availability. The high abundance of *Bifidobacterium* in breastfed infants has been correlated with the enrichment of genes required for the degradation of HMOs [135,136]. In recent years, several infant gut-associated *Bifidobacterium* species have been reported to harbor genes encoding enzymes necessary for the degradation of milk oligosaccharides, including *Bifidobacterium breve, Bifidobacterium bifidum, Bifidobacterium longum* subsp. *longum* (*B. longum*), *B. longum* subsp. *infantis* (*Bifidobacterium infantis*), and more rarely *Bifidobacterium pseudocatenulatum* [137–140]. Strategies for HMO degradation vary among *Bifidobacterium* species, and the metabolic capabilities for HMOs within the species are strain specific. Overall, conserved consumption of HMOs has been attributed to *B. bifidum* and *B. infantis,* while more variable and strain-dependent degradation of these carbohydrates has been demonstrated in *B. breve* and *B. longum* [132,141].

Based on the recent experimental evidence, it has been suggested that two main strategies for HMO metabolism have evolved in bifidobacteria [132,151]. The first one relies on the presence of functional membrane associated extracellular glycosyl hydrolases (GHs), namely lacto-*N*-biosidases. These enzymes hydrolyze HMOs, particularly lacto-*N*-tetraose (LNT, Fig. 9.1), into mono- and disaccharides, which are then imported into the cell by solute-binding proteins (SBP) [141]. This strategy seems to be widely employed by *B. bifidum,* but also by *B. longum* strains harboring the gene encoding lacto-*N*-biosidase (LnbX) [132,142]. The second strategy relies on the SBP-mediated import of intact HMOs inside the cell and their degradation by intracellular GHs, including α-fucosidases, sialidases, β-galactosidases, and β-*N*-acetylhexosaminidases with affinities for a wide spectrum of milk oligosaccharides [141]. This system has been demonstrated in *B. infantis, B. breve,* and *B. longum* strains that are LnbX-negative [132,142].

Previous in vitro studies have demonstrated that certain *Bifidobacterium* members, for example, *B. bifidum,* do not fully degrade HMOs, which raised a question of possible cross-feeding within infant-associated *Bifidobacterium* communities [143]. Recent work of Gotoh et al. [132] reported that wild-type strain of *B. longum* 105-A (LnbX-positive), which showed very limited growth on LNT in mono-culture, grew very well in coculture with an extracellular HMO degrader *B. bifidum.* In addition, supplementation of *B. bifidum* in HMO-containing stool cultures stimulated the growth of other *Bifidobacterium* species, especially in samples which did not initially contain *B. bifidum.* Most recently, results suggest similar cooperation between nonextracellular HMO degraders and other members of *Bifidobacterium* community [144]. Using conditioned media from *B. longum* grown on 2′-FL as a carbon source for other non-HMO-degrading *Bifidobacterium* strains, it was demonstrated that this could support the growth of other isolates within the same (infant) host. These results were further confirmed in coculture experiments. Altogether, these results suggest that bifidobacterial HMO degradation products may exert an effect on the acquisition and abundance of *Bifidobacterium* in an ecosystem [132,144].

Acquisition of specific strains and species of *Bifidobacterium* also correlates with a critical period of immune maturation and programming, with previous studies indicating that this genus strongly modulates specific immune cells and pathways. *Bifidobacterium* metabolism of these maternal-derived glycans, that is, HMOs, results in increased gut levels of SCFAs [145,146]. These compounds play regulatory roles within the gut-associated mucosal immune system, providing energy to epithelial cells, which in turn helps maintain the coherence of tight junctions and the intestinal barrier, thus providing mechanisms of defense against pathogens. SCFAs have also been implicated in regulating mucosal cells of the innate immune system such as macrophages or dendritic cells, as well as neutrophils, and the antigen-triggered adaptive immunity activities are mediated by T and B lymphocytes [147–150].

The formula effect: impact on the infant gut microbiome

The composition of early-life microbiome and the resulting gastrointestinal function are determined by a number of factors including mode of birth (i.e., vaginal delivery vs Cesarean section), antibiotic treatment, and skin-to-skin contact [97,151]. However, the infant diet is proposed to induce the most significant changes, and crucially notable differences in the composition of intestinal microbiome between breast- and formula-fed infants have been widely reported [152–157]. In contrast to the gut microbiome of breastfed infants, formula-fed infants are characterized by higher microbial diversity and elevated levels of potentially pathogenic organisms, such as *E. coli* and *Clostridium difficile* and reduced levels of beneficial *Bifidobacterium* [135,158,159].

Compared to breastfeeding, formula feeding has been associated with increased risk for several diseases in adulthood, which has been linked to microbiome profiles, and the previously mentioned development of the immune system in early life [67]. Numerous epidemiological studies have suggested that feeding practiced in infancy may contribute to obesity or the development of type 2 diabetes later in life [160–163]. Some studies have also suggested an increased risk of high blood pressure [164] and high cholesterol [165,166].

Over the years, improved formulations of infant formulas have been designed and produced as an alternative to breast milk. These products must contain water, carbohydrate, protein, fat, vitamins, and minerals in adequate proportions [167]. The composition of infant formulas falls under strict regulations, and manufacturers around the world are obliged to follow guidelines set by government agencies [167]. All major components (protein, lipids, carbohydrates) added to the formula must have a proven history of safe use [168]. The numerical values for the effectiveness of major formula components are defined, and the required range of each nutrient must be maintained throughout the shelf life of the product [167,169]. In terms of carbohydrate source, the use of fructose should be avoided, as it may lead to fructose intolerance in infants. Only the L forms of amino acids are approved as ingredients, as D forms may cause D-lactic acidosis, and are thus not allowed [170]. Hydrogenated fats are also not permitted in formulas, nor is using ionizing radiation to sterilize the product [169]. According to the Food and Agricultural Organization Codex Alimentarius [171], the consumption-ready infant formula should contain no less than 60 kcal (250 kJ) and no more than 70 kcal (295 kJ) of energy per 100 mL. Currently, infant formulas can be categorized in three major classes: cow milk-based formula, soy-based formula, and specialized formula. They differ in nutritional and caloric values, taste, digestion, and cost. Special formulas are available to meet the needs of infants with intolerances. Some of the cow's milk substitutes are rice-based or amino acid-based or contain extensively hydrolyzed whey or casein proteins [167].

HMOs are characteristic of human milk and are not present in the same composition in animal milk nor plant proteins that are normally used as the basis for formula [172]. As synthesizing HMOs is expensive, supplementing formulas with compounds that might mimic some of the HMO functions described earlier has been widely investigated in the recent years. Affordable NDCs that have been used for this purpose include GOS, FOS, and pectin oligomers (pectic oligosaccharides [POS]) [173].

"Optimizing" infant formulas with pre- and probiotics

Galacto-oligosaccharides

GOS are carbohydrates comprising galactosyl residues ranging from 2 to 10 units, and a terminal glucose linked by β-glycosidic bonds, whose type is determined by the source of enzyme (Fig. 9.2) [174]. The β-glycosidic linkages are resistant to the action of gastrointestinal enzymes, which allows GOS to reach the colon and undergo fermentation

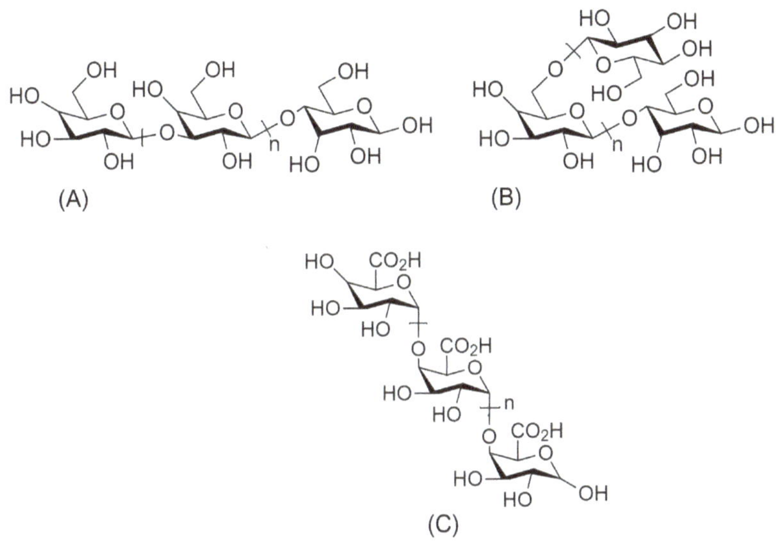

FIGURE 9.2 Structures of prebiotics commonly added to infant formulas. (A) β-1,3 galacto-oligosaccharide, (B) β-1,6 galacto-oligosaccharide, and (C) pectin.

by bacteria [175]. GOS are produced by microbial GHs by the transglycosylation of lactose [174].

A number of studies have shown a positive effect of adding GOS to the formula on the numbers of bifidobacteria and lactobacilli. Ben et al. [176] and Fanaro et al. [177] have established that feeding formula supplemented with GOS to infants increases the numbers of *Bifidobacterium* spp. after 3 months, in comparison to infants fed with formula without GOS. Using an in vitro model of the proximal colon, Maathuis et al. [178] demonstrated that representatives of several species of *Bifidobacterium* and *Lactobacillus*, namely *B. longum*, *B. bifidum*, *Bifidobacterium catenulatum*, *Lactobacillus gasseri*, and *L. salivarius* directly metabolized GOS. Other bacteria, including members of the Enterobacteria, Bacteroidetes, and Firmicutes could also ferment these carbohydrates, but to a lesser degree [178]. Another study [179] has shown that strains of bifidobacteria and lactobacilli can selectively metabolize GOS in vitro. Sierra et al. [180] conducted a clinical trial, whose results have demonstrated a significant increase in the percentage of acetic acid and the decrease in butyric acid in infants fed with formula supplemented with GOS, a similar pattern to that described in breastfed infants, correlating with higher abundance of bifidobacteria. In addition, the same study has shown a significant increase in proportion of infants fed the GOS-containing formula colonized with *Bifidobacterium adolescentis* at 4 months of age [180].

Fructans

Fructans are polymeric carbohydrates, whose structure consists of a linear chain of fructose units with a terminal glucose. The chains are linked by β-glycosidic bonds, which make fructans resistant to digestive enzymes. Chains vary in length between specific fructans and can contain between 2 and 200 residues. Fructans are a heterogeneous family of carbohydrates and have been organized into subclasses based on the length of chain: short-chain fructo-oligosaccharides (scFOS) with 3–5 residues per chain, FOS containing 6–10 residues, and structurally similar inulin and levan with a degree of polymerization ranging from below 10 to 200 residues [174]. scFOS and FOS are produced by transfructosylation of sucrose from sugar beet. They can also be directly isolated from plants such as onions, garlic, wheat, and bananas. In addition, scFOS, FOS, and inulin can be commercially produced by enzymatic digestion of long-chain inulin [174].

In vitro studies have demonstrated that bifidobacteria and lactobacilli can selectively metabolize scFOS and FOS as a sole carbon source. Marx et al. [181] have shown that different strains of bifidobacteria, *B. adolescentis*, *B. longum*, *B. breve*, and *B. pseudocatenulatum*, are capable of metabolizing FOS. Interestingly, *B. adolescentis* demonstrated the best growth and the highest degree of acidification. It was also the only strain capable of fermenting both short- and long-chain FOS; growth of other strains slowed or stopped once short-chain oligosaccharides were consumed. Rossi et al. [182] screened 55 human and animal *Bifidobacterium* strains for growth on FOS, and concluded that all strains fermented FOS, but most of them failed to grow on inulin. Consistent with previous studies, tested bifidobacteria preferred scFOS as a substrate. Similarly, Kaplan and Hutkins [183] have assessed the ability of different strains of lactobacilli to ferment scFOS and found that out of the 16 strains tested, 12 were able to ferment scFOS to SCFAs. Studies looking at the effects of fructan supplementation on microbiome composition in both infants and adults have linked metabolism of FOS, or mixtures of FOS and GOS, to the increase in the numbers of beneficial bacteria, such as bifidobacteria and lactobacilli, and the decrease in abundance of potentially pathogenic bacteria, including *Clostridium perfringens* and *E. coli* [184–187]. This has been explained by the increased production of SCFAs as a result of fructan fermentation. SCFAs lower pH in the gut, which in turn inhibits the growth of a number of pH-sensitive pathogenic members of Enterobacteriaceae and Clostridia [188,189]. Furthermore, it has been suggested that feeding infants formula supplemented with fructans correlates with higher abundance of beneficial bacteria;

Paineau et al. [190] demonstrated increased numbers of bifidobacteria in infants fed formula containing scFOS at 2 and 3 months of age compared to the placebo group.

Pectins

Pectins are structurally complex polysaccharides with backbone mainly composed of galacturonic acid linked by α-1,4 glycosidic bonds (Fig. 9.2). The backbone, as well as side chains, can incorporate other monosaccharides, including arabinose, rhamnose, and galactose [191]. Over 67 biosynthetic transferases have been estimated to be involved in the biosynthesis of pectins, due to the complexity of these polysaccharides. Pectins are classified according to the degree of methyl esterification, which affects the plant cell wall structure and properties [191].

The potential prebiotic effects of five POS produced by enzymatic treatment and acid hydrolysis of citrus peel have recently be evaluated by Di et al. [192]. The study demonstrated that POS showed bifidogenic effect in in vitro batch fermentation experiment, resulting in the increase in the numbers of *Bifidobacterium* over time. The prebiotic activity of POS was evaluated to be structure dependent and linked to arabinose-rich oligosaccharides. Apart from members of *Bifidobacterium*, several species of *Bacteroides*, as well as the Firmicutes, *Eubacterium eligens*, and *Faecalibacterium prausnitzii* have been demonstrated to degrade pectins [193,194]. The ability of prebiotic formulations containing pectin-derived acidic oligosaccharides to mimic the effects of the acidic fraction of HMOs has recently been shown [174]. These formulations have been proposed to have immune modulating effects and influence the development of immune responses in infants [174]. Stam et al. [195] have shown that a formulation composed of GOS, long-chain FOS, and acidic pectic oligosaccharides had no adverse effects on vaccine-specific antibody levels in healthy term infants. Furthermore, authors postulated that this formulation could promote Th1- and Treg-dependent immune responses and induce a downregulation of IgE allergic responses. Studies in mice have indicated that the addition of a similar oligosaccharide mixture with acidic pectic oligosaccharides exerted similar immunologic effects. One study [196] demonstrated that, in comparison to GOS/FOS formulation, a mixture of GOS, FOS, and arabino-oligosaccharides gave an optimal Th1-dependent delayed-type hypersensitivity response and reduced Th2 cytokine production in young adult mice, while the other [197] found that the same combination of oligosaccharides decreased symptoms of allergic asthma. Some pectins have been indicated to exert undesired effects in in vitro models [174]. Bosscher et al. [198] found that supplementing infant formulas with either esterified pectin or locust bean gum resulted in decreased availability of calcium, iron, and zinc compared to standard infant formula and human milk. This finding may suggest a potential change in the bioavailability of these nutrients in vivo.

Human milk oligosaccharides

Supplementation of infant formulas with NDC has been a widely used as a cost-effective method to provide health benefits associated with complex glycans to infants not consuming maternal breast milk. However, prebiotic plant-derived carbohydrates cannot substitute all the beneficial functions of HMOs [175]. Thus, much effort globally has been directed toward production of HMOs on a commercial scale. One method with a potential to deliver well-defined carbohydrate structures is chemical synthesis, but reliable methods for large-scale production are not available [175]. Instead, enzymatic and chemoenzymatic HMO synthesis methods have been researched. In vitro production of HMOs was successfully accomplished using bacterial transferases in a number of studies [199,200]. The use of engineered bacteria for HMO production has been approved by FDA and led to the commercialization of large-scale production of one of the major HMOs, 2′-fucosyllactose [175] (Fig. 9.1). Currently, a number of expression systems are available in *E. coli* and *Saccharomyces cerevisiae* [201–203]. Alternatively, bacterial GHs, widely used to produce GOS on an industrial scale, have been researched for the production of HMOs [204,205]. These enzymes can perform the reaction of transglycosylation, which can be used to attach carbohydrate moieties to existing sugars, resulting in the formation of fucosylated or sialylated glycans mimicking HMOs [206]. This approach can potentially be used to manufacture rare complex polysaccharides that are currently commercially unavailable [175].

To date, several companies have implemented these innovative methods to produce synthetic HMOs. In 2016, Glycom A/S (DK) received authorization from the European Commission to introduce two of their products obtained through fermentation by engineered bacteria, 2′-*O*-fucosyllactose and lacto-*N*-neotetraose (Fig. 9.1), as new food ingredients on the European market [207,208]. This was followed in 2017 by an authorization for sialic acid manufactured by enzymatic synthesis to be used in infant nutrition, follow-on formulas, and baby foods [209]. Similarly, Jennewein Biotechnology GmbH (DE) and FrieslandCampina (NL) received the Novel Food authorization for 2′-FL [210]. This technology has already been implemented in infant formulas. Abbot was the first company to launch a product (Similac Pro-Advance) containing one of these HMOs, 2′-FL, in the United States in 2016. Nestlé followed with Nan Optipro Supreme containing both 2′-FL and LNnT, which was launched in Spain, formulas containing 2′-FL introduced in Asia under the Illuma brand, and SMA Nutrition products containing 2′-FL and LNnT available in the United Kingdom and Ireland.

Probiotics

Probiotics have been defined by the WHO and the FAO as "live microorganisms (bacteria or yeasts), which when ingested or locally applied in sufficient numbers confer one or more specified demonstrated health benefits for the host" [211]. Based on the recent scientific and clinical developments, the ISAPP have reviewed and updated this definition, with a minor grammatical correction, to "live microorganisms that, when administered in adequate amounts, confer a health benefit on the host" [212]. The published consensus statement has proposed a set of benchmark standards for the differentiation of probiotic products and clarified the proper scope and appropriate use of the term probiotic [212]. Typically, probiotic bacteria belong to genus *Lactobacillus* and *Bifidobacterium* [213,214]. Health benefits associated with these organisms resulted in their widespread use as ingredients in functional foods, including a broad range of dairy and nondairy products [215,216].

Supplementation of infant formulas with probiotics has been explored as a method to replicate the health benefits associated with breastfeeding (and the transfer of breast milk microbial members) and to establish an intestinal microbiome similar to that of breastfed infants not consuming mother's milk [151,217]. In their position paper, the European Society for Paediatric Gastroenterology, Hepatology, and Nutrition has recommended that infant formulas with added probiotics should not be marketed unless their health benefits and safety have been fully documented [218]. Production of infant formulas is strictly regulated globally, and the probiotics aimed to be used as supplements in infant foods are subject to a range of regulatory requirements [151]. In Europe, the European Food Safety Authority has introduced a premarket system for safety assessment of microorganisms used in food and feed [219], while the U.S. Food and Drug Administration (FDA) has implemented Evidence-Based Review System for the Scientific Evaluation of Health Claims to provide guidelines for producers [220].

Beneficial effects of probiotics are strain specific; thus, food producers have to scientifically prove probiotic function and cell viability when making a health claim related to a specific product [151,217,218,221]. According to Kent and Doherty [151] large formula producers, such as

Nestle and Heinz, routinely use four probiotic strains whose properties have been documented in numerous scientific studies, *Bifidobacterium animalis* subsp. *lactis* Bb12, *Lactobacillus rhamnosus* GG, *Lactobacillus acidophilus* NCFM (along with Bb12), and *Lactobacillus reuteri* DSM 17938. In addition, dried probiotic formulations containing cocktails of lactobacilli and bifidobacteria that can be added to breast milk, human milk fortifiers, and formulas are commercially available; however, their health claims often prove difficult to validate due to the lack of strain designations and scientific evidence for health benefits [151,222].

Concerns regarding the safety of administering probiotics to infants have been raised in terms of their immunogenic effects and their impact on infant growth, as well as potential bacterial infection due to the underdeveloped infant immune system [151,223]. The widespread use of probiotic formulations in infants requires careful and comprehensive surveillance to monitor for unintended consequences. A systematic analysis of 57 clinical trials and eight follow-up studies by van den Nieuwboer et al. [224] has indicated that the use of probiotics in infants between 0 and 2 years of age is safe with regard to the evaluated strains in infants with a particular health status or susceptibility. However, long-term effects of probiotic administration to infants have rarely been measured. In addition, questions remain about the ability to identify infants at risk at the time of birth [223].

Life after milk: the crucial weaning window

Transition from breastfeeding to a more diversified diet and the introduction of solid foods is considered to cause an overall increase in the bacterial diversity in the infant gut, leading to the establishment of Firmicutes and Bacteroidetes as dominant phyla members of the infant microbiome [225–227]. Thompson et al. [228] examined the microbiomes of exclusively breastfed infants and nonexclusively breastfed infants during weaning and reported differences in bacterial composition between these two groups. Species such as *Veillonella*, *Roseburia*, and members of the Lachnospiraceae family were identified in exclusively breastfed infants after the introduction of complementary feeding, whereas *Streptococcus* and *Coprobacillus* were found present after solid foods were introduced in nonexclusively breastfed infants. Interestingly, the authors reported an increase in the relative abundance of *Bifidobacterium* during complementary feeding in nonexclusively breastfed infants, whereas in the exclusively breastfed infants the relative abundance of this genus showed a decrease. These findings may suggest differential effects of carbohydrates and starches introduced in diet during weaning [228].

Several studies have reported changes in the composition of infant microbiome during the complementary feeding period in late infancy (+9 months). Consistent with previous studies [135,154], Laursen et al. [226] have shown an increase in the relative abundance of families Lachnospiraceae, Ruminococcaceae, Eubacteriaceae, Rikenellaceae, Sutterellaceae, and a decrease in Bifidobacteriaceae, Actinomycetaceae, Veillonellaceae, Enterobacteriaceae, Lactobacillaceae, Enterococcaceae, Clostridiales incertae sedis XI, Carnobacteriaceae, and Fusobacteriaceae in infants over time. The overall increase in the bacterial diversity during weaning contributes to functional changes in the microbiome. The analysis of metabolic pathways between exclusively breastfed and nonexclusively breastfed infants showed an increased representation of sugar transporters in breastfed infants compared to those who were not exclusively breastfed, which can be explained by the abundance of HMOs in the diet of breastfed infants [228]. In contrast, nonexclusively breastfed infants were enriched in genes involved in nitrogen and methane metabolism, as well as peptidases, which could be explained by the higher protein content of formula [228].

The transition from milk-based diet to solid foods initiates the development of a more stable, and functionally more complex, adult-like microbiome with genes responsible for degradation of complex carbohydrates, starches, and xenobiotics, as well as production of vitamins [225]. Several studies reported an increase in total levels of SCFAs, especially butyrate and propionate, during weaning, which has been linked to an increase on abundance of Firmicutes and Bacteroidetes [154,225]. Widely reported increase in the members of family Lachnospiraceae and a decrease in saccharolytic bacteria, namely bifidobacteria, during complementary feeding has been correlated with the introduction of proteins into the diet, while higher numbers of Prevotellaceae have been associated with the presence of fiber in complementary foods [97].

While it is generally appreciated that the numbers of bifidobacteria in humans decrease over time, the reports of bifidobacterial species composition during infancy have been divergent [229]. Many studies have suggested that bifidobacteria are dominant members of the infant gut microbiome [230,231]; however, other studies have described fluctuations in the numbers of *Bifidobacterium*, including very low abundance or absence in many subjects [135,229,232]. These results could be explained by variation in cohort age, size, geographical location, or methodology between studies, including differences in experimental protocols, sequencing technology, and data analysis. For the studies based on the identification of the 16S rRNA gene, the accuracy of this approach strongly depends on the primer choice and the balance between efficiency, specificity, and sensitivity in the amplification of the different bacterial 16S rRNA sequences in samples [233]. In general,

B. bifidum and *B. longum* have been indicated as the dominant species in infants, whereas *B. adolescentis* and *B. longum* have mostly been associated with the "adult" gut microbiome [234]. A longitudinal survey of Japanese infants over the first 3 years of life revealed that at the age of 3, the bifidobacterial community was dominated by *B. longum* and *B. breve* followed by *B. catenulatum* and *B. bifidum* [229]. Another study on healthy Japanese adults reported *B. longum*, *B. catenulatum* group, and *B. adolescentis* as predominant in the adult microbiome [235]. Based on these results, it has been proposed that the bifidobacterial population is in transition at the age of 3 and stabilizes later in life [229].

The influence of additional complex dietary components on the early-life gut microbiome

The composition of complementary foods plays a crucial role in this transition to adult-like microbiome, providing essential nutrients to the infant and the developing intestinal microbiome while stimulating the immune development [236]. Starchy foods are common complementary foods due to their texture and palatability [236]. It has been proposed that plant-derived complex carbohydrates in complementary foods may exert beneficial health effects through their bifidogenic and prebiotic properties, resulting in modulation of the intestinal microbiome and metabolic end products [237,238]. Some examples of such NDCs with prebiotic effects include poly- and oligosaccharides containing fructose and a terminal glucose, for example, inulin-type fructans (ITF), or arabinose and xylose, for example, arabinoxylan (AX) or AXOS [239,240].

Inulin

Inulin is widely distributed in nature and can be found in fruits and plants such as chicory roots, onions, bananas, and leeks [240]. It is composed of a backbone of fructose monomers linked by β-(2-1) bonds with a degree of polymerization (DP) between 2 and 65, often linked to a terminal glucose monomer by an α-(1–2) bond (Fig. 9.3). Oligofructose is a subgroup of inulin, consisting of polymers with a (DP) below 10 [241]. Inulin has been shown to be degraded by β-fructofuranosidases, which act by cleaving terminal fructose residues from the nonreducing ends of the fructose polymers [242]. Several of these enzymes have been characterized in several bifidobacteria, including *B. longum* subsp. *infantis* [243], *B. animalis* subsp. *lactis* [244], and *B. adolescentis* [245].

The ability of bifidobacteria to degrade inulin is species- and strain-specific. Falony et al. [246] identified 4 different clusters of strains

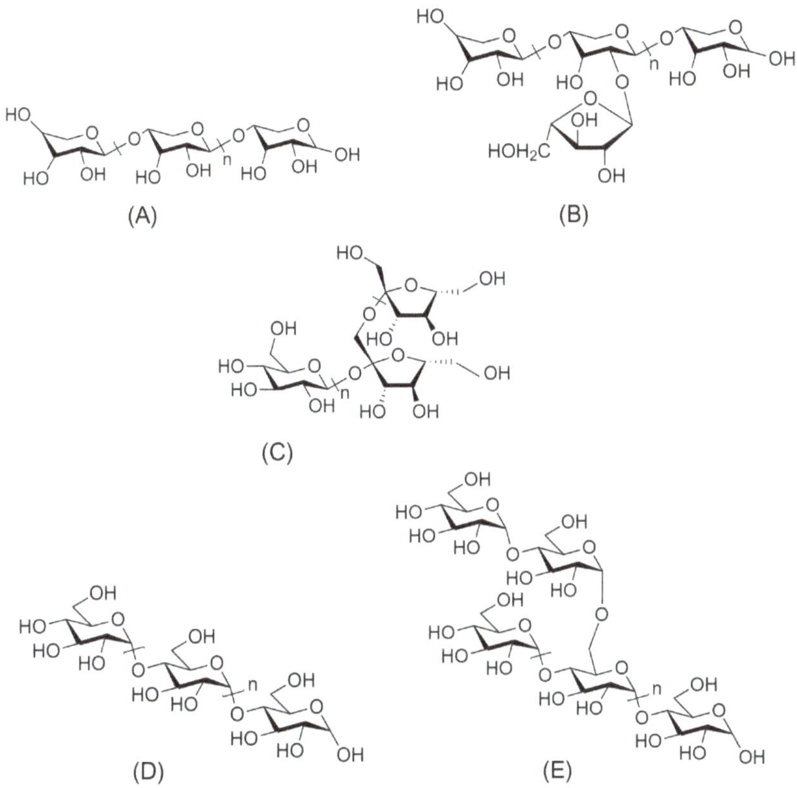

FIGURE 9.3 Structures of selected nondigestible carbohydrates found in complementary foods. Amylose and amylopectin constitute building blocks of starch. (A) xylan/XOS, (B) arabinoxylan, (C) inulin-type fructan (β-2,1), (D) amylose, and (E) amylopectin.

representing 10 different *Bifidobacterium* species based on their inulin degradation capabilities. Cluster A could only metabolize fructose, whereas cluster B was able to degrade both fructose and oligofructose after importing the oligosaccharide inside the cell. Clusters C and D were found to ferment both inulin and oligofructose extracellularly, followed by the release of fructose into the medium. Selak et al. [247] examined the ITF degradation capabilities of 190 bifidobacterial strains representing five species and originating from different regions of the intestine and found that the fermentation abilities of different strains were not correlated to the region in the intestine. Interestingly, strains with different ITF metabolism capabilities were found coexisting in the same intestinal region, which suggested bacterial cooperation in the degradation of ITF, and a potential for cross-feeding interactions between strains or on species level [247]. Cross-feeding between bifidobacteria and butyrate-producing bacteria grown simultaneously on

ITF was previously reported, where the consumption of ITF by bifidobacteria resulted in production of acetate, which was in turn used as a cosubstrate to produce butyrate by colonic butyrate producers *Roseburia intestinalis* DSM 14610 [248], *Roseburia inulinivorans* DSM 16841 [249], and *F. prausnitzii* DSM 17677 [250]. Proposed effects of the consumption of ITF on human health include increased absorption of minerals in the colon (calcium and magnesium), increased secretion of satiety hormones, increased stool frequency, and a decrease in proteolytic fermentation in distant colon [251].

Arabinoxylan and arabinoxylo-oligosaccharides

AX and AXOS have also received scientific attention in recent years due to their prebiotic and bifidogenic properties. AX is a hemicellulose consisting of a linear backbone of β-1,4 xylose residues with arabinose substitution (Fig. 9.3). The cleavage of the internal β-xylosidic glycosidic linkages of AX by endo-β1, 4-xylanases to xylo-oligosaccharides results in production of AXOS [252]. Complete degradation of complex structures of AX and AXOS requires a cooperation of a number of enzymes with debranching and depolymerizing actions, including aforementioned β-endoxylanases, as well as β-xylosidases, exo-oligoxylanases and α-arabinofuranosidases [238]. AXs have been found in all major cereal grains, with the highest content found in rye followed by wheat, barley, oats, rice, and sorghum [253].

Bioinformatic analyses of genome sequences have identified genes encoding enzymes involved in the cleavage of the AX and AXOS backbones in several bifidobacterial strains, some out of which have been isolated and characterized [254]. Examples include α-arabinofuranosidases in *B. adolescentis* ATCC 15703, *B. adolescentis* DSM 20083, *B. longum* B667, and *B. longum* NCC2705; β-xylosidases in *B. adolescentis* ATCC 15703 and *B. animalis* subsp. *lactis* BB12; and exo-oligoxylanases in *B. adolescentis* LMG 10502 [255–257]. According to Riviere et al. [238], no β-endoxylanases have been identified in bifidobacteria, so it is likely that members of this genus require cooperation with other bacterial species, for example, *Roseburia* or *Bacteroides*, for complete degradation of AX and AXOS. The bifidogenic properties of AX and AXOS have been reported in a number of studies. Hughes et al. [258] demonstrated in vitro that fermentation of wheat-derived AX resulted in proliferation of bifidobacteria, lactobacilli, and eubacteria. In an in vivo study with rats, Van Craeyveld et al. [259] showed that the bifidogenic effect is only caused by AXOS with low average DPs ≤ 5 and $A/X \leq 0.27$, while another study reported a 60-fold increase in bifidobacteria in the cecum of rats fed long-chain AX with average $DP = 60$ and A/X of 0.70 [260]. The results of the latter study were further validated by another study which reported the presence of two different *B. longum* species during the fermentation of long-chain AX in an in vitro model of the proximal colon [261]. Riviere et al. [262] demonstrated that the ability of bifidobacteria to metabolize AX and AXOS is strain specific. The authors identified five bacterial clusters capable of differential degradation of AXOS. Cluster I contained 15 strains, representatives of 7 different bifidobacterial species (*B. adolescentis, Bifidobacterium angulatum, B. bifidum, Bifidobacterium breve, Bifidobacterium dentium, B. longum,* and *Bifidobacterium thermophilum*), which were not able to metabolize neither arabinose substitutions nor the xylan backbone of AXOS, although some strains could utilize arabinose and xylose. Members of the identified Cluster II, eight *B. longum* strains, could not utilize the xylan backbone of AXOS, but were able to degrade the arabinose substitutions. Cluster III was comprised of 10 strains spanning six bifidobacterial species (*B. adolescentis, B. angulatum, B. longum, B. animalis, Bifidobacterium gallicum,* and *Bifidobacterium pseudolongum*), which were capable of utilizing the xylan backbone of AXOS up to xylotetraose but could not metabolize the arabinose substitutions or showed limited capabilities to do so. Cluster IV contained only two strains of *B. longum,* which shared the ability to degrade AXOS, whereas Cluster V was comprised of a single strain of *B. catenulatum* LMG 11043, which could use all XOS fractions in a nonpreferential way and displayed a broad degradation capabilities for arabinose substitutions [262].

AX and AXOS are slow-fermenting carbohydrates, and as such they have been associated with a number of health benefits. They have been indicated to affect the production of toxic metabolites by protein and lipid-fermenting bacteria in distal colon through stimulation of saccharolytic bacterial species, such as bifidobacteria, resulting in increased production of SCFAs and a lowering of the luminal pH [259,263,264]. In mice, AXOS consumption and the resulting increase in bifidobacteria have been proposed to help restore gut barrier functions and cure endotoxemia [265]. In addition, consumption of AX and AXOS has been associated with increased colonic absorption of dietary minerals, increased antioxidant capacity, reduced postprandial glycaemic response, and reduced blood cholesterol [239,266–268].

Resistant starches

RSs are homopolysaccharides of glucose that are not digested in the upper intestine [269]. They have been grouped into four main types. RS1 is physically indigestible, for example, starches in whole grains. RS2 is a granular starch that is present in foods such as green bananas, uncooked potato, or high-amylose maize. RS3 is retrograded starch produced by cooking and cooling processes, whereas RS4 is a starch that has been chemically modified by esterification, crosslinking, or transglycosylation and is not found in nature [269,276] (Fig. 9.3).

Numerous in vivo studies reported modulatory effects of RSs on the host intestinal microbiome. Tachon et al. [270] demonstrated that mice fed diets containing RS2-type starch from high-amylose maize were colonized by higher numbers of Bacteroidetes and species of *Bifidobacterium*, *Akkermansia*, and *Allobaculum* in proportions dependent on the starch concentration. For example, mice fed 18% RS were mostly colonized by close relatives of *B. pseudolongum* subsp. *pseudolongum* and *globosum*, whereas those fed 36% RS were enriched with *Bifidobacterium* most closely related to *B. animalis*. In addition, the levels of *Bifidobacterium* and *Akkermansia* were correlated with expression levels of proglucagon, the precursor of the gut antiobesity/diabetic hormone GLP-1, mice feeding responses, and the weight of the gut [270]. The prebiotic effects of RSs were also demonstrated in adult humans. Martinez et al. [271] reported differential effects of starch types on human microbiome composition. Type RS4 was associated with phylum-level changes in the microbial composition, significantly increasing Actinobacteria and Bacteroidetes while decreasing Firmicutes. Moreover, at the species level, consumption of RS4 was correlated with an increase in *B. adolescentis* and *Parabacteroides distasonis*, whereas RS2 significantly raised the numbers of *Ruminococcus bromii* and *Eubacterium rectale* [271].

Strain abilities of *Bifidobacterium* to metabolize starch and starch derivatives were tested in vitro by Duranti et al. [272] using representatives of 47 known bifidobacterial taxa. The results showed that strains representing *B. adolescentis*, *B. angulatum*, *Bifidobacterium boum*, *B. breve*, *Bifidobacterium cuniculi*, *B. dentium*, *B. longum* subsp. *longum*, *B. longum* subsp. *infantis*, *Bifidobacterium merycicum*, *Bifidobacterium minimum*, *Bifidobacterium reuteri*, *B. pseudocatenulatum*, *Bifidobacterium ruminantium*, and *B. thermophilum* species grew substantially (OD $>$ 0.5) on RS2 as sole carbon source. The study identified two strains that exhibited very good growth among representatives of *B. adolescentis*, out of which *B. adolescentis* 22L could grow on starch and all starch derivatives (maltodextrin, maltotriose, pullulan, and glycogen). Genomic and proteomic analyses of *B. adolescentis* 22L revealed the presence of a number of GHs putatively implicated in degradation of starch, as well as starch-derived glycans [272]. They included four proteins with predicted α-amylase activity, a glycogen/starch phosphorylase, putative phosphoglucomutase, and 4-α-glucanotransferase, as well as two genes predicted to encode amylopullulanases, similar to the starch-degrading enzyme previously characterized in *B. breve* UCC2003 [273]. In addition, genomic analyses identified two genes predicted to encode α-1,4- and α-1,6-glucosidases, an ABC-type system, and a gene encoding a LacI-type regulator flanking the gene encoding an amylopullulanase, whose products may also be involved in starch utilization using mechanisms similar to those described for *B. breve* UCC2003 [272,273]. Recent work of Jung et al. [274] has also identified enzymes implicated in direct or indirect degradation of RS and its derivatives in silico in another bifidobacterial strain, *Bifidobacterium choerinum* FMB-1. The putative enzymes included α-amylase, maltosyltransferase, pullulanase type I, glycogen phosphorylase, glycogen-debranching enzyme, 4-α-glucanotransferase, and glucan-branching enzyme. Interestingly, the aforementioned study by Duranti et al. [272] did not identify the representative strain of *B. choerinum* as one that could utilize starch, which could further suggest that RS utilization in bifidobacteria is strain specific.

A number of studies reported the potential of RSs to exert health benefits, among which maintenance of healthy blood glucose levels was widely mentioned [275–277]. Bacterial fermentation of RS and associated changes in microbiome have also been associated with the increase in the production SCFAs and a reduction in secondary bile acids, phenol, and ammonia [271,278].

Conclusions

Human milk is considered the "optimal" source of nutrients in early life. Exclusive breastfeeding is encouraged as it promotes proper infant growth and development. Milk-associated microorganisms are transmitted to the infant via breastfeeding, and HMOs, which are characteristic components of breast milk and have unique features, beneficially modulate microbiome composition through stimulation of *Bifidobacterium*, modulation of the early immune system, and defense against pathogens. Changes in feeding practices, such as introduction of formula milk, have been shown to drastically influence the early-life microbiome, leading to a significant reduction in *Bifidobacterium* and an increase in potentially detrimental microbes, for example, Enterobacteriaceae. Moreover, formula-fed babies have been shown to have increased risk of inflammatory and atopic diseases in later life. Therefore, addition of NDCs to formulas to mimic HMO functions is often used to exert beneficial health effects that can extend to later stages of life. Complementary feeding during weaning can further support infant health and development through balanced maturation and modulation of the intestinal microbiome. Complementary foods provide essential nutrients to the developing infant and contain prebiotic compounds that promote the acquisition of commensal gut bacteria and support bacterial production of beneficial metabolites, which may exert lasting effects on health.

References

[1] Langley-Evans SC. Nutrition in early life and the programming of adult disease: a review. J Hum Nutr Diet 2015;28(Suppl 1):1–14.
[2] Lucas A. Programming by early nutrition in man. Ciba Found Symposia 1991;156:38–55.
[3] Hanson M, Fall C, Robinson S, Baird J. Early life nutrition and lifelong health, British Medical Association Board of Science. 2009. Online: http://bmaopac.hosted.exlibrisgroup.com/exlibris/aleph/a25_1/apache_media/HLFT6AEIM5Q6R9S8RH50MVLPGXRHLS.pdf.
[4] Scientific Advisory Committee on Nutrition. Dietary reference values for energy. Scientific Advisory Committee on Nutrition; 2011. http://bmaopac.hosted.exlibrisgroup.com/exlibris/aleph/a23_1/apache_media/HLFT5AEIM4Q5R8S7RH48MVLPGXRHLS.pdf.
[5] World Health Organisation. Comprehensive implementation plan on maternal, infant and young child nutrition. Geneva: World Health Organisation; 2014.
[6] UNICEF and World Health Organisation. Capture the moment: early initiation of breastfeeding – the best start for every newborn. New York/Geneva: UNICEF/World Health Organisation; 2018.
[7] Olszak T, An DD, Zeissig S, Vera MP, Richter J, Franke A, et al. Microbial exposure during early life has persistent effects on natural killer T cell function. Science 2012;336:489–93.
[8] Renz H, Brandtzaeg P, Hornef M. The impact of perinatal immune development on mucosal homeostasis and chronic inflammation. Nat Rev Immunol 2012;12:9–23.
[9] Turnbaugh PJ, Ley RE, Mahowald MA, Magrini V, Mardis ER, Gordon JI. An obesity-associated gut microbiome with increased capacity for energy harvest. Nature 2006;444:1027–31.
[10] Bokulich NA, Chung J, Battaglia T, Henderson N, Jay M, Li HL, et al. Antibiotics, birth mode, diet shape microbiome maturation early life. Sci Transl Med 2016;8.
[11] Burcelin R. Gut microbiota and immune crosstalk in metabolic disease. Mol Metab 2016;5:771–81.
[12] Pedersen HK, Gudmundsdottir V, Nielsen HB, Hyotylainen T, Nielsen T, Jensen BAH, et al. Human gut microbes impact host serum metabolome and insulin sensitivity. Nature 2016;535:376.
[13] Tang WHW, Kitai T, Hazen SL. Gut microbiota in cardiovascular health and disease. Circ Res 2017;120:1183–96.
[14] de Martel C, Ferlay J, Franceschi S, Vignat J, Bray F, Forman D, et al. Global burden of cancers attributable to infections in 2008: a review and synthetic analysis. Lancet Oncol 2012;13.
[15] Feng Q, Liang SS, Jia HJ, Stadlmayr A, Tang LQ, Lan Z, et al. Gut microbiome development along the colorectal adenoma-carcinoma sequence. Nat Commun 2015;6.
[16] Voreades N, Kozil A, Weir TL. Diet and the development of the human intestinal microbiome. Front Microbiol 2014;5.
[17] Singh RK, Chang HW, Yan D, Lee KM, Ucmak D, Wong K, et al. Influence of diet on the gut microbiome and implications for human health. J Transl Med 2017;15.
[18] Gibson GR, Roberfroid MB. Dietary modulation of the human colonic microbiota - introducing the concept of prebiotics. J Nutr 1995;125:1401–12.
[19] Sanders ME. International Scientific Association for Probiotics and Prebiotics 2010 Meeting Report. International Scientific Association for Probiotics and Prebiotic. 2010. https://4cau4jsaler1zglkq3wnmje1-wpengine.netdna-ssl.com/wp-content/uploads/2018/11/sanders-2010-isapp-meeting-report-101.pdf.
[20] Bindels LB, Delzenne NM, Cani PD, Walter J. Towards a more comprehensive concept for prebiotics. Nat Rev Gastroenterol Hepatol 2015;12:303–10.

[21] Hutkins RW, Krumbeck JA, Bindels LB, Cani PD, Fahey G, Goh YJ, et al. Prebiotics: why definitions matter. Curr Opin Biotechnol 2016;37:1–7.
[22] Gibson GR, Hutkins R, Sanders ME, Prescott SL, Reimer RA, Salminen SJ, et al. The International Scientific Association for Probiotics and Prebiotics (ISAPP) consensus statement on the definition and scope of prebiotics. Nat Rev Gastroenterol Hepatol 2017;14:491–502.
[23] Gibson GR, Scott KP, Rastall RA, Tuohy KM, Hotchkiss A, Dubert-Ferrandon A, et al,. Dietary prebiotics: Current status and new definition. IFIS Functional Foods Bulletin 2010;7(1):1–19.
[24] Gibson GR, Probert HM, Van Loo J, Rastall RA, Roberfroid MB. Dietary modulation of the human colonic microbiota: updating the concept of prebiotics. Nutr Res Rev 2004;17:259–75.
[25] Stowell J. Calorie control and weight management. In: Mitchell H, editor. Sweeteners and sugar alternatives in food technology. Oxford: Blackwell Publishing Ltd; 2007.
[26] Barile D, Rastall RA. Human milk and related oligosaccharides as prebiotics. Curr Opin Biotechnol 2013;24:214–19.
[27] Bode L. Human milk oligosaccharides: prebiotics and beyond. Nutr Rev 2009;67: S183–91.
[28] Coppa GV, Bruni S, Morelli L, Soldi S, Gabrielli O. The first prebiotics in humans - Human milk oligosaccharides. J Clin Gastroenterol 2004;38:S80–3.
[29] Collins SL, McMillan A, Seney S, Van Der Veer C, Kort R, Sumarah MW, et al. Promising prebiotic candidate established by evaluation of lactitol, lactulose, raffinose, and oligofructose for maintenance of a *Lactobacillus*-dominated vaginal microbiota. Appl Environ Microbiol 2018;84.
[30] do Carmo MMR, Walker JCL, Novello D, Caselato VM, Sgarbieri VC, Ouwehand AC, et al. Polydextrose: physiological function, and effects on health. Nutrients 2016;8.
[31] Markowiak P, Slizewska K. Effects of probiotics, prebiotics, and synbiotics on human health. Nutrients 2017;9.
[32] Wu Y, Pan L, Shang QH, Ma XK, Long SF, Xu YT, et al. Effects of isomalto-oligosaccharides as potential prebiotics on performance, immune function and gut microbiota in weaned pigs. Anim Feed Sci Technol 2017;230:126–35.
[33] Sabater-Molina M, Larque E, Torrella F, Zamora S. Dietary fructooligosaccharides and potential benefits on health. J Physiol Biochem 2009;65:315–28.
[34] Xu BH, Wang YB, Li JR, Lin Q. Effect of prebiotic xylooligosaccharides on growth performances and digestive enzyme activities of allogynogenetic crucian carp (*Carassius auratus gibelio*). Fish Physiol Biochem 2009;35:351–7.
[35] Mandal V, Sen SK, Mandal NC. Effect of prebiotics on bacteriocin production and cholesterol lowering activity of *Pediococcus acidilactici* LAB 5. World J Microbiol Biotechnol 2009;25:1837–47.
[36] Vamanu E, Vamanu A. The influence of prebiotics on bacteriocin synthesis using the strain *Lactobacillus paracasei* CMGB16. Afr J Microbiol Res 2010;4:534–7.
[37] Yeo SK, Liong MT. Effect of prebiotics on viability and growth characteristics of probiotics in soymilk. J Sci Food Agric 2010;90:267–75.
[38] Carlson JL, Erickson JM, Lloyd BB, Slavin JL. Health effects and sources of prebiotic dietary fiber. Curr Dev Nutr 2018;2.
[39] Fuentes-Zaragoza E, Sanchez-Zapata E, Sendra E, Sayas E, Navarro C, Fernandez-Lopez J, et al. Resistant starch as prebiotic: a review. Starch- tarke 2011;63:406–15.
[40] Vaidya RH, Sheth MK. Processing and storage of Indian cereal and cereal products alters its resistant starch content. J Food Sci Technol Mysore 2011;48:622–7.
[41] Ballard O, Morrow AL. Human milk composition: nutrients and bioactive factors. Pediatr Clin North Am 2013;60:49–74.

[42] Garwolinska D, Namiesnik J, Kot-Wasik A, Hewelt-Belka W. Chemistry of human breast milk-a comprehensive review of the composition and role of milk metabolites in child development. J Agric Food Chem 2018;66:11881–96.
[43] World Health Organisation. Global strategy for infant and young child feeding. Geneva: World Health Organisation; 2003.
[44] American Academy of Pediatrics. Breastfeeding and the use of human milk. Pediatrics 2012;129:e827–41.
[45] Gomez-Gallego C, Garcia-Mantrana I, Salminen S, Collado MC. The human milk microbiome and factors influencing its composition and activity. Semin Fetal Neonatal Med 2016;21:400–5.
[46] Le Huerou-Luron I, Blat S, Boudry G. Breast- v. formula-feeding: impacts on the digestive tract and immediate and long-term health effects. Nutr Res Rev 2010;23:23–36.
[47] Vieira Borba V, Sharif K, Shoenfeld Y. Breastfeeding and autoimmunity: programing health from the beginning. Am J Reprod Immunol 2018;79.
[48] Boix-Amoros, A., Collado, M.C., Van't Land, B., Calvert, A., Le Doare, K., Garssen, J., et al., 2019. Reviewing the evidence on breast milk composition and immunological outcomes. *Nutr Rev.* **77** (8), 2019, 541–556.
[49] Hennet T, Borsig L. Breastfed at Tiffany's. Trends Biochem Sci 2016;41:508–18.
[50] Munblit D, Treneva M, Korsunskiy I, Asmanov A, Pampura A, Warner JO. A national survey of Russian physicians' knowledge of diagnosis and management of food-induced anaphylaxis. BMJ Open 2017;7:e015901.
[51] Bravi F, Wiens F, Decarli A, Dal Pont A, Agostoni C, Ferraroni M. Impact of maternal nutrition on breast-milk composition: a systematic review. Am J Clin Nutr 2016;104:646–62.
[52] Gomez-Gallego C, Kumar H, Garcia-Mantrana I, Du Toit E, Suomela JP, Linderborg KM, et al. Breast milk polyamines and microbiota interactions: impact of mode of delivery and geographical location. Ann Nutr Metab 2017;70:184–90.
[53] Lackey KA, Williams JE, Meehan CL, Zachek JA, Benda ED, Price WJ, et al. What's normal? Microbiomes in human milk and infant feces are related to each other but vary geographically: the INSPIRE study. Front Nutr 2019;6:45.
[54] Ruiz L, Espinosa-Martos I, Garcia-Carral C, Manzano S, McGuire MK, Meehan CL, et al. What's normal? Immune profiling of human milk from healthy women living in different geographical and socioeconomic settings. Front Immunol 2017;8:696.
[55] Dorea JG. Selenium and breast-feeding. Br J Nutr 2002;88:443–61.
[56] Novak EM, Innis SM. Impact of maternal dietary n-3 and n-6 fatty acids on milk medium-chain fatty acids and the implications for neonatal liver metabolism. Am J Physiol Endocrinol Metab 2011;301:E807–17.
[57] Innis SM. Impact of maternal diet on human milk composition and neurological development of infants. Am J Clin Nutr 2014;99:734S–741SS.
[58] Lonnerdal B. Effects of maternal dietary intake on human milk composition. J Nutr 1986;116:499–513.
[59] Ares Segura S, Arena Ansotegui J, Diaz-Gomez NM, en representacion del comite de Lactancia Materna de la Asociacion Espanola de Pediatria. The importance of maternal nutrition during breastfeeding: do breastfeeding mothers need nutritional supplements? An Pediatr (Barc) 2016;84:347 e1–7.
[60] Hermoso M, Vollhardt C, Bergmann K, Koletzko B. Critical micronutrients in pregnancy, lactation, and infancy: considerations on vitamin D, folic acid, and iron, and priorities for future research. Ann Nutr Metab 2011;59:5–9.
[61] de Waard M, Brands B, Kouwenhoven SMP, Lerma JC, Crespo-Escobar P, Koletzko B, et al. Optimal nutrition in lactating women and its effect on later health of

offspring: a systematic review of current evidence and recommendations (EarlyNutrition project). Crit Rev Food Sci Nutr 2017;57:4003–16.
[62] Lassek WD, Gaulin SJ. Maternal milk DHA content predicts cognitive performance in a sample of 28 nations. Matern Child Nutr 2015;11:773–9.
[63] Bernard JY, Armand M, Peyre H, Garcia C, Forhan A, Agostini DE, et al. Breastfeeding, polyunsaturated fatty acid levels in colostrum and child intelligence uotient at age 5-6 years. J Pediatr 2017;183:43–50 e3.
[64] Reynolds RM. Excess maternal weight gain during pregnancy is associated with overweight/obesity in offspring at age 16 years, but maternal pre-pregnancy obesity has a greater effect. Evid Based Nurs 2013;16:43–4.
[65] Obanewa O, Newell ML. Maternal nutritional status during pregnancy and infant immune response to routine childhood vaccinations. Future Virol 2017;12:525–36.
[66] Munblit D, Treneva M, Peroni DG, Colicino S, Chow LY, Dissanayeke S, et al. Immune components in human milk are associated with early infant immunological health outcomes: a prospective three-country analysis. Nutrients 2017;9.
[67] Stuebe A. The risks of not breastfeeding for mothers and infants. Rev Obstet Gynecol 2009;2:222–31.
[68] Forchielli ML, Walker WA. The role of gut-associated lymphoid tissues and mucosal defence. Br J Nutr 2005;93(Suppl 1):S41–8.
[69] Peterson JA, Patton S, Hamosh M. Glycoproteins of the human milk fat globule in the protection of the breast-fed infant against infections. Biol Neonate 1998;74:143–62.
[70] Lehmann S, Hiller J, Van Bergenhenegouwen J, Knippels LM, Garssen J, Traidl-Hoffmann C. In vitro evidence for immune-modulatory properties of non-digestible oligosaccharides: direct effect on human monocyte derived dendritic cells. PLoS One 2015;10:e0132304.
[71] Naarding MA, Ludwig IS, Groot F, Berkhout B, Geijtenbeek TB, Pollakis G, et al. Lewis X component in human milk binds DC-SIGN and inhibits HIV-1 transfer to CD4+ T lymphocytes. J Clin Invest 2005;115:3256–64.
[72] Oddy WH, Rosales F. A systematic review of the importance of milk TGF-beta on immunological outcomes in the infant and young child. Pediatr Allergy Immunol 2010;21:47–59.
[73] Jones CA, Holloway JA, Popplewell EJ, Diaper ND, Holloway JW, Vance GH, et al. Reduced soluble CD14 levels in amniotic fluid and breast milk are associated with the subsequent development of atopy, eczema, or both. J Allergy Clin Immunol 2002;109:858–66.
[74] Zdolsek HA, Jenmalm MC. Reduced levels of soluble CD14 in atopic children. Clin Exp Allergy 2004;34:532–9.
[75] Munblit D, Peroni DG, Boix-Amoros A, Hsu PS, Van't Land B, Gay MCL, et al. Human milk and allergic diseases: an unsolved puzzle. Nutrients 2017;9.
[76] Aydin S, Ozkan Y, Erman F, Gurates B, Kilic N, Colak R, et al. Presence of obestatin in breast milk: relationship among obestatin, ghrelin, and leptin in lactating women. Nutrition 2008;24:689–93.
[77] van Rossem L, Wijga AH, De Jongste JC, Koppelman GH, Oldenwening M, Postma DS, et al. Blood pressure in 12-year-old children is associated with fatty acid composition of human milk: the prevention and incidence of asthma and ite allergy birth cohort. Hypertension 2012;60:1055.
[78] Martin R, Jimenez E, Heilig H, Fernandez L, Marin ML, Zoetendal EG, et al. Isolation of bifidobacteria from breast milk and assessment of the bifidobacterial population by PCR-denaturing gradient gel electrophoresis and quantitative real-time PCR. Appl Env Microbiol 2009;75:965–9.
[79] Heikkila MP, Saris PE. Inhibition of *Staphylococcus aureus* by the commensal bacteria of human milk. J Appl Microbiol 2003;95:471–8.

[80] Jost T, Lacroix C, Braegger C, Chassard C. Assessment of bacterial diversity in breast milk using culture-dependent and culture-independent approaches. Br J Nutr 2013;110:1253–62.
[81] Gueimonde M, Laitinen K, Salminen S, Isolauri E. Breast milk: a source of bifidobacteria for infant gut development and maturation? Neonatology 2007;92:64–6.
[82] Khodayar-Pardo P, Mira-Pascual L, Collado MC, Martinez-Costa C. Impact of lactation stage, gestational age and mode of delivery on breast milk microbiota. J Perinatol 2014;34:599–605.
[83] Collado MC, Delgado S, Maldonado A, Rodriguez JM. Assessment of the bacterial diversity of breast milk of healthy women by quantitative real-time PCR. Lett Appl Microbiol 2009;48:523–8.
[84] Delgado S, Arroyo R, Martin R, Rodriguez JM. PCR-DGGE assessment of the bacterial diversity of breast milk in women with lactational infectious mastitis. BMC Infect Dis 2008;8:51.
[85] Filleron A, Simon M, Hantova S, Jacquot A, Cambonie G, Marchandin H, et al. tuf-PCR-temporal temperature gradient gel electrophoresis for molecular detection and identification of staphylococci: application to breast milk and neonate gut microbiota. J Microbiol Methods 2014;98:67–75.
[86] Cabrera-Rubio R, Collado MC, Laitinen K, Salminen S, Isolauri E, Mira A. The human milk microbiome changes over lactation and is shaped by maternal weight and mode of delivery. Am J Clin Nutr 2012;96:544–51.
[87] Hunt KM, Foster JA, Forney LJ, Schutte UM, Beck DL, Abdo Z, et al. Characterization of the diversity and temporal stability of bacterial communities in human milk. PLoS One 2011;6:e21313.
[88] Moossavi S, Sepehri S, Robertson B, Bode L, Goruk S, Field CJ, et al. Composition and variation of the human milk microbiota are influenced by maternal and early-life factors. Cell Host Microbe 2019;25:324–335.e4.
[89] Ward TL, Hosid S, Ioshikhes I, Altosaar I. Human milk metagenome: a functional capacity analysis. BMC Microbiol 2013;13:116.
[90] Patel SH, Vaidya YH, Patel RJ, Pandit RJ, Joshi CG, Kunjadiya AP. Culture independent assessment of human milk microbial community in lactational mastitis. Sci Rep 2017;7:7804.
[91] Ruiz L, Garcia-Carral C, Rodriguez JM. Unfolding the human milk microbiome landscape in the Omics era. Front Microbiol 2019;10:1378.
[92] Fitzstevens JL, Smith KC, Hagadorn JI, Caimano MJ, Matson AP, Brownell EA. Systematic review of the human milk microbiota. Nutr Clin Pract 2017;32:354–64.
[93] Fernandez L, Langa S, Martin V, Jimenez E, Martin R, Rodriguez JM. The microbiota of human milk in healthy women. Cell Mol Biol (Noisy-le-grand) 2013;59:31–42.
[94] Fernandez L, Langa S, Martin V, Maldonado A, Jimenez E, Martin R, et al. The human milk microbiota: origin and potential roles in health and disease. Pharmacol Res 2013;69:1–10.
[95] Biagi E, Quercia S, Aceti A, Beghetti I, Rampelli S, Turroni S, et al. The bacterial ecosystem of mother's milk and infant's mouth and gut. Front Microbiol 2017;8:1214.
[96] Jost T, Lacroix C, Braegger CP, Rochat F, Chassard C. Vertical mother-neonate transfer of maternal gut bacteria via breastfeeding. Env Microbiol 2014;16:2891–904.
[97] Milani C, Duranti S, Bottacini F, Casey E, Turroni F, Mahony J, et al. The first microbial colonizers of the human gut: composition, activities, and health implications of the infant gut microbiota. Microbiol Mol Biol Rev 2017;81.
[98] Yassour M, Jason E, Hogstrom LJ, Arthur TD, Tripathi S, Siljander H, et al. Strain-level analysis of mother-to-child bacterial transmission during the first few months of life. Cell Host Microbe 2018;24:146–154.e4.

[99] Benito D, Lozano C, Jimenez E, Albujar M, Gomez A, Rodriguez JM, et al. Characterization of *Staphylococcus aureus* strains isolated from faeces of healthy neonates and potential mother-to-infant microbial transmission through breastfeeding. FEMS Microbiol Ecol 2015;91.

[100] Makino H, Kushiro A, Ishikawa E, Kubota H, Gawad A, Sakai T, et al. Mother-to-infant transmission of intestinal bifidobacterial strains has an impact on the early development of vaginally delivered infant's microbiota. PLoS One 2013;8:e78331.

[101] Makino H, Kushiro A, Ishikawa E, Muylaert D, Kubota H, Sakai T, et al. Transmission of intestinal *Bifidobacterium longum* subsp. *longum* strains from mother to infant, determined by multilocus sequencing typing and amplified fragment length polymorphism. Appl Env Microbiol 2011;77:6788–93.

[102] Milani C, Mancabelli L, Lugli GA, Duranti S, Turroni F, Ferrario C, et al. Exploring vertical transmission of bifidobacteria from mother to child. Appl Env Microbiol 2015;81:7078–87.

[103] Dzidic M, Boix-Amoros A, Selma-Royo M, Mira A, Collado MC. Gut microbiota and mucosal immunity in the neonate. Med Sci (Basel) 2018;6.

[104] Sprockett D, Fukami T, Relman DA. Role of priority effects in the early-life assembly of the gut microbiota. Nat Rev Gastroenterol Hepatol 2018;15:197–205.

[105] Gensollen T, Iyer SS, Kasper DL, Blumberg RS. How colonization by microbiota in early life shapes the immune system. Science 2016;352:539–44.

[106] Perez-Cano FJ, Dong H, Yaqoob P. In vitro immunomodulatory activity of *Lactobacillus fermentum* CECT5716 and *Lactobacillus salivarius* CECT5713: two probiotic strains isolated from human breast milk. Immunobiology 2010;215:996–1004.

[107] Olivares M, Diaz-Ropero MP, Martin R, Rodriguez JM, Xaus J. Antimicrobial potential of four *Lactobacillus* strains isolated from breast milk. J Appl Microbiol 2006;101:72–9.

[108] Maldonado J, Canabate F, Sempere L, Vela F, Sanchez AR, Narbona E, et al. Human milk probiotic *Lactobacillus fermentum* CECT5716 reduces the incidence of gastrointestinal and upper respiratory tract infections in infants. J Pediatr Gastroenterol Nutr 2012;54:55–61.

[109] Boix-Amoros A, Martinez-Costa C, Querol A, Collado MC, Mira A. Multiple approaches detect the presence of fungi in human breastmilk samples from healthy mothers. Sci Rep 2017;7:13016.

[110] Boix-Amoros A, Puente-Sanchez F, Toit DU, Linderborg E, Zhang KM, Yang Y, et al. Mycobiome profiles in breast milk from healthy women depend on mode of delivery, geographic location, and interaction with bacteria. Appl Env Microbiol 2019;85.

[111] Garcia-Loygorri MC, De Luis D, Torreblanca B, March GA, Bachiller MR, Eiros JM. Breast milk as a vehicle of transmission of virus. Nutr Hosp 2015;32:4–10.

[112] Duranti S, Lugli GA, Mancabelli L, Armanini F, Turroni F, James K, et al. Maternal inheritance of bifidobacterial communities and bifidophages in infants through vertical transmission. Microbiome 2017;5:66.

[113] Pannaraj PS, Ly M, Cerini C, Saavedra M, Aldrovandi GM, Saboory AA, et al. Shared and distinct features of human milk and infant stool viromes. Front Microbiol 2018;9:1162.

[114] Sakwinska O, Moine D, Delley M, Combremont S, Rezzonico E, Descombes P, et al. Microbiota in breast milk of Chinese lactating mothers. PLoS One 2016;11: e0160856.

[115] Kunz C, Meyer C, Collado MC, Geiger L, Garcia-Mantrana I, Bertua-Rios B, et al. Influence of gestational age, secretor, and Lewis blood group status on the oligosaccharide content of human milk. J Pediatr Gastroenterol Nutr 2017;64:789–98.

[116] Sprenger N, Lee LY, De Castro CA, Steenhout P, Thakkar SK. Longitudinal change of selected human milk oligosaccharides and association to infants' growth, an observatory, single center, longitudinal cohort study. PLoS One 2017;12:e0171814.

[117] Hoashi M, Meche L, Mahal LK, Bakacs E, Nardella D, Naftolin F, et al. Human milk bacterial and glycosylation patterns differ by delivery Mode. Reprod Sci 2016;23:902–7.

[118] Gonzalez R, Maldonado A, Martin V, Mandomando I, Fumado V, Metzner KJ, et al. Breast milk and gut microbiota in African mothers and infants from an area of high HIV prevalence. PLoS One 2013;8:e80299.

[119] Olivares M, Albrecht S, De Palma G, Ferrer MD, Castillejo G, Schols HA, et al. Human milk composition differs in healthy mothers and mothers with celiac disease. Eur J Nutr 2015;54:119–28.

[120] Collado MC, Laitinen K, Salminen S, Isolauri E. Maternal weight and excessive weight gain during pregnancy modify the immunomodulatory potential of breast milk. Pediatr Res 2012;72:77–85.

[121] Gronlund MM, Gueimonde M, Laitinen K, Kociubinski G, Gronroos T, Salminen S, et al. Maternal breast-milk and intestinal bifidobacteria guide the compositional development of the *Bifidobacterium* microbiota in infants at risk of allergic disease. Clin Exp Allergy 2007;37:1764–72.

[122] Angelopoulou A, Field D, Ryan CA, Stanton C, Hill C, Ross RP. The microbiology and treatment of human mastitis. Med Microbiol Immunol 2018;207:83–94.

[123] Jimenez E, De Andres J, Manrique M, Pareja-Tobes P, Tobes R, Martinez-Blanch JF, et al. Metagenomic analysis of milk of healthy and mastitis-suffering women. J Hum Lact 2015;31:406–15.

[124] Williams JE, Carrothers JM, Lackey KA, Beatty NF, York MA, Brooker SL, et al. Human milk microbial community structure is relatively stable and related to variations in macronutrient and micronutrient intakes in healthy lactating women. J Nutr 2017;147:1739–48.

[125] Cabrera-Rubio R, Mira-Pascual L, Mira A, Collado MC. Impact of mode of delivery on the milk microbiota composition of healthy women. J Dev Orig Health Dis 2016;7:54–60.

[126] Hermansson H, Kumar H, Collado MC, Salminen S, Isolauri E, Rautava S. Breast milk microbiota is shaped by mode of delivery and intrapartum antibiotic exposure. Front Nutr 2019;6:4.

[127] Cabrera-Rubio R, Collado MC, Laitinen K, Salminen S, Isolauri E, Mira A. The human milk microbiome changes over lactation and is shaped by maternal weight and mode of delivery. Am J Clin Nutr 2012;96(3):544–551.

[128] Urbaniak C, Angelini M, Gloor GB, Reid G. Human milk microbiota profiles in relation to birthing method, gestation and infant gender. Microbiome 2016;4:1.

[129] Soto A, Martin V, Jimenez E, Mader I, Rodriguez JM, Fernandez L. Lactobacilli and bifidobacteria in human breast milk: influence of antibiotherapy and other host and clinical factors. J Pediatr Gastroenterol Nutr 2014;59(1):78–88.

[130] Kumar H, Du Toit E, Kulkarni A, Aakko J, Linderborg KM, Zhang Y, et al. Distinct patterns in human milk microbiota and fatty acid profiles across specific geographic locations. Front Microbiol 2016;7:1619.

[131] Li SW, Watanabe K, Hsu CC, Chao SH, Yang ZH, Lin YJ, et al. Bacterial composition and diversity in Breast Milk Samples from Mothers Living in Taiwan and Mainland China. Front Microbiol 2017;8:965.

[132] Gotoh A, Katoh T, Sakanaka M, Ling YW, Yamada C, Asakuma S, et al. Sharing of human milk oligosaccharides degradants within bifidobacterial communities in faecal cultures supplemented with *Bifidobacterium bifidum*. Sci Rep 2018;8.

[133] Marcobal A, Sonnenburg JL. Human milk oligosaccharide consumption by intestinal microbiota. Clin Microbiol Infect 2012;18:12–15.

[134] Ninonuevo MR, Perkins PD, Francis J, Lamotte LM, Locascio RG, Freeman SL, et al. Daily variations in oligosaccharides of human milk determined by microfluidic chips and mass spectrometry. J Agric Food Chem 2008;56:618–26.
[135] Backhed F, Roswall J, Peng Y, Feng Q, Jia H, Kovatcheva-Datchary P, et al. Dynamics and stabilization of the human gut microbiome during the first year of life. Cell Host Microbe 2015;17:852.
[136] Yamada C, Gotoh A, Sakanaka M, Hattie M, Stubbs KA, Katayama-Ikegami A, et al. Molecular insight into evolution of symbiosis between breast-fed infants and a member of the human gut microbiome *Bifidobacterium longum*. Cell Chem Biol 2017;24:515.
[137] Garrido D, Ruiz-Moyano S, Kirmiz N, Davis JC, Totten SM, Lemay DG, et al. A novel gene cluster allows preferential utilization of fucosylated milk oligosaccharides in *Bifidobacterium longum* subsp. *longum* SC596. Sci Rep 2016;6.
[138] James K, Motherway MO, Bottacini F, Van Sinderen D. *Bifidobacterium breve* UCC2003 metabolises the human milk oligosaccharides lacto-N-tetraose and lacto-N-neo-tetraose through overlapping, yet distinct pathways. Sci Rep 2016;6:38560.
[139] Katayama T. Host-derived glycans serve as selected nutrients for the gut microbe: human milk oligosaccharides and bifidobacteria. Biosci Biotechnol Biochem 2016;80:621–32.
[140] Sela DA, Chapman J, Adeuya A, Kim JH, Chen F, Whitehead TR, et al. The genome sequence of *Bifidobacterium longum* subsp *infantis* reveals adaptations for milk utilization within the infant microbiome. Proc Natl Acad Sci U S A 2008;105:18964–9.
[141] Thomson P, Medina DA, Garrido D. Human milk oligosaccharides and infant gut bifidobacteria: molecular strategies for their utilization. Food Microbiol 2018;75:37–46.
[142] Odamaki T, Horigome A, Sugahara H, Hashikura N, Minami J, Xiao JZ, et al. Comparative genomics revealed genetic diversity and species/strain-level differences in carbohydrate metabolism of three probiotic bifidobacterial species. Int J Genomics 2015;2015:567809.
[143] Asakuma S, Hatakeyama E, Urashima T, Yoshida E, Katayama T, Yamamoto K, et al. Physiology of consumption of human milk oligosaccharides by infant gut-associated bifidobacteria. J Biol Chem 2011;286:34583–92.
[144] Lawson MAE, O'Neill IJ, Kujawska M, et al. Breast milk-derived human milk oligosaccharides promote *Bifidobacterium* interactions within a single ecosystem. ISME J 2020;14:635–648. https://doi.org/10.1038/s41396-019-0553-2.
[145] LeBlanc JG, Chain F, Martin R, Bermudez-Humaran LG, Courau S, Langella P. Beneficial effects on host energy metabolism of short-chain fatty acids and vitamins produced by commensal and probiotic bacteria. Microb Cell Fact 2017;16.
[146] Hemalatha R, Ouwehand AC, Saarinen MT, Prasad UV, Swetha K, Bhaskar V. Effect of probiotic supplementation on total lactobacilli, bifidobacteria and short chain fatty acids in 2-5-year-old children. Microb Ecol Health Dis 2017;28:1298340.
[147] Li M, Van Esch B, Wagenaar GTM, Garssen J, Folkerts G, Henricks PAJ. Pro- and anti-inflammatory effects of short chain fatty acids on immune and endothelial cells. Eur J Pharmacol 2018;831:52–9.
[148] Luu M, Weigand K, Wedi F, Breidenbend C, Leister H, Pautz S, et al. Regulation of the effector function of CD8(+) T cells by gut microbiota-derived metabolite butyrate. Sci Rep 2018;8:14430.
[149] Vitetta L, Vitetta G, Hall S. Immunological tolerance and function: associations between intestinal bacteria, probiotics, prebiotics, and phages. Front Immunology 2018;9.
[150] O'Neil I, Schofield Z, Hall LJ. Exploring the role of the microbiota member Bifidobacterium in modulating immune-linked diseases. Emerg Top Life Sci 2017;1 (4):333–349.

[151] Kent RM, Doherty SB. Probiotic bacteria in infant formula and follow-up formula: microencapsulation using milk and pea proteins to improve microbiological quality. Food Res Int 2014;64:567–76.
[152] Harmsen HJ, Wildeboer-Veloo AC, Raangs GC, Wagendorp AA, Klijn N, Bindels JG, et al. Analysis of intestinal flora development in breast-fed and formula-fed infants by using molecular identification and detection methods. J Pediatr Gastroenterol Nutr 2000;30:61–7.
[153] Lu CY, Ni YH. Gut microbiota and the development of pediatric diseases. J Gastroenterol 2015;50:720–6.
[154] Martin R, Makino H, Cetinyurek Yavuz A, Ben-Amor K, Roelofs M, Ishikawa E, et al. Early-life events, including mode of delivery and type of feeding, siblings and gender, shape the developing gut microbiota. PLoS One 2016;11:e0158498.
[155] O'Sullivan A, Farver M, Smilowitz JT. The influence of early infant-feeding practices on the intestinal microbiome and body composition in infants. Nutr Metab Insights 2015;8:1–9.
[156] Stiemsma LT, Michels KB. The role of the microbiome in the developmental origins of health and disease. Pediatrics 2018;141.
[157] Yoshioka H, Iseki K, Fujita K. Development and differences of intestinal flora in the neonatal period in breast-fed and bottle-fed infants. Pediatrics 1983;72:317–21.
[158] Bezirtzoglou E, Tsiotsias A, Welling GW. Microbiota profile in feces of breast- and formula-fed newborns by using fluorescence in situ hybridization (FISH). Anaerobe 2011;17:478–82.
[159] Stewart CJ, Ajami NJ, O'brien JL, Hutchinson DS, Smith DP, Wong MC, et al. Temporal development of the gut microbiome in early childhood from the TEDDY study. Nature 2018;562:583.
[160] Das UN. Breastfeeding prevents type 2 diabetes mellitus: but, how and why? Am J Clin Nutr 2007;85:1436–7.
[161] Harder T, Bergmann R, Kallischnigg G, Plagemann A. Duration of breastfeeding and risk of overweight: a meta-analysis. Am J Epidemiol 2005;162:397–403.
[162] Ip S, Chung M, Raman G, Chew P, Magula N, Devine D, et al. Breastfeeding and maternal and infant health outcomes in developed countries. Evid Rep Technol Assess (Full Rep) 2007;1–186.
[163] Ortega-Garcia JA, Kloosterman N, Alvarez L, Tobarra-Sanchez E, Carceles-Alvarez A, Pastor-Valero R, et al. Full breastfeeding and obesity in children: a prospective study from birth to 6 years. Child Obes 2018;14:327–37.
[164] Martin RM, Gunnell D, Smith GD. Breastfeeding in infancy and blood pressure in later life: systematic review and meta-analysis. Am J Epidemiol 2005;161:15–26.
[165] Mercer KE, Bhattacharyya S, Diaz-Rubio ME, Piccolo BD, Pack LM, Sharma N, et al. Infant formula feeding increases hepatic cholesterol 7 alpha hydroxylase (CYP7A1) expression and fecal bile acid loss in neonatal piglets. J Nutr 2018;148:702–11.
[166] Owen CG, Whincup PH, Odoki K, Gilg JA, Cook DG. Infant feeding and blood cholesterol: a study in adolescents and a systematic review. Pediatrics 2002;110:597–608.
[167] Martin CR, Ling PR, Blackburn GL. Review of infant feeding: key features of breast milk and infant formula. Nutrients 2016;8.
[168] Koletzko B, Baker S, Cleghorn G, Neto UF, Gopalan S, Hernell O, et al. Global standard for the composition of infant formula: recommendations of an ESPGHAN Coordinated International Expert Group. J Pediatr Gastroenterol Nutr 2005;41:584–99.
[169] Cook DA. Nutrient levels in infant formulas: technical considerations. J Nutr 1989;119:1773–7.

[170] Papagaroufalis K, Fotiou A, Egli D, Tran LA, Steenhout P. A Randomized double blind controlled safety trial evaluating d-lactic acid production in healthy infants fed a *Lactobacillus reuteri*-containing formula. Nutr Metab Insights 2014;7:19–27.
[171] FAO. Standard for infant formula and formulas for special medical purposes intended for infants. Rome: FAO; 2016.
[172] Kunz C, Rudloff S, Baier W, Klein N, Strobel S. Oligosaccharides in human milk: structural, functional, and metabolic aspects. Annu Rev Nutr 2000;20:699–722.
[173] Vandenplas Y, Zakharova I, Dmitrieva Y. Oligosaccharides in infant formula: more evidence to validate the role of prebiotics. Br J Nutr 2015;113:1339–44.
[174] Ackerman DL, Craft KM, Townsend SD. Infant food applications of complex carbohydrates: structure, synthesis, and function. Carbohydr Res 2017;437:16–27.
[175] Verkhnyatskaya S, Ferrari M, De Vos P, Walvoort MTC. Shaping the infant microbiome with non-digestible carbohydrates. Front Microbiol 2019;10.
[176] Ben XM, Zhou XY, Zhao WH, Yu WL, Wei P, Zhang WL, et al. Supplementation of milk formula with galacto-oligosaccharides improves intestinal micro-flora and fermentation in term infants. Chin Med J 2004;117:927–31.
[177] Fanaro S, Boehm G, Garssen J, Knol J, Mosca F, Stahl B, et al. Galacto-oligosaccharides and long-chain fructo-oligosaccharides as prebiotics in infant formulas: a review. Acta Paediatr 2005;94:22–6.
[178] Maathuis AJH, Van Den Heuvel EC, Schoterman MHC, Venema K. Galacto-oligosaccharides have prebiotic activity in a dynamic in vitro colon model using a C-13-labeling technique. J Nutr 2012;142:1205–12.
[179] Watson D, Motherway MO, Schoterman MHC, Van Neerven RJJ, Nauta A, Van Sinderen D. Selective carbohydrate utilization by lactobacilli and bifidobacteria. J Appl Microbiol 2013;114:1132–46.
[180] Sierra C, Bernal MJ, Blasco J, Martinez R, Dalmau J, Ortuno I, et al. Prebiotic effect during the first year of life in healthy infants fed formula containing GOS as the only prebiotic: a multicentre, randomised, double-blind and placebo-controlled trial. Eur J Nutr 2015;54:89–99.
[181] Marx SP, Winkler S, Hartmeier W. Metabolization of beta-(2,6)-linked fructose-oligosaccharides by different bifidobacteria. FEMS Microbiol Lett 2000;182 163–9.
[182] Rossi M, Corradini C, Amaretti A, Nicolini M, Pompei A, Zanoni S, et al. Fermentation of fructooligosaccharides and inulin by bifidobacteria: a comparative study of pure and fecal cultures. Appl Environ Microbiol 2005;71:6150–8.
[183] Kaplan H, Hutkins RW. Fermentation of fructooligosaccharides by lactic acid bacteria and bifidobacteria. Appl Environ Microbiol 2000;66:2682–4.
[184] Bouhnik Y, Achour L, Paineau D, Riottot M, Attar A, Bornet F. Four-week short chain fructo-oligosaccharides ingestion leads to increasing fecal bifidobacteria and cholesterol excretion in healthy elderly volunteers. Nutr J 2007;6.
[185] Buddington RK, Williams CH, Chen SC, Witherly SA. Dietary supplement of neosugar alters the fecal flora and decreases activities of some reductive enzymes in human subjects. Am J Clin Nutr 1996;63:709–16.
[186] Costalos C, Kapiki A, Apostolou M, Papathoma E. The effect of a prebiotic supplemented formula on growth and stool microbiology of term infants. Early Hum Dev 2008;84:45–9.
[187] Guigoz Y, Rochat F, Perruisseau-Carrier G, Rochat I, Schiffrin EJ. Effects of oligosaccharide on the faecal flora and non-specific immune system in elderly people. Nutr Res 2002;22:13–25.
[188] Cherrington CA, Hinton M, Pearson GR, Chopra I. Short-chain organic-acids at Ph 5.0 Kill *Escherichia coli* and *Salmonella* spp. without causing membrane perturbation. J Appl Bacteriol 1991;70:161–5.

[189] Duncan SH, Louis P, Thomson JM, Flint HJ. The role of pH in determining the species composition of the human colonic microbiota. Environ Microbiol 2009;11:2112–22.
[190] Paineau D, Respondek F, Menet V, Sauvage R, Bornet F, Wagner A. Effects of short-chain fructooligosaccharides on faecal bifidobacteria and specific immune response in formula-fed term infants: a randomized, double-blind, placebo-controlled trial. J Nutr Sci Vitaminol 2014;60:167–75.
[191] Mohnen D. Pectin structure and biosynthesis. Curr Opin Plant Biol 2008;11:266–77.
[192] Di R, Vakkalanka MS, Onumpai C, Chau HK, White A, Rastall RA, et al. Pectic oligosaccharide structure-function relationships: prebiotics, inhibitors of *Escherichia coli* O157:H7 adhesion and reduction of Shiga toxin cytotoxicity in HT29. Cell Food Chem 2017;227:245–54.
[193] Lopez-Siles M, Khan TM, Duncan SH, Harmsen HJM, Garcia-Gil LJ, Flint HJ. Cultured representatives of two major phylogroups of human colonic *Faecalibacterium prausnitzii* can utilize pectin, uronic acids, and host-derived substrates for growth. Appl Environ Microbiol 2012;78:420–8.
[194] Louis P, Flint HJ, Michel C. How to manipulate the microbiota: prebiotics. Microbiota Hum Body Implic Health Dis 2016;902:119–42.
[195] Stam J, Van Stuijvenberg M, Garssen J, Knipping K, Sauer PJ. A mixture of three prebiotics does not affect vaccine specific antibody responses in healthy term infants in the first year of life. Vaccine 2011;29:7766–72.
[196] Vos AP, Haarman M, Van Ginkel JW, Knol J, Garssen J, Stahl B, et al. Dietary supplementation of neutral and acidic oligosaccharides enhances Th1-dependent vaccination responses in mice. Pediatr Allergy Immunol 2007;18:304–12.
[197] Vos AP, Van Esch BC, Stahl B, M'rabet L, Folkerts G, Nijkamp FP, et al. Dietary supplementation with specific oligosaccharide mixtures decreases parameters of allergic asthma in mice. Int Immunopharmacol 2007;7:1582–7.
[198] Bosscher D, Van Caillie-Bertrand M, Van Cauwenbergh R, Deelstra H. Availabilities of calcium, iron, and zinc from dairy infant formulas is affected by soluble dietary fibers and modified starch fractions. Nutrition 2003;19:641–5.
[199] Chen CC, Zhang Y, Xue MY, Liu XW, Li YH, Chen X, et al. Sequential one-pot multienzyme (OPME) synthesis of lacto-N-neotetraose and its sialyl and fucosyl derivatives. Chem Commun 2015;51:7689–92.
[200] Yu H, Li YH, Wu ZG, Li L, Zeng J, Zhao C, et al. H-pylori alpha 1-3/4-fucosyltransferase (Hp3/4FT)-catalyzed one-pot multienzyme (OPME) synthesis of Lewis antigens and human milk fucosides. Chem Commun 2017;53:11012–15.
[201] Chin YW, Kim JY, Kim JH, Jung SM, Seo JH. Improved production of 2′-fucosyllactose in engineered *Escherichia coli* by expressing putative alpha-1,2-fucosyltransferase, Wcf from Bacteroides fragilis. J Biotechnol 2017;257:192–8.
[202] Qin HM, Li ST, Zhang YF, Wang JW, Li JX, Song SY, et al. Multienzymatic cascade synthesis of fucosyloligosaccharide via a two-step fermentation strategy in *Escherichia coli*. Biotechnol Lett 2016;38:1747–52.
[203] Yu S, Liu JJ, Yun EJ, Kwak S, Kim KH, Jin YS. Production of a human milk oligosaccharide 2′-fucosyllactose by metabolically engineered *Saccharomyces cerevisiae*. Microb Cell Factories 2018;17.
[204] Manas NHA, Illias RM, Mahadi NM. Strategy in manipulating transglycosylation activity of glycosyl hydrolase for oligosaccharide production. Crit Rev Biotechnol 2018;38:272–93.
[205] Danby PM, Withers SG. Advances in enzymatic glycoside synthesis. Acs Chem Biol 2016;11:1784–94.
[206] Zeuner B, Jers C, Mikkelsen JD, Meyer AS. Methods for improving enzymatic transglycosylation for synthesis of human milk oligosaccharide biomimetics. J Agric Food Chem 2014;62:9615–31.

[207] European Commision. Commission Implementing Decision (EU) 2016/375 of 11 March 2016 authorising the placing on the market of lacto-N-neotetraose as a novel food ingredient under Regulation (EC) No 258/97 of the European Parliament and of the Council (notified under document C(2016) 1419). Brussels: European Commision; 2016.
[208] European Commision. Commission Implementing Decision (EU) 2016/376 of 11 March 2016 authorising the placing on the market of 2′-O-fucosyllactose as a novel food ingredient under Regulation (EC) No 258/97 of the European Parliament and of the Council (notified under document C(2016) 1423). Brussels: European Commision; 2016.
[209] European Commision. Commission Implementing Decision (EU) 2017/2375 of 15 December 2017 authorising the placing on the market of N-acetyl-D-neuraminic acid as a novel food ingredient under Regulation (EC) No 258/97 of the European Parliament and of the Council (notified under document C(2017) 8431). Brussels: European Commision; 2017.
[210] European Commision. Commission Implementing Decision (EU) 2017/2201 of 27 November 2017 authorising the placing on the market of 2′-fucosyllactose produced with *Escherichia coli* strain BL21 as a novel food ingredient under Regulation (EC) No 258/97 of the European Parliament and of the Council (notified under document C(2017) 7662). Brussels: European Commision; 2017.
[211] WHO/FAO. Report of a joint FAO/WHO expert consultation on evaluation of health and nutritional properties of probiotics in food including powder milk with live lactic acid bacteria. Geneva/Rome: WHO/FAO; 2001.
[212] Hill C, Guarner F, Reid G, Gibson GR, Merenstein DJ, Pot B, et al. The International Scientific Association for Probiotics and Prebiotics consensus statement on the scope and appropriate use of the term probiotic. Nat Rev Gastroenterol Hepatol 2014;11:506–14.
[213] Fuller R. Probiotics in man and animals. J Appl Bacteriol 1989;66:365–78.
[214] Veld JHJH, Havenaar R. Probiotics and health in man and animal. J Chem Technol Biotechnol 1991;51:562–7.
[215] Anal AK, Singh H. Recent advances in microencapsulation of probiotics for industrial applications and targeted delivery. Trends Food Sci Technol 2007;18:240–51.
[216] de Vos P, Faas MM, Spasojevic M, Sikkema J. Encapsulation for preservation of functionality and targeted delivery of bioactive food components. Int Dairy J 2010;20:292–302.
[217] Mugambi MN, Musekiwa A, Lombard M, Young T, Blaauw R. Synbiotics, probiotics or prebiotics in infant formula for full term infants: a systematic review. Nutr J 2012;11.
[218] Braegger C, Chmielewska A, Decsi T, Kolacek S, Mihatsch W, Moreno L, et al. Supplementation of infant formula with probiotics and/or prebiotics: a systematic review and comment by the ESPGHAN Committee on Nutrition. J Pediatr Gastroenterol Nutr 2011;52:238–50.
[219] European Food Safety Authority. Opinion of the Scientific Committee on a request from EFSA on the introduction of a QualifiedPresumption of Safety (QPS) approach for assessment of selected microorganisms referred to EFSA. The EFSA Journal 2007;587:1–16. Online: https://efsa.onlinelibrary.wiley.com/doi/epdf/10.2903/j.efsa.2007.587
[220] U. S. Food and Drug Administration. Evidence-based review system for the scientific evaluation of health claim. Silver Spring, MD: U. S. Food and Drug Administration; 2009.
[221] European Parliament. Regulation (EC) No 1924/2006 of the European Parliament and of the Council of 20 December 2006 on nutrition and health claims made on foods. Brussels: European Parliament; 2006.

[222] Sanders ME, Merenstein D, Merrifield CA, Hutkins R. Probiotics for human use. Nutr Bull 2018;43:212–25.
[223] Sanders ME, Akkermans LM, Haller D, Hammerman C, Heimbach J, Hormannsperger G, et al. Safety assessment of probiotics for human use. Gut Microbes 2010;1:164–85.
[224] van den Nieuwboer M, Claassen E, Morelli L, Guarner F, Brummer RJ. Probiotic and synbiotic safety in infants under two years of age. Benef Microbes 2014;5:45–60.
[225] Koenig JE, Spor A, Scalfone N, Fricker AD, Stombaugh J, Knight R, et al. Succession of microbial consortia in the developing infant gut microbiome. Proc Natl Acad Sci U S A 2011;108(Suppl 1):4578–85.
[226] Laursen MF, Andersen LBB, Michaelsen KF, Molgaard C, Trolle E, Bahl MI, et al. Infant gut microbiota development is driven by transition to family foods independent of maternal obesity. Msphere 2016;1.
[227] Fallani M, Amarri S, Uusijarvi A, Adam R, Khanna S, Aguilera M, et al. Determinants of the human infant intestinal microbiota after the introduction of first complementary foods in infant samples from five European centres. Microbiology 2011;157:1385–92.
[228] Thompson AL, Monteagudo-Mera A, Cadenas MB, Lampl ML, Azcarate-Peril MA. Milk- and solid-feeding practices and daycare attendance are associated with differences in bacterial diversity, predominant communities, and metabolic and immune function of the infant gut microbiome. Front Cell Infect Microbiol 2015;5:3.
[229] Nagpal R, Kurakawa T, Tsuji H, Takahashi T, Kawashima K, Nagata S, et al. Evolution of gut *Bifidobacterium* population in healthy Japanese infants over the first three years of life: a quantitative assessment. Sci Rep 2017;7:10097.
[230] Turroni F, Marchesi JR, Foroni E, Gueimonde M, Shanahan F, Margolles A, et al. Microbiomic analysis of the bifidobacterial population in the human distal gut. ISME J 2009;3:745–51.
[231] Zoetendal EG, Rajilic-Stojanovic M, De Vos WM. High-throughput diversity and functionality analysis of the gastrointestinal tract microbiota. Gut 2008;57:1605–15.
[232] Tannock GW, Lawley B, Munro K, Gowri Pathmanathan S, Zhou SJ, Makrides M, et al. Comparison of the compositions of the stool microbiotas of infants fed goat milk formula, cow milk-based formula, or breast milk. Appl Env Microbiol 2013;79:3040–8.
[233] Sambo F, Finotello F, Lavezzo E, Baruzzo G, Masi G, Peta E, et al. Optimizing PCR primers targeting the bacterial 16S ribosomal RNA gene. BMC Bioinforma 2018;19.
[234] Turroni F, Peano C, Pass DA, Foroni E, Severgnini M, Claesson MJ, et al. Diversity of bifidobacteria within the infant gut microbiota. PLoS One 2012;7.
[235] Matsuki T, Watanabe K, Fujimoto J, Kado Y, Takada T, Matsumoto K, et al. Quantitative PCR with 16S rRNA-Gene-targeted species-specific primers for analysis of human intestinal bifidobacteria. Appl Environ Microbiol 2004;70:167–73.
[236] McKeen S, Young W, Mullaney J, Fraser K, McNabb WC, Roy NC. Infant complementary feeding of prebiotics for the microbiome and immunity. Nutrients 2019;11.
[237] Hald S, Schioldan AG, Moore ME, Dige A, Laerke HN, Agnholt J, et al. Effects of arabinoxylan and resistant starch on intestinal microbiota and short-chain fatty acids in subjects with metabolic syndrome: a randomised crossover study. PLoS One 2016;11.
[238] Riviere A, Selak M, Lantin D, Leroy F, De Vuyst L. Bifidobacteria and butyrate-producing colon bacteria: importance and strategies for their stimulation in the human gut. Front Microbiol 2016;7:979.
[239] Broekaert WF, Courtin CM, Verbeke K, Van de Wiele T, Verstraete W, Delcour JA. Prebiotic and other health-related effects of cereal-derived arabinoxylans,

arabinoxylan-oligosaccharides, and xylooligosaccharides. Crit Rev Food Sci Nutr 2011;51:178–94.

[240] Roberfroid MB. Inulin-type fructans: functional food ingredients. J Nutr 2007;137:2493S–502S.

[241] Niness KR. Inulin and oligofructose: what are they? J Nutr 1999;129:1402s–6s.

[242] Scott KP, Martin JC, Chassard C, Clerget M, Potrykus J, Campbell G, et al. Substrate-driven gene expression in *Roseburia inulinivorans*: importance of inducible enzymes in the utilization of inulin and starch. Proc Natl Acad Sci U S A 2011;108:4672–9.

[243] Warchol M, Perrin S, Grill JP, Schneider F. Characterization of a purified beta-fructofuranosidase from *Bifidobacterium infantis* ATCC 15697. Lett Appl Microbiol 2002;35:462–7.

[244] Ehrmann MA, Korakli M, Vogel RF. Identification of the gene for beta-fructofuranosidase of *Bifidobacterium lactis* DSM10140(T) and characterization of the enzyme expressed in *Escherichia coli*. Curr Microbiol 2003;46:391–7.

[245] Omori T, Ueno K, Muramatsu K, Kikuchi M, Onodera S, Shiomi N. Characterization of recombinant beta-fructofuranosidase from *Bifidobacterium adolescentis* G1. Chem Cent J 2010;4.

[246] Falony G, Lazidou K, Verschaeren A, Weckx S, Maes D, De Vuyst L. In vitro kinetic analysis of fermentation of prebiotic inulin-type fructans by *Bifidobacterium* species reveals four different phenotypes. Appl Environ Microbiol 2009;75:454–61.

[247] Selak M, Riviere A, Moens F, Van Den Abbeele P, Geirnaert A, Rogelj I, et al. Inulin-type fructan fermentation by bifidobacteria depends on the strain rather than the species and region in the human intestine. Appl Microbiol Biotechnol 2016;100:4097–107.

[248] Falony G, Vlachou A, Verbrugghe K, De Vuyst L. Cross-feeding between *Bifidobacterium longum* BB536 and acetate-converting, butyrate-producing colon bacteria during growth on oligofructose. Appl Environ Microbiol 2006;72:7835–41.

[249] Falony G, Verschaeren A, De Bruycker F, De Preter V, Verbeke K, Leroy F, et al. In vitro kinetics of prebiotic inulin-type fructan fermentation by butyrate-producing colon bacteria: implementation of online gas chromatography for quantitative analysis of carbon dioxide and hydrogen gas production. Appl Environ Microbiol 2009;75:5884–92.

[250] Moens F, Weckx S, De Vuyst L. Bifidobacterial inulin-type fructan degradation capacity determines cross-feeding interactions between bifidobacteria and *Faecalibacterium prausnitzii*. Int J Food Microbiol 2016;231:76–85.

[251] Schaafsma G, Slavin JL. Significance of inulin fructans in the human diet. Compr Rev Food Sci Food Saf 2015;14:37–47.

[252] Morgan NK, Keerqin C, Wallace A, Wu SB, Choct M. Effect of arabinoxylo-oligosaccharides and arabinoxylans on net energy and nutrient utilization in broilers. Anim Nutr 2019;5:56–62.

[253] Izydorczyk MS, Biliaderis CG. Cereal arabinoxylans: advances in structure and physicochemical properties. Carbohydr Polym 1995;28:33–48.

[254] Van Den Broek LAM, Voragen AGJ. *Bifidobacterium* glycoside hydrolases and (potential) prebiotics. Innovat Food Sci Emerg Technol 2008;9:401–7.

[255] Lagaert S, Pollet A, Courtin CM, Volckaert G. beta-Xylosidases and alpha-L-arabinofuranosidases: accessory enzymes for arabinoxylan degradation. Biotechnol Adv 2014;32:316–32.

[256] Lagaert S, Pollet A, Delcour JA, Lavigne R, Courtin CM, Volckaert G. Substrate specificity of three recombinant alpha-L-arabinofuranosidases from *Bifidobacterium adolescentis* and their divergent action on arabinoxylan and arabinoxylan oligosaccharides. Biochemical Biophysical Res Commun 2010;402:644–50.

[257] Lagaert S, Pollet A, Delcour JA, Lavigne R, Courtin CM, Volckaert G. Characterization of two beta-xylosidases from *Bifidobacterium adolescentis* and their

contribution to the hydrolysis of prebiotic xylooligosaccharides. Appl Microbiol Biotechnol 2011;92:1179–85.

[258] Hughes SA, Shewry PR, Li L, Gibson GR, Sanz ML, Rastall RA. In vitro fermentation by human fecal microflora of wheat arabinoxylans. J Agric Food Chem 2007;55:4589–95.

[259] Van Craeyveld V, Swennen K, Dornez E, Van De Wiele T, Marzorati M, Verstraete W, et al. Structurally different wheat-derived arabinoxylooligosaccharides have different prebiotic and fermentation properties in rats. J Nutr 2008;138:2348–55.

[260] Van den Abbeele P, Gerard P, Rabot S, Bruneau A, El Aidy S, Derrien M, et al. Arabinoxylans and inulin differentially modulate the mucosal and luminal gut microbiota and mucin-degradation in humanized rats. Environ Microbiol 2011;13:2667–80.

[261] Truchado P, Van Den Abbeele P, Riviere A, Possemiers S, De Vuyst L, Van De Wiele T. *Bifidobacterium longum* D2 enhances microbial degradation of long-chain arabinoxylans in an in vitro model of the proximal colon. Benef Microbes 2015;6:849–60.

[262] Riviere A, Moens F, Selak M, Maes D, Weckx S, Vuyst LD. The ability of bifidobacteria to degrade arabinoxylan oligosaccharide constituents and derived oligosaccharides is strain dependent. Appl Environ Microbiol 2014;80:204–17.

[263] Grootaert C, Van Den Abbeele P, Marzorati M, Broekaert WF, Courtin CM, Delcour JA, et al. Comparison of prebiotic effects of arabinoxylan oligosaccharides and inulin in a simulator of the human intestinal microbial ecosystem. FEMS Microbiol Ecol 2009;69:231–42.

[264] Sanchez JI, Marzorati M, Grootaert C, Baran M, Van Craeyveld V, Courtin CM, et al. Arabinoxylan-oligosaccharides (AXOS) affect the protein/carbohydrate fermentation balance and microbial population dynamics of the simulator of human intestinal microbial ecosystem. Microb Biotechnol 2009;2:101–13.

[265] Neyrinck AM, Pachikian B, Taminiau B, Daube G, Frederick R, Cani PD, et al. Intestinal sucrase as a novel target contributing to the regulation of glycemia by prebiotics. PLoS One 2016;11.

[266] Mendis M, Simsek S. Arabinoxylans and human health. Food Hydrocoll 2014;42:239–43.

[267] Damen B, Verspreet J, Pollet A, Broekaert WF, Delcour JA, Courtin CM. Prebiotic effects and intestinal fermentation of cereal arabinoxylans and arabinoxylan oligosaccharides in rats depend strongly on their structural properties and joint presence. Mol Nutr Food Res 2011;55:1862–74.

[268] Grootaert C, Delcour JA, Courtin CM, Broekaert WF, Verstraete W, Van De Wiele T. Microbial metabolism and prebiotic potency of arabinoxylan oligosaccharides in the human intestine. Trends Food Sci Technol 2007;18:64–71.

[269] Lattimer JM, Haub MD. Effects of dietary fiber and its components on metabolic health. Nutrients 2010;2:1266–89.

[270] Tachon S, Zhou JN, Keenan M, Martin R, Marco ML. The intestinal microbiota in aged mice is modulated by dietary resistant starch and correlated with improvements in host responses. Fems Microbiol Ecol 2013;83:299–309.

[271] Martinez I, Kim J, Duffy PR, Schlegel VL, Walter J. Resistant starches types 2 and 4 have differential effects on the composition of the fecal microbiota in human subjects. PLoS One 2010;5.

[272] Duranti S, Turroni F, Lugli GA, Milani C, Viappiani A, Mangifesta M, et al. Genomic characterization and transcriptional studies of the starch-utilizing strain *Bifidobacterium adolescentis* 22L. Appl Environ Microbiol 2014;80:6080–90.

[273] Motherway MO, Fitzgerald GF, Neirynck S, Ryan S, Steidler L, Van Sinderen D. Characterization of ApuB, an extracellular type II amylopullulanase from *Bifidobacterium breve* UCC2003. Appl Environ Microbiol 2008;74:6271–9.

[274] Jung DH, Chung WH, Seo DH, Nam YD, Yoon S, Park CS. Complete genome sequence of *Bifidobacterium choerinum* FMB-1, a resistant starch-degrading bacterium. J Biotechnol 2018;274:28–32.

[275] Behall KM, Scholfield DJ, Hallfrisch JG, Liljeberg-Elmstahl HGM. Consumption of both resistant starch and beta-glucan improves postprandial plasma glucose and insulin in women. Diabetes Care 2006;29:976–81.

[276] Maziarz MP, Preisendanz S, Juma S, Imrhan V, Prasad C, Vijayagopal P. Resistant starch lowers postprandial glucose and leptin in overweight adults consuming a moderate-to-high-fat diet: a randomized-controlled trial. Nutr J 2017;16.

[277] Nilsson AC, Ostman EM, Hoist JJ, Bjorck IME. Including indigestible carbohydrates in the evening meal of healthy subjects improves glucose tolerance, lowers inflammatory markers, and increases satiety after a subsequent standardized breakfast. J Nutr 2008;138:732–9.

[278] Louis P, Scott KP, Duncan SH, Flint HJ. Understanding the effects of diet on bacterial metabolism in the large intestine. J Appl Microbiol 2007;102:1197–208.

CHAPTER

10

The early gut microbiome and the risk of chronic disease

Ehud Rinott[1] *and Ilan Youngster*[2,3]

[1]Faculty of Health Sciences, Ben-Gurion University of the Negev, Beer-Sheva, Israel, [2]Pediatric Infectious Diseases Unit and Center for Microbiome Research, Shamir Medical Center, Be'er Ya'akov, Israel, [3]Sackler School of Medicine, Tel-Aviv University, Tel-Aviv, Israel

The early gastrointestinal microbiome plays a critical role in nutritional, metabolic, and immune functions in infants and young children and has implications for future health status. A growing understanding of the role of intestinal dysbiosis in chronic disease development and progression allows implementation of new approaches to prevention and treatment of chronic diseases that originate in early childhood but affect the patient throughout life. In this chapter, we will use autoimmune diseases and inflammatory bowel disease (IBD) as representative examples and discuss central pertinent aspects of the intriguing connection between early gut microbiome and later development of chronic diseases.

The early gut microbiome and autoimmune disease

A central feature of the human immune system is its ability to generate an inflammatory response to foreign materials while avoiding harm to self-tissues. Congruently, an autoimmune disease is characterized by tissue injury, caused by the immunologic reaction of the human body against its own tissues. Though autoimmunity is present in all human individuals, emergence of an autoimmune disease is a result of aberrant immune tolerance regulation [1].

The Human Microbiome in Early Life
DOI: https://doi.org/10.1016/B978-0-12-818097-6.00010-9

The pathogenesis of autoimmune diseases is dependent on interactions between genetic and environmental determinants. In the last few decades, the developed world has seen a marked increase in immune-mediated disorders. As this trend was preceded by a decline in infectious diseases, a "hygiene theory" has emerged, linking modern environment to a decreased microbial exposure, harming immune tolerance regulation and resulting in self-reactivity and autoimmunity [2,3].

The human gut forms a hub of extensive contact with the external environment, being exposed to a myriad of foreign organisms, pathogenic, commensal or beneficial, as well as xenobiotics and food antigens [4,5]. In the healthy individual, the intestinal tract preserves homeostasis of balanced interactions between the gut microbiome and the host's immune system [6,7]. Any breakdown of intestinal homeostasis can lead to a pro-inflammatory state in a response to otherwise innocuous antigens, leading to the development of chronic inflammation [8,9]. However, data regarding the mechanisms that link the gut microbiome, specific environmental factors and the host immune response are still limited, with most evidence acquired through distinct associations and epidemiological studies. Two such immunologically mediated disorders, type 1 diabetes (T1D) and celiac disease (CeD), have massively increased in prevalence in last few decades, especially in young children, and have been reported to be associated with gut dysbiosis [10]. Here, we focus on current evidence related to the gut microbiome and its development, and the role of gut microbiome–environmental factors interaction, in the pathogenesis of T1D and CeD.

Type 1 diabetes mellitus

T1D is an autoimmune disorder, characterized by the destruction of insulin-producing β cells in the pancreas by a T cell-mediated response [1]. Although 70% of T1D patients are considered genetically predisposed by having human leukocyte antigen (HLA) risk alleles, only some 3%–7% of children with this genetic predisposition will go on to develop T1D [11], suggesting environmental factors have a central role in disease pathogenesis. Furthermore, the incidence of T1D has been rapidly increasing over the past few decades, particularly in early life (0–4 years) [12], again supporting the hypothesis that certain environmental factors are associated with T1D risk.

Recent evidence linked intestinal homeostasis and T1D susceptibility, as dysbiosis and increased gut barrier permeability were observed in T1D subjects [13,14]. In other studies, the microbiome of prediagnosed T1D children was noted to contain fewer genes related to fermentation and biosynthesis of short-chain fatty acids (SCFAs). As the cases and controls did not contain distinct consistent taxonomic patterns, it was suggested that overall, T1D-associated microbiomes encompass taxonomically diffuse but functionally coherent traits [15]. Although some features of the gut microbiome in pediatric T1D subjects are distinguishable only at the time of diagnosis (specific T1D associated taxa), gut alpha diversity tends to undergo a gradual decline preceding diagnosis, starting as soon as seroconversion is apparent [16]. Changes occuring in the infant gut microbiome prior to T1D diagnosis include increases in abundance of inflammation-favoring organisms and distinct changes in gene functions, and stool metabolites [16].

The role of microbial exposures in T1D pathogenesis has been well established in diabetes mice models. The knockout of MyD88, an adaptor downstream of multiple Toll-like receptors involved in microbial sensing, resulted in complete protection from diabetes [17]. However, heterozygous $MyD88^{KO/+}$ mice, which normally develop diabetes, are protected from T1D when exposed to the gut microbiome of a MyD88-KO donor from birth [17]. Therefore it is hypothesized that the disease progression is at least partially mediated by an innate immune response to the gut microbiome, and that modulation of the microbiome might mitigate this response and reduce T1D risk.

In light of the establishment of the gut microbiome's pivotal role in T1D development [17], it has become imperative to pinpoint specific gut microbiome modulating factors, allowing both a better understanding of the disease pathogenesis, along with the identification of novel treatment avenues. To date, several factors contributing to intestinal homeostasis modification have been identified, including delivery mode, infant feeding modality, antibiotic use, geography, and hygiene (Fig. 10.1) [19].

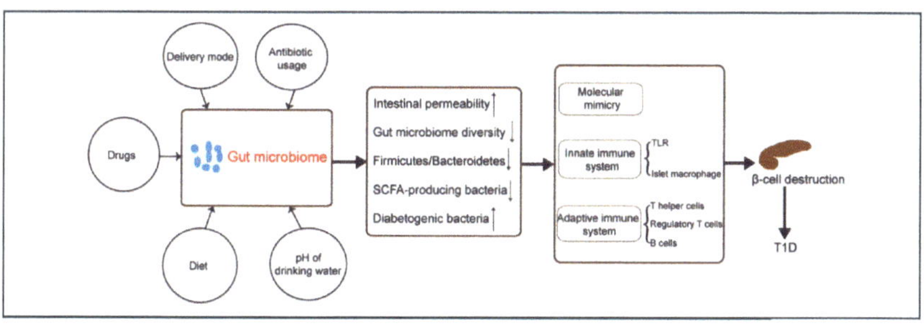

FIGURE 10.1 The role of gut microbiome in type 1 diabetes. *Source: From Zheng P, Li Z, Zhou Z. Gut microbiome in type 1 diabetes: a comprehensive review. Diabetes Metab Res Rev 2018;34(7):1–9 [18].*

Early colonization

A recent study discovered that early-life colonization by a distinct microbiome could have "immunologically silencing" effects, compromising immune education and inducing autoimmunity risk [20]. The investigators were able to recognize a different pattern of gram-negative bacterial colonization between infants from countries with high autoimmunity incidence (characterized by the dominance of *Bacteroides* species) and a country with low autoimmunity incidence (characterized by low levels of *Bacteroides* species). It was also demonstrated that lipopolysaccharide structure is bacteria-specific and has a differential effect on endotoxin tolerance, with *Bacteroides*-associated lipopolysaccharide inducing higher diabetes risk when injected to mice, possibly explaining the higher autoimmunity incidence in certain human populations [20].

Delivery mode

It has been established that a Cesarean section (C-section) delivery confers prominent, lasting, effects on the gut microbiome composition and function, presumably due to a priority effect of the inoculating microbes during birth (as described in previous chapters) [21–23]. This hypothesis has been validated by the identification of significantly higher relative abundance of vagina-associated bacteria, such as *Lactobacillus*, in the gut microbiome of children delivered vaginally [23,24].

C-section has been associated with slightly increased T1D incidence. Several observational studies have evaluated the association between delivery mode and T1D, concluding, in a reproducible manner, that C-sections increase T1D risk in several distinct cohorts from different countries, including Germany [25] Taiwan [26], Scotland [27], and Denmark [28].

These observational data further indicated that the increased T1D risk finding is consistent even when assessed only in planned, prelabor surgery [27,28]. Interestingly, scheduled surgeries were associated with a greater T1D risk as compared to intrapartum C-sections [28].

Altogether, it could be concluded that early colonization by the birth canal's microbiome, as well as maternal gut microbiome, is of importance and could promote T1D protection in later life.

Infant feeding

Breast milk could be regarded as the most common type of nutrition introduced to the newborn gut, with the World Health Organization recommending that infants should be exclusively breastfed until 6 months of age [29]. Breast milk has a key role in the establishment of a healthy gut barrier through its bioactive components [30], with some evidence indicating that breast milk might contain the mother's intestinal bacteria, translocated through an "entero-mammary" pathway [31].

Breast milk continuously changes in composition throughout infancy [30], as described in the previous chapter, and is generally rich in human milk oligosaccharides, a prebiotic substrate fermented by the infant gut microbiome. Due to the fermentation, SCFAs are released, modifying the gut barrier integrity and limiting the growth of pathogenic organisms [32]. The association between T1D and breast milk was established as human milk oligosaccharides introduced to a mice model provided protection against autoimmune diabetes [33].

An observational study, based on two Scandinavian cohorts with 155,392 children, found that children who were never breastfed had a twofold increased risk of T1D, compared with those who were breastfed (to any extent), and that duration of breastfeeding had no significant effect on T1D risk [34]. A different analysis, facilitating a pooled analysis of 43 observational studies, concluded that exclusive breastfeeding for over 2 weeks was the most prominent and consistent T1D protective factor across populations and studies [35]. It was also suggested that perturbation of the infant gut microbiome, caused by C-section, could be partially restored by breastfeeding [36].

The timing of the infants' first exposure to solid foods was also found to be associated with T1D risk, with results suggesting that the safest age to introduce solid foods in children at increased genetic risk for T1DM is between 4 and 5 months of age [37].

In conclusion, several feeding strategies are known to affect T1D development, including breastfeeding initiation, extent and duration, formula feeding, and first exposure to solid foods.

Antibiotic use

Due to the extensive effect of antibiotic treatment on the gut microbiome, it was presumed that when applied in early life, antibiotics might modulate the development of the immune system, eventually affecting T1D risk. A mice study testing this hypothesis concluded that a single course of antibiotics had pronounced and consistent effects on the gut microbiome composition, while modifying ileal gene immune expression, inducing accelerated T1D development [38].

The antibiotics-T1D hypothesis was also tested epidemiologically, as a study surveying data from about 1 million subjects under the age of 1 concluded that beta-lactam-based antibiotics are associated with increased T1D risk [39]. A different study reproduced this results with broad-spectrum antibiotics, indicating that antibiotic treatment until age of 2 is associated with an increased T1D risk later in life [40].

Celiac disease

CeD is a chronic immune-mediated enteropathy driven by dietary gluten ingestion in genetically predisposed individuals. Though the etiology of CeD is largely unknown, environmental, immunologic, and genetic factors are known to contribute to disease pathogenesis. The immune system constitutes a pivotal component in CeD pathogenesis, characterized by both an adaptive and innate immune response, with serum antibodies considered a hallmark of the disease, whether as a primary response or secondary to the tissue damage [1].

Exposure to dietary gluten and the presence of HLA DQ2 or DQ8 are necessary, though insufficient, conditions for CeD development. As very few HLA DQ2/DQ8 carrying individuals tend to develop CeD (Approximately 2%–5%), it was presumed that additional factors affect disease development. Genome-wide association studies have identified some non-HLA CeD risk loci, but those have been estimated to contribute up to 14% of the genetic variance of CeD, suggesting that environmental factors should be considered when assessing CeD pathogenesis [41,42].

Several studies demonstrated that CeD is associated with gut microbiome composition alterations, when compared to healthy control's microbiome. For example, it was reported that CeD is associated with reduced *Bifidobacterium* and increased *Bacteroides* abundance, a change that could denote a pro-inflammatory shift in the gut composition [20,43–45]. CeD was also associated with changes in abundance of Firmicutes and Proteobacteria, both in pediatric and adult active patients [46,47]. A distinct signature of microbial fermentation products (SCFAs) have also been reported in patients with CeD [45].

As CeD tends to have various clinical manifestations, a study examined whether the microbiome is associated with different CeD symptoms. It was demonstrated that the gut microbiome of patients experiencing gastrointestinal symptoms was different from the microbiome of patients experiencing dermatological symptoms or dyspepsia [46].

To date, several gut microbiome-affected environmental factors have been described to be associated with CeD pathogenesis (Fig. 10.2). Shedding light on their role in disease pathogenesis, while establishing the role of the gut microbiome, could ultimately clarify some of the unknowns in CeD etiology.

Early gut colonization

The early colonization of the gut has a strong influence on the development of systemic immunity. In a germ-free mice model, immunoglobulin levels were reduced, spleens were smaller, and gut-associated lymphoid tissue was immature as compared to postcolonization features [48]; with

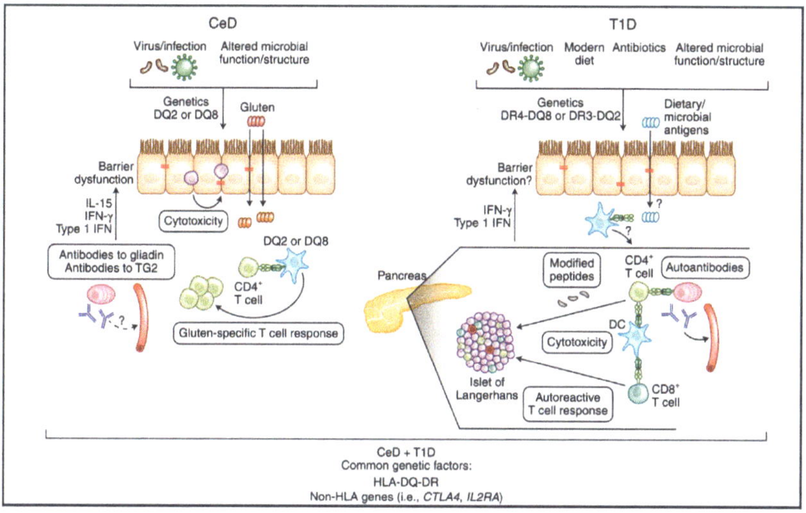

FIGURE 10.2 The role of gut microbiome in celiac disease (CeD) and the common features in pathogenesis with type 1 diabetes. *Source: From Verdu EF, Danska JS. Common ground: shared risk factors for type 1 diabetes and celiac disease. Nat Immunol 2018;19(7):685–95.*

similar evidence observed in human observational studies [49,50]. A study based on 22 infants with either high or low genetic risk of CeD (HLA-DQ2 carriers and non HLA-DQ2/8 carriers, respectively), all having at least one first-degree relative with CeD, evaluated early gut composition in genetically predisposed individuals. High genetic risk infants were colonized with a significantly higher proportions of Firmicutes and Proteobacteria and lower proportions of Actinobacteria, as compared to low-risk infants, with an observed difference down to the genus level. The authors suggested that a disease-risk-associated host genotype may also select for the first gut colonizers, ultimately affecting disease risk [51].

Antibiotic use

In a trial using a high CeD-risk mice model, a "healthy microbiome," devoid of pathogens, was found to protect against gluten immunopathology. Moreover, perinatal exposure to vancomycin was associated with a Proteobacteria bloom in the small intestine, with dysbiosis persisting following antibiotics discontinuation, causing increased gluten sensitization in the antibiotic-treated mice [52].

A recent nationwide observational study from Denmark and Norway evaluated the association between antibiotic use in the first year of life and CeD ($n > 1.7$ million; age at end of follow-up: 1–20.3 years). The researchers showed that exposure to systemic antibiotics in the first year of life was positively associated with CeD in both cohorts (pooled

OR = 1.26). They were also able to identify a dose-response trend between the number of dispensed antibiotic doses and the risk of CeD (pooled OR for each antibiotic = 1.08) [53]. These findings suggest that microbiome alteration through early antibiotic exposure might affect CeD risk in later life.

Early feeding

Recent studies established the role of breastfeeding in CeD development and progression. A retrospective study performed in Serbia, using medical records of 89 infants diagnosed with CeD, evaluated the association between breastfeeding duration and disease severity. Breastfeeding significantly delayed CeD onset (OR 0.655). Moreover, breastfeeding duration was most significantly associated with CeD development, as compared to other covariates. Infants still breastfed at the time of first gluten introduction developed the disease later than infants that stopped breastfeeding prior to first gluten introduction [54]. It was also shown that exclusively breastfed children with CeD had a delayed onset, accompanied by a reduction in symptoms severity, as compared to children with CeD who had not been exclusively breastfed [55]. However, in a recent 10-year follow-up study among high-risk newborns, neither breastfeeding nor delayed introduction of gluten modified the risk of CeD [56].

Early-life gut microbiome and inflammatory bowel disease

IBD is an immune-mediated chronic disease state characterized by intestinal inflammation, which can be divided into two major types: ulcerative colitis and Crohn's disease [1]. IBD incidence is higher in developed countries, with developing countries showing rising incidence in the past years, emphasizing the central role of environmental factors in disease pathogenesis [57]. Interestingly enough, while immigration to a western country is associated with increased IBD risk, returning to the country of birth does not appear to ameliorate the risk [58], suggesting that some western-specific environmental factors confer lasting changes. Several genetic and environmental factors have been associated with IBD [59,60], some of which, including maternal exposures during pregnancy, early feeding (breastfeeding vs bottle feeding), drug exposure, and diet, were found to be associated with the gut microbiome [59].

The role of the gut microbiome in IBD garnered attention, as it was shown that the underlying uninflamed mucosa in IBD patients harbors a different microbiome than a healthy mucosa [61,62], raising the hypothesis that the gut microbiome affects IBD susceptibility. The gut microbiome was also found to have a causal role in IBD, as it was shown that the introduction of an IBD-associated *Escherichia coli* strain induced chronic colitis in a murine model [63].

In order to assess the effect of IBD on the gut microbiome beyond environmental and genetic factors, family-based studies were performed, comparing the gut microbiome between family relatives and twin pairs with IBD discordancy. The trials suggested that dysbiosis is in an independent association with IBD [64,65].

Associations between the gut microbiome and IBD were reported in multiple studies, describing microbiome composition differences between IBD patients and healthy controls. It was shown that the IBD-associated microbiome is not only characterized by reduced diversity [66], but exhibit specific changes in bacterial taxa, virome, and metabolome (Fig. 10.3) [67].

For example, *Faecalibacterium prausnitzii*, a bacterial species with reported antiinflammatory features [68], is often reduced in relative abundance in the stool of patients with Crohn's disease, although studies using mucosal biopsies suggested a more dynamic role for this organism in IBD pathophysiology [69]. As *F. prausnitzii* is a dominant butyrate-producing

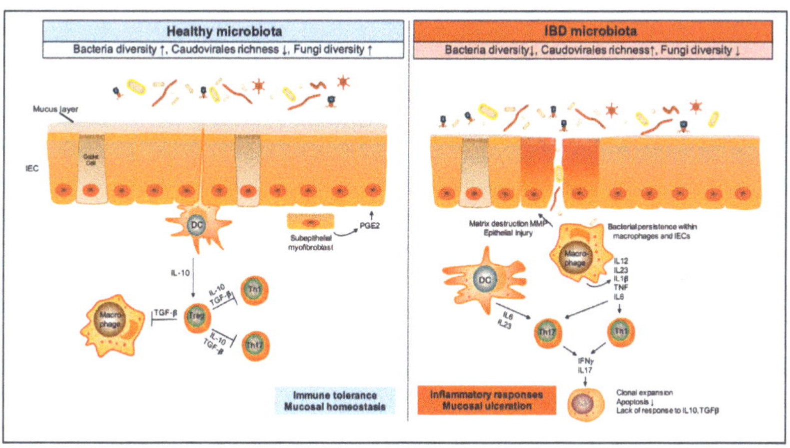

FIGURE 10.3 Gut microbiota alteration and immune responses in IBD. The gut induces a dysfunctional immune response, engaging Tregs, T-helper 1 (Th1), and Th17, that are implicated in IBD pathogenesis. During homeostasis, gut microbes induce an immune tolerance phenotype in the host, while in inflammatory conditions such as IBD, antigens from dysbiotic microbes activate Th1 and Th17 cells, resulting in tissue injury, decreased mucus layer, and microbial penetration and persistence in the intestinal tissues. This mucosal injury results in further uptake of microbial antigens, TLR ligands, and viable organisms that perpetuate the immune responses. *DC*, dendritic cell; *MMP*, matrix metalloproteinase; *TGF*, transforming growth factor. *Source: From T. Zuo and S.C. Ng, The gut microbiota in the pathogenesis and therapeutics of inflammatory bowel disease, Front Microbiol [Internet] 9 (September), 2018, 2247 [67].*

organism, its reduction suggested an accordant decrease in colonic butyrate levels, a hypothesis that was established in later reports [70,71]. Conversely, specific taxa from the Proteobacteria phylum, including the family Enterobacteriaceae and more specifically *E. coli* were increased in IBD patients [72].

Several environmental factors have been suggested to affect IBD susceptibility by the mediation of gut microbiome modification in early life.

Maternal exposures

It was demonstrated, using a murine model of colitis, that maternal antibiotic exposure is associated with increased IBD susceptibility in their offspring. Germ-free pregnant mice, inoculated with a postantibiotic gut microbiome, transmitted its perturbed microbiome to their offspring, causing markedly increased colitis [73]. A different study examined, using a mouse model, whether peripartum antibiotic administration affects IBD risk in the offspring. The investigators showed that offspring of a cefoperazone-exposed mother in the peripartum timeframe develop persistent gut dysbiosis into adulthood, affecting the immune system, and is associated with increased susceptibility to colitis [74]. Thus it could be suggested that exposure to antibiotic-induced maternal dysbiosis, before birth or peripartum, could affect the gut microbiome assemblage and immune programming, causing an increased IBD risk in genetically susceptible offspring.

Maternal smoking during pregnancy was also found to affect IBD, possibly by vertical transmission of the gut microbiome. In an observational study examining disease progression in Crohn's disease patients, maternal smoking during pregnancy was associated with an increased hospitalization risk, with an odds ratio of 1.75 for hospitalization during a 3-year follow up [75]. This warrants further research, as many possible confounders could have a role in this observed association.

Early-life antibiotics

Several studies have established the effect of antibiotics use on the human gut microbiome. A short-term study revealed that a 5-day course of oral ciprofloxacin induced a decrease in species diversity, evenness, and overall richness of the gut microbiome [76]. A study based on new-onset cases, before treatment, allowed to compare the gut microbiome in patients with Crohn's disease, with and without antibiotic exposure. The results indicated that the use of antibiotics amplifies microbial dysbiosis associated with IBD [62]. This evidence is in agreement with observational studies, demonstrating a link between early-life antibiotic use (within the first year of life) and later IBD development [77,78], with a gradual reduction of antibiotics-associated risk by the age of 15 at time of prescription [78]. A recent large observational study, following 827,339 children, aimed to estimate the association between antibiotic exposure during pregnancy or infancy and early onset IBD, defined as IBD diagnosis <6 years of age. The investigators found that exposure to antibiotics during pregnancy, but not in infancy, was associated with an increased risk of early IBD [79]. This finding calls for further exploration, assessing whether this effect is due to an antibiotic effect on the fetus during pregnancy, possibly mediated by the maternal gut microbiome, or, possibly, due to vertical transmission of maternal dysbiosis during labor.

Early feeding

The gut microbiome is strongly affected by early feeding type. While Bifidobacteria are the most dominant species in both breastfed and formula-fed infants [21,80], breastfed infants tend to harbor twofold higher Bifidobacteria levels than formula-fed infants [81] and to have lower bacterial diversity [82].

A recent review of 53 meta-analyses demonstrated a strong inverse association between breastfeeding and Crohn's disease (OR 0.71, 95% CI 0.59–0.85), ulcerative colitis (OR 0.78, 95% CI 0.67–0.91), and total IBD (OR 0.74, 95% CI 0.66–0.83) [82]. In a different study based on a diagnosed cohort of 1119 patients with Crohn's disease, breastfeeding was inversely associated with complicated disease [75]. The accumulated observational data suggest a possible association between a feeding modality-dependent gut microbiome configuration and future-life risk of IBD. However, this model is yet to be validated in a comprehensive study.

Other environmental factors

Several other environmental factors have been established to be associated with IBD risk, possibly by gut microbiome modification, including living near farm animals (OR, 0.46; 0.20–0.72), home sharing (OR, 0.49; 0.25–0.75), bed sharing (OR, 0.54; 0.43–0.65), having pets (OR, 0.77; 0.59–0.94), and having two or more siblings (OR, 0.93; 0.88–0.98) [75].

References

[1] Jameson JL, Fauci AS, Kasper DL, Hauser SL, Longo DL, Loscalzo J. Harrison's principles of internal medicine. 20th ed. New York: McGraw Hill Education; 2018.
[2] Strachan DP. Hay fever, hygiene, and household size. BMJ 1989;299(6710):1259–60.
[3] Bach J-F. The effect of infections on susceptibility to autoimmune and allergic diseases. N Engl J Med 2002;347(12):911–20.

[4] Lynch SV, Pedersen O. The human intestinal microbiome in health and disease. N Engl J Med 2016;375(24):2369–79.

[5] Spanogiannopoulos P, Bess EN, Carmody RN, Turnbaugh PJ. The microbial pharmacists within us: a metagenomic view of xenobiotic metabolism. Nat Rev Microbiol 2016;14(5):273–87.

[6] Pabst O, Mowat AM. Oral tolerance to food protein. Mucosal Immunol 2012;5 (3):232–9.

[7] Thaiss CA, Zmora N, Levy M, Elinav E. The microbiome and innate immunity. Nature 2016;535(7610):65–74.

[8] Belkaid Y, Hand TW. Role of the microbiota in immunity and inflammation. Cell 2014;157(1):121–41.

[9] Sommer F, Bäckhed F. The gut microbiota - masters of host development and physiology. Nat Rev Microbiol 2013;11(4):227–38.

[10] Verdu EF, Danska JS. Common ground: shared risk factors for type 1 diabetes and celiac disease. Nat Immunol 2018;19(7):685–95.

[11] Achenbach P, Bonifacio E, Koczwara K, Ziegler A-G. Natural history of type 1 diabetes. Diabetes 2005;54(Suppl 2):S25–31.

[12] Harjutsalo V, Sjöberg L, Tuomilehto J. Time trends in the incidence of type 1 diabetes in Finnish children: a cohort study. Lancet 2008;371(9626):1777–82.

[13] Pellegrini S, Sordi V, Bolla AM, et al. Duodenal mucosa of patients with type 1 diabetes shows distinctive inflammatory profile and microbiota. J Clin Endocrinol Metab 2017;102(5):1468–77.

[14] Gavin PG, Mullaney JA, Loo D, et al. Intestinal metaproteomics reveals host-microbiota interactions in subjects at risk for type 1 diabetes. Diabetes Care 2018;41(10):2178–86.

[15] Vatanen T, Franzosa EA, Schwager R, et al. The human gut microbiome in early-onset type 1 diabetes from the TEDDY study. Nature 2018;562(7728):589–94.

[16] Kostic AD, Gevers D, Siljander H, et al. The dynamics of the human infant gut microbiome in development and in progression toward type 1 diabetes. Cell Host Microbe 2015;17(2):260–73.

[17] Wen L, Ley RE, Volchkov PY, et al. Innate immunity and intestinal microbiota in the development of Type 1 diabetes. Nature 2008;455:1109.

[18] Zheng P, Li Z, Zhou Z. Gut microbiome in type 1 diabetes: a comprehensive review. Diabetes Metab Res Rev 2018;34(7):1–9.

[19] Boerner BP, Sarvetnick NE. Type 1 diabetes: role of intestinal microbiome in humans and mice. Ann N Y Acad Sci 2011;1243(1):103–18.

[20] Vatanen T, Kostic AD, D'Hennezel E, et al. Variation in microbiome LPS immunogenicity contributes to autoimmunity in humans. Cell 2016;165(4):842–53.

[21] Azad MB, Konya T, Maughan H, et al. Gut microbiota of healthy Canadian infants: profiles by mode of delivery and infant diet at 4 months. CMAJ 2013;185(5):385–94.

[22] Jakobsson HE, Abrahamsson TR, Jenmalm MC, et al. Decreased gut microbiota diversity, delayed Bacteroidetes colonisation and reduced Th1 responses in infants delivered by caesarean section. Gut 2014;63(4):559–66.

[23] Dominguez-Bello MG, Costello EK, Contreras M, et al. Delivery mode shapes the acquisition and structure of the initial microbiota across multiple body habitats in newborns. Proc Natl Acad Sci U S A 2010;107(26):11971–5.

[24] Pandey PK, Verma P, Kumar H, Bavdekar A, Patole MS, Shouche YS. Comparative analysis of fecal microflora of healthy full-term Indian infants born with different methods of delivery (vaginal vs cesarean): *Acinetobacter* sp. prevalence in vaginally born infants. J Biosci 2012;37(6):989–98.

[25] Bonifacio E, Warncke K, Winkler C, Wallner M, Ziegler A-G. Cesarean section and interferon-induced helicase gene polymorphisms combine to increase childhood type 1 diabetes risk. Diabetes 2011;60:3300–6.

[26] Lee H-Y, Lu C-L, Chen H-F, Su H-F, Li C-Y. Perinatal and childhood risk factors for early-onset type 1 diabetes: a population-based case-control study in Taiwan. Eur J Public Health 2015;25(6):1024–9.

[27] Black M, Bhattacharya S, Philip S, Norman JE, McLernon DJ. Planned cesarean delivery at term and adverse outcomes in childhood health. JAMA 2015;314(21):2271.

[28] Clausen TD, Bergholt T, Eriksson F, Rasmussen S, Keiding N, Løkkegaard EC. Prelabor cesarean section and risk of childhood type 1 diabetes. Epidemiology 2016;27(4):547–55.

[29] WHO/UNICEF. Global strategy for infant and young child feeding [Internet]. Geneva: WHO/UNICEF; 2003.

[30] Mosca F, Giannì ML. Human milk: composition and health benefits. La Pediatr Medica e Chir 2017;39(2).

[31] Rodríguez JM. The origin of human milk bacteria: is there a bacterial entero-mammary pathway during late pregnancy and lactation? Adv Nutr 2014;5 (6):779–84.

[32] Davis JCC, Lewis ZT, Krishnan S, et al. Growth and morbidity of gambian infants are influenced by maternal milk oligosaccharides and infant gut microbiota. Sci Rep 2017;7(1):40466.

[33] Xiao L, van't Land B, Engen PA, et al. Human milk oligosaccharides protect against the development of autoimmune diabetes in NOD-mice. Sci Rep 2018;8(1):3829.

[34] Lund-Blix NA, Dydensborg Sander S, Størdal K, et al. Infant feeding and risk of type 1 diabetes in two large Scandinavian birth cohorts. Diabetes Care 2017;40(7):920–7.

[35] Cardwell CR, Stene LC, Ludvigsson J, et al. Breast-feeding and childhood-onset type 1 diabetes: a pooled analysis of individual participant data from 43 observational studies. Diabetes Care 2012;35(11):2215–25.

[36] Liu Y, Qin S, Song Y, et al. The perturbation of infant gut microbiota caused by cesarean delivery is partially restored by exclusive breastfeeding. Front Microbiol 2019;10:598.

[37] Frederiksen B, Kroehl M, Lamb MM, et al. Infant exposures and development of type 1 diabetes mellitus: the Diabetes Autoimmunity Study in the Young (DAISY). JAMA Pediatr 2013;167(9):808–15.

[38] Zhang X-S, Li J, Krautkramer KA, et al. Antibiotic-induced acceleration of type 1 diabetes alters maturation of innate intestinal immunity. Elife 2018;7:1–37.

[39] Boursi B, Mamtani R, Haynes K, Yang Y-X. The effect of past antibiotic exposure on diabetes risk. Eur J Endocrinol 2015;172(6):639–48.

[40] Clausen TD, Bergholt T, Bouaziz O, et al. Broad-spectrum antibiotic treatment and subsequent childhood type 1 diabetes: a Nationwide Danish Cohort Study. PLoS One 2016;11(8):e0161654.

[41] Trynka G, Wijmenga C, van Heel DA. A genetic perspective on coeliac disease. Trends Mol Med 2010;16(11):537–50.

[42] Trynka G, Hunt KA, Bockett NA, et al. Dense genotyping identifies and localizes multiple common and rare variant association signals in celiac disease. Nat Genet 2011;43(12):1193–201.

[43] Collado M, Donat E, Ribes-Koninckx C, Calabuig M, Sanz Y. Imbalances in faecal and duodenal Bifidobacterium species composition in active and non-active coeliac disease. BMC Microbiol 2008;8(1):232.

[44] Collado MC, Donat E, Ribes-Koninckx C, Calabuig M, Sanz Y. Specific duodenal and faecal bacterial groups associated with paediatric coeliac disease. J Clin Pathol 2009;62(3):264–9.

[45] Di Cagno R, De Angelis M, De Pasquale I, et al. Duodenal and faecal microbiota of celiac children: molecular, phenotype and metabolome characterization. BMC Microbiol 2011;11(1):219.

[46] Wacklin P, Kaukinen K, Tuovinen E, et al. The duodenal microbiota composition of adult celiac disease patients is associated with the clinical manifestation of the disease. Inflamm Bowel Dis 2013;19(5):934–41.
[47] Sánchez E, Donat E, Ribes-Koninckx C, Fernández-Murga ML, Sanz Y. Duodenal-mucosal bacteria associated with celiac disease in children. Appl Environ Microbiol 2013;79(18):5472–9.
[48] Tlaskalová-Hogenová H, Štěpánková R, Hudcovic T, et al. Commensal bacteria (normal microflora), mucosal immunity and chronic inflammatory and autoimmune diseases. Immunol Lett 2004;93(2–3):97–108.
[49] Grönlund MM, Arvilommi H, Kero P, Lehtonen OP, Isolauri E. Importance of intestinal colonisation in the maturation of humoral immunity in early infancy: a prospective follow up study of healthy infants aged 0-6 months. Arch Dis Child Fetal Neonatal Ed 2000;83(3):F186–92.
[50] Rautava S, Arvilommi H, Isolauri E. Specific probiotics in enhancing maturation of iga responses in formula-fed infants. Pediatr Res 2006;60(2):221–4.
[51] Olivares M, Neef A, Castillejo G, et al. The HLA-DQ2 genotype selects for early intestinal microbiota composition in infants at high risk of developing coeliac disease. Gut 2015;64(3):406–17.
[52] Galipeau HJ, McCarville JL, Huebener S, et al. Intestinal microbiota modulates gluten-induced immunopathology in humanized mice. Am J Pathol 2015;185(11):2969–82.
[53] Dydensborg Sander S, Nybo Andersen AM, Murray JA, Karlstad Ø, Husby S, Størdal K. Association between antibiotics in the first year of life and celiac disease. Gastroenterology 2019;156(8):2217–29.
[54] Radlovic NP, Mladenovic MM, Lekovic ZM, Stojsic ZM, Radlovic VN. Influence of early feeding practices on celiac disease in infants. Croat Med J 2010;51 (5):417–22.
[55] D'Amico MA, Holmes J, Stavropoulos SN, et al. Presentation of pediatric celiac disease in the united states: prominent effect of breastfeeding. Clin Pediatr (Phila) 2005;44(3):249–58.
[56] Lionetti E, Castellaneta S, Francavilla R, et al. Introduction of gluten, HLA Status, and the risk of celiac disease in children. N Engl J Med 2014;371(14):1295–303.
[57] Ng SC, Bernstein CN, Vatn MH, et al. Geographical variability and environmental risk factors in inflammatory bowel disease. Gut 2013;62(4):630–49. http://dx.doi.org/10.1136/gutjnl-2012-303661.
[58] Barreiro-de Acosta M, Alvarez Castro A, Souto R, Iglesias M, Lorenzo A, Dominguez-Muñoz JE. Emigration to western industrialized countries: A risk factor for developing inflammatory bowel disease. J Crohn's Colitis [Internet] 2011;5 (6):566–9. http://www.ncbi.nlm.nih.gov/pubmed/22115376.
[59] Ananthakrishnan AN, Bernstein CN, Iliopoulos D, et al. Environmental triggers in IBD: a review of progress and evidence. Nat Rev Gastroenterol Hepatol [Internet] 2017;15(1):39–49. http://www.nature.com/articles/nrgastro.2017.136.
[60] McGovern DPB, Kugathasan S, Cho JH. Genetics of inflammatory bowel diseases. Gastroenterology [Internet] 2015;149(5):1163–1176.e2. https://linkinghub.elsevier.com/retrieve/pii/S0016508515010926.
[61] Kostic A D, Xavier RJ, Gevers D. The microbiome in inflammatory bowel disease: Current status and the future ahead. Gastroenterology 146(6), 2014,1489–1499.
[62] Gevers D, Kugathasan S, Denson LA, et al. The treatment-naive microbiome in new-onset Crohn's disease. Cell Host Microbe [Internet] 2014;15(3):382–92. http://www.ncbi.nlm.nih.gov/pubmed/24629344.
[63] Carvalho FA, Koren O, Goodrich JK, et al. Transient inability to manage proteobacteria promotes chronic gut inflammation in TLR5-deficient mice. Cell Host Microbe [Internet] 2012;12:139–52. http://dx.doi.org/10.1016/j.chom.2012.07.004.

[64] Willing BP, Dicksved J, Halfvarson J, et al. A pyrosequencing study in twins shows that gastrointestinal microbial profiles vary with inflammatory bowel disease phenotypes. Gastroenterology [Internet] 2010;139(6):1844–1854.e1. http://www.ncbi.nlm.nih.gov/pubmed/20816835.

[65] Joossens M, Huys G, Cnockaert M, et al. Dysbiosis of the faecal microbiota in patients with Crohn's disease and their unaffected relatives. Gut [Internet] 2011;60(5):631–7. http://www.ncbi.nlm.nih.gov/pubmed/21209126.

[66] Manichanh C, Rigottier-Gois L, Bonnaud E, et al. Reduced diversity of faecal microbiota in Crohn's disease revealed by a metagenomic approach. Gut [Internet] 2006;55(2):205–11. http://www.ncbi.nlm.nih.gov/pubmed/16188921.

[67] Zuo T, Ng SC. The gut microbiota in the pathogenesis and therapeutics of inflammatory bowel disease. Front Microbiol [Internet] 2018;9(September):2247. http://www.ncbi.nlm.nih.gov/pubmed/30319571%0Ahttp://www.pubmedcentral.nih.gov/articlerender.fcgi?artid = PMC6167487.

[68] Sokol H, Pigneur B, Watterlot L, et al. *Faecalibacterium prausnitzii* is an anti-inflammatory commensal bacterium identified by gut microbiota analysis of Crohn disease patients. Proc Natl Acad Sci [Internet] 2008;105(43):16731–6. http://www.ncbi.nlm.nih.gov/pubmed/18936492.

[69] Hansen R, Russell RK, Reiff C, et al. Microbiota of de-novo pediatric IBD: increased *Faecalibacterium prausnitzii* and reduced bacterial diversity in Crohn's but not in ulcerative colitis. Am J Gastroenterol [Internet] 2012;107(12):1913–22. http://insights.ovid.com/crossref?an = 00000434-201212000-00026.

[70] Parada Venegas D, De la Fuente MK, Landskron G, et al. Short chain fatty acids (SCFAs)-mediated gut epithelial and immune regulation and its relevance for inflammatory bowel diseases. Front Immunol [Internet] 2019;10:277. http://www.ncbi.nlm.nih.gov/pubmed/30915065.

[71] Erickson AR, Cantarel BL, Lamendella R, et al. Integrated metagenomics/metaproteomics reveals human host-microbiota signatures of Crohn's Disease. PLoS One [Internet] 2012;7(11):e49138. http://www.ncbi.nlm.nih.gov/pubmed/23209564.

[72] Baumgart M, Dogan B, Rishniw M, et al. Culture independent analysis of ileal mucosa reveals a selective increase in invasive *Escherichia coli* of novel phylogeny relative to depletion of Clostridiales in Crohn's disease involving the ileum. ISME J [Internet] 2007;1(5):403–18. http://www.ncbi.nlm.nih.gov/pubmed/18043660.

[73] Schulfer AF, Battaglia T, Alvarez Y, et al. Intergenerational transfer of antibiotic-perturbed microbiota enhances colitis in susceptible mice. Nat Microbiol [Internet] 2018;3(2):234–42. http://www.nature.com/articles/s41564-017-0075-5.

[74] Miyoshi J, Bobe AM, Miyoshi S, et al. Peripartum antibiotics promote gut dysbiosis, loss of immune tolerance, and inflammatory bowel disease in genetically prone offspring. Cell Rep [Internet] 2017;20(2):491–504. http://www.ncbi.nlm.nih.gov/pubmed/28700948.

[75] Lindoso L, Mondal K, Venkateswaran S, et al. The effect of early-life environmental exposures on disease phenotype and clinical course of Crohn's disease in children. Am J Gastroenterol [Internet] 2018;113(10):1524–9. http://dx.doi.org/10.1038/s41395-018-0239-9.

[76] Dethlefsen L, Relman DA. Incomplete recovery and individualized responses of the human distal gut microbiota to repeated antibiotic perturbation. Proc Natl Acad Sci [Internet] 2011;108(Supplement_1):4554–61. http://www.ncbi.nlm.nih.gov/pubmed/20847294.

[77] Shaw SY, Blanchard JF, Bernstein CN. Association between the use of antibiotics in the first year of life and pediatric inflammatory bowel disease. Am J Gastroenterol [Internet] 2010;105(12):2687–92. http://insights.ovid.com/crossref?an = 00000434-201012000-00029.

[78] Kronman MP, Zaoutis TE, Haynes K, Feng R, Coffin SE. Antibiotic exposure and IBD development among children: a population-based cohort study. Pediatrics [Internet] 2012;130(4):e794–803. http://www.ncbi.nlm.nih.gov/pubmed/23008454.
[79] Örtqvist AK, Lundholm C, Halfvarson J, Ludvigsson JF, Almqvist C. Fetal and early life antibiotics exposure and very early onset inflammatory bowel disease: a population-based study. Gut [Internet] 2019;68(2):218–25. http://www.ncbi.nlm.nih.gov/pubmed/29321166.
[80] Guaraldi F, Salvatori G. Effect of breast and formula feeding on gut microbiota shaping in newborns. Front Cell Infect Microbiol [Internet] 2012;2. http://www.frontiersin.org.
[81] Bezirtzoglou E, Tsiotsias A, Welling GW. Microbiota profile in feces of breast- and formula-fed newborns by using fluorescence in situ hybridization (FISH). Anaerobe [Internet] 2011;17(6):478–82. http://www.ncbi.nlm.nih.gov/pubmed/21497661.
[82] Piovani D, Danese S, Peyrin-Biroulet L, Nikolopoulos GK. Environmental risk factors for inflammatory bowel diseases: an umbrella review of meta-analyses. Gastroenterology 2019;157(3):647–659.e4. https://doi.org/10.1053/j.gastro.2019.04.016.

SECTION IV

Modifying early microbial contact

CHAPTER

11

Postbiotics: defining the impact of inactivated microbes and their metabolites on promotion of health

Collado Maria Carmen[1], *Salminen Seppo*[2] *and Gabriel Vinderola*[3]

[1]Institute of Agrochemistry and Food Technology-National Research Council, Valencia, Spain, [2]Functional Foods Forum, Faculty of Medicine, University of Turku, Turku, Finland, [3]Dairy Products Institute (INLAIN, UNL-CONICET), Faculty of Chemical Engineering, National University of Litoral, Santa Fe, Argentina

Introduction

Probiotics, prebiotics, and synbiotics have been a part of the human diet for a long time, and several definitions have evolved to describe them. Postbiotics, a novel term that comprises nonviable microbes and their metabolites that are able to confer a health benefit when administered in adequate amounts, have been used in fermented foods. However, as their significance as bioactive compounds is increasingly being acknowledged, their characterization and nomenclature need to be defined in the same way as other members of the "biotics" family.

The Human Microbiome in Early Life
DOI: https://doi.org/10.1016/B978-0-12-818097-6.00011-0

Existing definitions and the need of consensus over the definition of postbiotics

Several position papers exist on the terminology used for probiotics, prebiotics, and synbiotics. Probiotics are described as "live microorganisms that, when administered in adequate amounts, confer a health benefit on the host" [1] and Prebiotics as "substrates that are selectively utilized by host microorganisms and that confer a health benefit" [2]. Synbiotics have recently been defined as "a mixture comprising live microorganisms and substrates selectively utilized by host microbes, which confer a health benefit on the host" [3].

The mechanisms explaining the beneficial health effects of probiotics, prebiotics, and synbiotics are not completely understood. The benefits are thought to be facilitated by interactions between specific probiotics, prebiotics, or a combination thereof as synbiotics via the intestinal microbiome and immunological system. The interactions can involve intestinal microbial changes and direct interfaces with the intestinal epithelial or immune cells. Thus definitions exist that cover viable microbes and the substrates that are available to expedite the beneficial effects, but a definition is lacking that covers the health benefits of nonviable microbial cells and the food fermentation metabolites produced by them. Tentatively, they are referred to as postbiotics as they describe microbes, mainly bacteria, and the products of fermentation, after inactivation. However, health benefits have been reported for inactivated microbes alone, and earlier studies have focused on relatively harmless bacteria such as lactic acid bacteria and bifidobacteria. Knowledge of the related health benefits continues to increase rapidly [4,5].

The need for a postbiotic definition has arisen from both the research findings on novel probiotics (i.e., novel genera, such as *Bacteroides* and *Clostridium* or groups, such as fructophilic lactic acid bacteria), and the regulatory processes some of these have undergone during a novel food safety assessment by the European Union. Following the safety evaluation, novel probiotics, prebiotics, and synbiotics need to be further evaluated for potential health benefits in accordance with their definition. Background material for the definition should be based on inactivated bacteria and the tightening of European Union legislation on health claims because the emphasis on probiotics, prebiotics, and synbiotics is the same. Several studies recognize the fact that several mechanisms mediate the health benefits without the requirement of the viability of specific probiotic bacteria. It is important to determine how to define these components, while recognizing the differences between probiotics, prebiotics, and synbiotics. To this end, a new definition needs to be created to denote how nonviable microbial cells, their metabolites, microbial fractions, or even cell lysates can confer physiological benefits on the host by providing additional bioactivity, provided that they are administered in adequate amounts.

In this chapter, an overview is provided of a suggested postbiotic definition, and existing evidence of proven health benefits and possible mechanisms responsible for the postbiotic effect in the promotion of health are deliberated. Information is also provided on the ability of postbiotics to modify human biological responses. Inactivation methods and the future potential of their application in foods and feeds are briefly discussed.

Based on a comprehensive view of studies that has evaluated inactivated probiotics and fermented foods in the absence of live bacteria, it is proposed that the definition for postbiotics could be: "Compounds produced by microorganisms, released from food components or microbial constituents, including nonviable cells that, when administered in adequate amounts, promote health and well-being" [6] (Fig. 11.1).

How to identify postbiotics and the roles of different components

Food fermentation has a significant impact on the stability and health effects of different foods, especially milk and vegetable products. Lactic

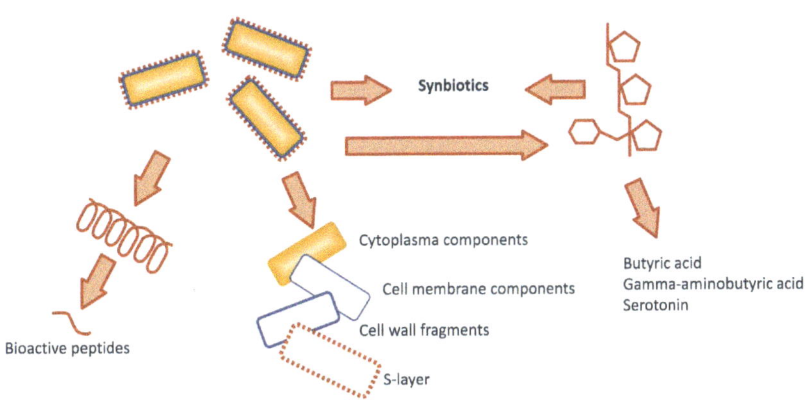

FIGURE 11.1 Graphic presentation of possible postbiotics [6]. Postbiotics could be produced by the fermentative action of microbes, probiotics or other fermentative microbes, or released from foods components by the enzymatic activity of microbes (e.g., the release of the antihypertensive peptide Val-Pro-Pro from milk proteins during milk fermentation).

acid bacteria produce a range of metabolites, some of which have been associated with health-promoting properties. The most notable of these are short-chain fatty acids (SCFAs) and bioactive peptides, which are released from milk proteins. Bioactive sequences released in fermented milks, together with the capacity of the cell wall of lactic acid bacteria to activate the mucosal immune system, have been associated with a decrease in blood pressure, for example.

Until now, the majority of postbiotics is derived from *Lactobacillus* and *Bifidobacterium* strains, which are commonly used as probiotics. However, recently, *Akkermansia*, *Streptococcus*, and *Faecalibacterium* species, as well as yeasts, have also been reported to be the potential sources of future postbiotics [7,8].

In general, fermented foods are a source of postbiotics as they may contain nonviable microbes and the products of fermentation. Postbiotics can also be obtained using cell disruption techniques, which include the application of heat, vibrations, or high pressure. However, other treatments such as ultraviolet light treatment and radiation may be used.

A large number of biological effects are associated with inactivated probiotics, and these are summarized in Table 11.1. Such effects constitute an important option for the enhancement of foods and feeds as a way of providing health benefits to the host. Although health benefits are associated with inactivated probiotics, data demonstrate that these effects are also linked to their metabolites or a combination of them.

Data on potential postbiotic components and their classification

Examples of potential postbiotic components are now provided, but consensus is currently lacking in this regard, and the need for a comprehensive definition is clear.

Case: *Bacteroides xylanisolvens*

Bacteroides xylanisolvens is a novel food ingredient that was introduced in the European Union regulatory processes following an evaluation by the European Food Safety Authority [9]. The food product produced with this microbe is a low-fat or skimmed-milk product and is manufactured using *B. xylanisolvens* DSM 23964 as a starter culture for milk fermentation. Thereafter, the product is heat treated and after this process, it does not contain viable *B. xylanisolvens* cells. The final product contains both inactivated bacteria and their metabolites [9]. Therefore, it could feasibly be classified as a postbiotic product.

TABLE 11.1 Potential postbiotics (depending on the definition) and their effects.

Examples	Activity linked to a potential health benefit
Heat-killed bacteria, UV-inactivated cells	Immunomodulation, antigastroenteritis
DNA, RNA, and proteins/peptides	Immunotherapy
S-layer proteins	Immunomodulation
Teichoic acids	Immunomodulation
Peptidoglycan-derived muropeptides	Immunomodulation
Polysaccharides	Immunomodulation
SCFAs (butyrate, acetate, lactate)	Antiinflammatory
Lactic acid	Antiinflammatory
Exopolysaccharides	Immunomodulation
Vitamins	Neuroprotection
Peptides	Immunomodulation
Gamma-aminobutyric acid (GABA)	Mood regulator
Bacteriocins	Antimicrobial activity
Antimicrobial peptides	Antimicrobial activity
Enzymes	Antimicrobial activity
Surfactants	Dermatological care
Antioxidants	Cancer prevention
Peptides released from milk proteins	Control of blood pressure
GOS produced by transgalactosylation	Prebiotic effect

Adapted from Collado MC, Vinderola G, Salminen S. Postbiotics: facts and open questions. A position paper on the need for a consensus definition. Benef Microbes 2019;10(7):711–19.

Case: Sorbiflore

Sobiflore, a microbial supplement product for animal feed, consists of *Lactobacillus rhamnosus* (CNCM-I-3698) and *L. farciminis* (CNCM-I-3699), both at a guaranteed concentration of 5×10^7 viable fluorescent units (VFUs) per gram (Section 2.1), giving a total of 1×10^{11} VFUs kg^{-1}. By using standard agar plating, it can be ensured that the product does not contain viable bacteria. Thus it is also a potential postbiotic. However, the use of a special method developed by the manufacturer ensures that the lactobacilli in the product are viable at the level of 1×10^{11} VFU/kg, which means that it can also be defined as a probiotic product (European Food Safety Authority 2009) [10].

Studies that support health effects

There is a scarcity of research on the classification of products as postbiotics. Accordingly, consensus regarding the definition of which products are considered to be postbiotics is required before classification can take place.

Postbiotics are attracting considerable scientific interest as a consequence of their potential benefits to human health [6]. Despite the lack of a consensus definition and limited available data, several postbiotics have been recognized to provide benefits through the long-standing and widely adopted consumption of fermented products [11,12]. The effects of postbiotics are mediated through interactions between the host and the microbial products; it is thought that these interactions trigger and activate the immune system and other systems via a systemic response [13]. Reports showing the potential for postbiotic molecular mechanisms have been described in in vitro studies, but the availability of human data remains limited [14].

The fermentation process also produces a range of secondary metabolites or peptides derived from the hydrolysis of food proteins that may exert effects on the human gut, as well as potential cell debris and inactivated bacteria. Other postbiotic effects are related to the interaction and inhibition of pathogens, viruses, and toxins; immunomodulation through interactions with toll-like receptors; antiinflammatory activities that modify the production of cytokines; and alterations to the gut microbiome. Modulation may take place by influencing intestinal homeostasis and the barrier function, controlling permeability and preventing bacterial translocation [14].

A recent systematic review reported the potential health-promoting benefits of fermented milk postbiotics, which included antioxidant activity due to peptides, folates, gamma-aminobutyric acid (GABA), and others; antihypertensive activity due the presence of angiotensin-converting enzyme inhibitory peptides and GABA; an increase in vitamin content; and an improvement in the digestibility and bioavailability of proteins. Following fermentation, some bacteria, primarily lactic acid bacteria, produce SCFAs, lactate, and other compounds that have been linked to antinflammatory and anticarcinogenic activities [15,16]. A recent in vitro study demonstrated the potential cytotoxicity impact of postbiotic metabolites produced by Lactobacillus *plantarum* strains on cancer cells [17]. Other studies have reported that heat-killed *Enterococcus faecium* BGPAS1-3 modulated tight junction protein expression and the immune function in Caco-2 cell models infected with *Listeria monocytogenes* [18].

In an animal model, a *L. rhamnosus* GG (LGG) supernatant was reported to accelerate immune maturation in neonatal rodents and prevented oral *Escherichia coli* K1 infection [19]. In addition, a postbiotic derived from

L. plantarum was demonstrated to improve ruminal epithelium growth, immune status, and intestinal barrier function in postweaning lambs [20].

Other studies have shown that heat-killed *Lactobacillus* strains had an impact on human microbiome composition and behavior [21]. Interestingly, an inactivated *Saccharomyces cerevisiae* strain was shown to have a positive effect on mice vagina as an antifungal and antiinflammatory agent in the treatment of vaginal candidiasis [22].

However, less is known on the effects of postbiotics on human health. It has been suggested that postbiotics can be used to promote health and decrease the symptoms of specific conditions, such as colic in infants or atopic dermatitis and diarrhea in adults [23]. It has also been reported that the administration of fermented milk that contains heat-inactivated *Lactobacillus gasseri* CP2305 (1×10^{10} cells/container) was able to regulate the intestinal function of patients with a tendency for constipation [24]. A specific heat-inactivated strain of *L. gasseri* was shown to ameliorate stress-related symptoms and improve the sleep quality of 32 healthy students [25]. A potentially beneficial effect of partially fermented infant formula on infant health has also been reported [26]. However, a limited number of trials have been performed ($n = 5$, on 1326 infants) [27].

Interestingly, one study demonstrated how fermented infant formula, supplemented with short-chain galacto-oligosaccharides and long-chain fructo-oligosaccharides, reduced overall crying time, lowered the incidence of infant colic, and resulted in a stool-softening effect in healthy term infants [28]. Further evidence is needed to determine the potential effects of postbiotics on infant health outcomes.

So far, most known postbiotics have been derived from well-known and commonly used probiotic strains, belonging to the lactic acid bacteria or bifidobacteria groups. However, other new potential bacterial species, such as *Akkermansia muciniphila, Eubacterium hallii, Streptococcus,* and *Faecalibacterium* spp., some fructophilic lactic acid bacteria and specific yeasts, have recently been reported to be potential sources of postbiotics [8,29]. For example, a recent study demonstrated the impact of inactivated *Bacillus coagulans* on endocrine, inflammatory, and performance responses in 16 male soldiers during training [30].

There is considerable interest in the potential probiotic and postbiotic effects of *A. muciniphila*. A recent mice model study suggests that *A. muciniphila*-derived extracellular vesicles could be used as a mucosal delivery vector to facilitate a reduction in obesity [31]. Furthermore, a recent human pilot study described the effect of pasteurized *A. muciniphila* on the levels of several blood markers for liver dysfunction and inflammation, and the effect does not appear to require the modification of microbiome in obese subjects [32]. Pasteurized *A. muciniphila* was also reported to improve insulin sensitivity, reduce insulinemia and plasma total cholesterol, decrease body weight, and reduce fat mass and hip circumference, compared to placebo [32].

Examples of postbiotics and their applications

Postbiotics constitute a heterogenic family of components that are diverse in nature and range from single molecules to an entire mixture of fermentation products, which include nonviable microbes and their cellular fragments. The molecules may be metabolites derived from cytoplasmic metabolic pathways, such as bacteriocins, exopolysaccharides, organic acids, enzymes, or other bioactive components or may be molecules released from food components by microbial fermentative action; bioactive peptides released from milk proteins are an example of this. Of the next-generation beneficial microbes that have been identified, *A. muciniphila* offers promises in the near future. Indeed, *A. muciniphila* administration in animal studies is inversely associated with obesity, diabetes mellitus, cardiometabolic disease, and low-grade inflammation [33]. This microbe or its membrane components could be a future postbiotic. Specifically, a purified membrane protein from *A. muciniphila* or the pasteurized bacterium improved the metabolism of obese and diabetic mice [34]. In addition, it was recently reported that supplementation with *A. muciniphila* in overweight and obese human volunteers reduced the levels of relevant blood markers for liver dysfunction and inflammation, thereby improving several metabolic parameters [32].

L. plantarum Tensia DSM 21380 is able to produce antihypertensive peptides, and their use is regulated by a US patent (No. 2011/0177198 A1). The effects of a food spread that contains tripeptides (isoleucine–proline–proline and valine–proline–proline) were studied in subjects with mild hypertension and elevated LDL cholesterol. The subjects consumed 20 g of the spread daily, which contained 4.2 mg of milk peptides. Beneficial effects were demonstrated in relation to two major cardiovascular risk factors (blood pressure and plasma lipids) in hypertensive, dyslipidemic subjects [35].

Besides their multifunctional properties, other potential postbiotics produced by lactic acid bacteria, such as exopolysaccharides, could also contribute to enhancing human health as prebiotics through antitumor, antiulcer, immunomodulatory, and cholesterol-lowering activities. However, EFSA and US Food and Drug Administration health claims for lactic acid bacteria exopolysaccarides have not yet been approved. For instance, dextran, an EPS that is produced by *Leuconostoc mesenteroides*, can be applied to confectionary, baked products, ice cream, and frozen or dried foods because of its capacity to confer stability and to improve the texture of the products [36].

In summary, so far, this chapter has described postbiotics based on membrane proteins, nonviable cells, exopolysaccharides, or peptides, that is to say, single molecules or microbes that are easily visualized. However, postbiotics may not be a completely defined mixture of components; for example, they could be the products left in a given culture medium after microbial fermentation. A mixture of thermophilic lactobacilli was used to ferment milk under pH-controlled conditions. The supernatant of the fermentation was administered to mice for 2 weeks, and the mice were then challenged with a single infective dose of *Salmonella typhimurium*. Significantly higher survival was observed in mice fed with the postbiotic mixture from the fermentation compared to that for the control animals [37]. An infant formula that contains milk fermented by *Bifidobacterium breve* C50 and *Streptococcus thermophilus* 065 was recently developed. The product was spray dried after fermentation; therefore it contained nonviable microbes together with the fermentation products. Thus, it could be regarded as a complex mix of postbiotics. In particular, a specific postbiotic was identified in the product, 3′-galactosyllactose (3′-GL), which is structurally similar to one of the human milk oligosaccharides (HMOs) found in breast milk. This fermented infant formula was shown to be well tolerated and also supported adequate infant growth [38].

Minimum criteria for a definition of postbiotics and how to assess their safety and health benefits

The minimum criteria for a definition of postbiotics should include (1) accurate and detailed identification of the bacteria used, (2) the chemical composition of the ferment or fermented food (cell free or with the cells included) to be used and (3) the identification of major metabolites that could be accumulated or reduced for specific properties.

Safety

Performing a safety assessment is an essential step in the development of postbiotics. In addition to a long history of the safe use of lactic acid bacteria, the oral administration of currently used probiotics in fermented foods or as food supplements in a viable or nonviable form has been well tolerated and proven to be safe. In the absence of viable active bacteria, the risk of bacterial translocation to extraintestinal sites is extremely small, and, therefore, postbiotics could be used in highly vulnerable populations, such as preterm infants or immunosuppressed persons. This issue constitutes an important and desirable safety characteristic of postbiotics.

Scientific evidence is accumulating on the beneficial effects that can be obtained from a whole cell-free fraction of fermented milk [39] without the need for the application of sophisticated and high-cost laboratory and industrial technologies to identify, separate, and purify each bioactive component. However, studying the multifunctional properties remains important. It is recognized that peptides can be more reactive than proteins because they have a lower molecular weight. Peptides present in cell-free fractions may react with other components, such as residual sugars or exopolysaccharides.

A most recent example of new postbiotics is a nonviable, heat inactivated *Bifidobacterium bifidum* MIMBb75 (SYN_HI001) which was reported to alleviate symptoms of irritable bowel disease in real life settings [40].

Conclusion

Health effects similar to those described for pro-, pre-, and synbiotics have been observed for nonviable microbes, their cell components, and their metabolites. It is suggested that the term "postbiotics" should be applied to these modulators until a definition on which consensus can be reached becomes available. Determining a proper definition for postbiotics will expedite their development, marketing, quality control, and regulation. The need for this is motivated by the potential safety of postbiotics, evidence of their stability, and increasing substantiation of their health-promoting capacities.

References

[1] Hill C, Guarner F, Reid G, Gibson GR, Merenstein DJ, Pot B, et al. Expert consensus document. The International Scientific Association for Probiotics and Prebiotics consensus statement on the scope and appropriate use of the term probiotic. Nat Rev Gastroenterol Hepatol 2014;11(8):506–14.

[2] Gibson GR, Hutkins R, Sanders ME, Prescott SL, Reimer RA, Salminen SJ, et al. Expert consensus document: The International Scientific Association for Probiotics and Prebiotics (ISAPP) consensus statement on the definition and scope of prebiotics. Nat Rev Gastroenterol Hepatol 2017;14(8):491–502.

[3] Swanson KS, Gibson GR, Sanders ME, Reid G, Verbeke K, Delzenne NM, et al. Expert consensus document: The International Scientific Association for Probiotics and Prebiotics (ISAPP) consensus statement on the definition and scope of synbiotics. Nat Rev Gastroenterol Nutr 2020; In press.

[4] Ouwehand A, Salminen S. The health effects of cultured milk products with viable and non-viable bacteria. Int Dairy J 1998;8(9):749–58.

[5] Lahtinen SJ. Probiotic viability - does it matter? Microb Ecol Health Dis 2012;23. Available from: https://doi.org/10.3402/mehd.v23i0.18567.

[6] Collado MC, Vinderola G, Salminen S. Postbiotics: facts and open questions. A position paper on the need for a consensus definition. Benef Microbes 2019;10(7):711–19.

[7] Aguilar-Toalá JE, Hall FG, Urbizo-Reyes UC, Garcia HS, Vallejo-Cordoba B, González-Córdova AF, et al. In silico prediction and in vitro assessment of multifunctional properties of postbiotics obtained from two probiotic bacteria. Probiotics Antimicrob Proteins 2019. Available from: https://doi.org/10.1007/s12602-019-09568-z.

[8] Tsilingiri K, Rescigno M. Postbiotics: what else? Benef Microbes 2013;4(1):101–7.

[9] Scientific opinion on the safety of 'heat-treated milk products fermented with *Bacteroides xylanisolvens* DSM 23964' as a novel food. EFSA Panel on Dietetic Products, Nutrition and Allergies (NDA). EFSA J 2015;13(1):3956.

[10] Scientific Opinion of the Panel on Additives and Products or Substances used in Animal Feed (FEEDAP) on a request from the European Commission on the safety and efficacy of the product Sorbiflore, a preparation of *Lactobacillus rhamnosus* and *Lactobacillus farciminis*, as feed additive for piglets. EFSA J 2008;771:1-13.

[11] Melini F, Melini V, Luziatelli F, Ficca AG, Ruzzi M. Health-promoting components in fermented foods: an up-to-date systematic review. Nutrients 2019;11(5):E1189.

[12] Granier A, Goulet O, Hoarau C. Fermentation products: immunological effects on human and animal models. Pediatr Res 2013;74(2):238–44.

[13] Gosálbez L, Ramón D. Probiotics in transition: novel strategies. Trends Biotechnol 2015;33(4):195–6.

[14] de Almada CM, Almada CN, Martinez RCR, Sant'Ana AS. Paraprobiotics: evidences on their ability to modify biological responses, inactivation methods and perspectives on their application in foods. Trends Food Sci Technol 2016;58:96–114.

[15] Iraporda C, Errea A, Romanin DE, Cayet D, Pereyra E, Pignataro O, et al. Lactate and short chain fatty acids produced by microbial fermentation downregulate proinflammatory responses in intestinal epithelial cells and myeloid cells. Immunobiology 2015;220(10):1161–9.

[16] Wu X, Wu Y, He L, Wu L, Wang X, Liu Z. Effects of the intestinal microbial metabolite butyrate on the development of colorectal cancer. J Cancer 2018;9(14):2510–17.

[17] Chuah LO, Foo HL, Loh TC, et al. Postbiotic metabolites produced by *Lactobacillus plantarum* strains exert selective cytotoxicity effects on cancer cells. BMC Complement Altern Med 2019;19(1):114.

[18] Popović N, Djokić J, Brdarić E, Dinić M, Terzić-Vidojević A, Golić N, et al. The influence of heat-killed *Enterococcus faecium* BGPAS1-3 on the tight junction protein expression and immune function in differentiated Caco-2 cells infected with *Listeria monocytogenes* ATCC 19111. Front Microbiol 2019;10:412.

[19] Gao J, Li Y, Wan Y, Hu T, Liu L, Yang S, et al. A novel postbiotic from *Lactobacillus rhamnosus* GG with a beneficial effect on intestinal barrier function. Front Microbiol 2019;10:477.

[20] Izuddin WI, Loh TC, Foo HL, Samsudin AA, Humam AM. Postbiotic *L. plantarum* RG14 improves ruminal epithelium growth, immune status and upregulates the intestinal barrier function in post-weaning lambs. Sci Rep 2019;9(1):9938.

[21] Warda AK, Rea K, Fitzgerald P, Hueston C, Gonzalez-Tortuero E, Dinan TG, et al. Heat-killed lactobacilli alter both microbiota composition and behaviour. Behav Brain Res 2019;362:213–23.

[22] Gabrielli E, Pericolini E, Ballet N, Roselletti E, Sabbatini S, Mosci P, et al. *Saccharomyces cerevisiae*-based probiotic as novel anti-fungal and anti-inflammatory agent for therapy of vaginal candidiasis. Benef Microbes 2018;9(2):219–30.

[23] Wegh CAM, Geerlings SY, Knol J, Roeselers G, Belzer C. Postbiotics and their potential applications in early life nutrition and beyond. Int J Mol Sci 2019;20(19):E4673.

[24] Sugawara T, Sawada D, Ishida Y, Aihara K, Aoki Y, Takehara I, et al. Regulatory effect of paraprobiotic *Lactobacillus gasseri* CP2305 on gut environment and function. Microb Ecol Health Dis 2016;27:30259.
[25] Nishida K, Sawada D, Kawai T, Kuwano Y, Fujiwara S, Rokutan K. Para-psychobiotic *Lactobacillus gasseri* CP2305 ameliorates stress-related symptoms and sleep quality. J Appl Microbiol 2017;123(6):1561–70.
[26] van de Heijning BJ, Berton A, Bouritius H, Goulet O. GI symptoms in infants are a potential target for fermented infant milk formulae: a review. Nutrients 2014;6(9):3942–67.
[27] Szajewska H, Skórka A, Pieścik-Lech M. Fermented infant formulas without live bacteria: a systematic review. Eur J Pediatr 2015;174(11):1413–20.
[28] Vandenplas Y, Ludwig T, Bouritius H, Alliet P, Forde D, Peeters S, et al. Randomised controlled trial demonstrates that fermented infant formula with short-chain galacto-oligosaccharides and long-chain fructo-oligosaccharides reduces the incidence of infantile colic. Acta Paediatr 2017;106(7):1150–8.
[29] Endo A, Maeno S, Tanizawa Y, Kneifel W, Arita M, Dicks L, et al. Fructophilic lactic acid bacteria, a unique group of fructose-fermenting microbes. Appl Env Microbiol 2018;84(19) e01290-18.
[30] Hoffman JR, Hoffman MW, Zelicha H, Gepner Y, Willoughby DS, Feinstein U, et al. The effect of 2 weeks of inactivated probiotic *Bacillus coagulans* on endocrine, inflammatory, and performance responses during self-defense training in soldiers. J Strength Cond Res 2019;33(9):2330–7.
[31] Ashrafian F, Shahriary A, Behrouzi A, Moradi HR, Keshavarz Azizi Raftar S, Lari A, et al. *Akkermansia muciniphila*-derived extracellular vesicles as a mucosal delivery vector for amelioration of obesity in mice. Front Microbiol 2019;10:2155.
[32] Depommier C, Everard A, Druart C, Plovier H, Van Hul M, Vieira-Silva S, et al. Supplementation with *Akkermansia muciniphila* in overweight and obese human volunteers: a proof-of-concept exploratory study. Nat Med 2019;25(7):1096–103.
[33] Cani PD, de Vos WM. Next-generation beneficial microbes: the case of *Akkermansia muciniphila*. Front Microbiol 2017;8:1765.
[34] Plovier H, Everard A, Druart C, Depommier C, Van Hul M, Geurts L, et al. A purified membrane protein from *Akkermansia muciniphila* or the pasteurized bacterium improves metabolism in obese and diabetic mice. Nat Med 2017;23(1):107–13.
[35] Turpeinen A, Kumpu M, Rönnback M, Seppo L, Kautiainen H, Jauhiainen T, et al. Antihypertensive and cholesterol-lowering effects of a spread containing bioactive peptides IPP and VPP and plant sterols. J Funct Foods 2009;1:260–5.
[36] Zannini E, Waters DM, Coffey A, Arendt EK. Production, properties, and industrial food application of lactic acid bacteria-derived exopolysaccharides. Appl Microbiol Biotechnol 2016;100(3):1121–35.
[37] Dunand E, Burns P, Binetti A, Bergamini C, Peralta GH, Forzani L, et al. Postbiotics produced at laboratory and industrial level as potential functional food ingredients with the capacity to protect mice against *Salmonella* infection. J Appl Microbiol 2019;127(1):219–29.
[38] Huet F, Abrahamse-Berkeveld M, Tims S, Simeoni U, Beley G, Savagner C, et al. Partly fermented infant formulae with specific oligosaccharides support adequate infant growth and are well-tolerated. J Pediatr Gastroenterol Nutr 2016;63(4):e43–53.
[39] Vinderola G. Dried cell-free fraction of fermented milks: new functional additives for the food industry. Trends Food Sci Technol 2008;19:40–6.
[40] Andresen V, Gschossmann J, Layer P. Heat-inactivated *Bifidobacterium bifidum* MIMBb75 (SYN-HI-001) in the treatment of irritable bowel syndrome: a multicentre, randomised, double-blind, placebo-controlled clinical trial. Lancet Gastroenterol Hepatol 2020; Apr 8.

CHAPTER

12

Modification of the gut microbiome in an attempt to reduce the risk of child disease: Clinical data from prenatal interventions

Erika Isolauri[1,2] *and Samuli Rautava*[3,4]

[1]Department of Paediatrics, Institute of Clinical Medicine, University of Turku, Turku, Finland, [2]Department of Paediatric and Adolescent Medicine, Turku University Hospital, Turku, Finland, [3]Department of Pediatrics, University of Helsinki, Helsinki, Finland, [4]New Children's Hospital, Helsinki University Hospital, Helsinki, Finland

Cutting the cord: Too late to initiate preventive measures?

According to the developmental origins of health and disease hypothesis, the child's health is regulated by events experienced in utero and during early infancy. Specifically, a synchronized transition from the intrauterine to the extrauterine environment ensures healthy adaptation to the postnatal environment, and thereby later health. Conversely, the later health is endangered if the environment after birth differs from the situation before birth, particularly in terms of nutrition supply and the nutritional state. The Barker hypothesis envisions that the human fetus, when adapting to limited supply of nutrients, programs its physiology and metabolism to survive in nutritional scarcity, and in so doing adjusts the key regulatory systems of the body to meager circumstances [1,2].

DOI: https://doi.org/10.1016/B978-0-12-818097-6.00012-2

Maternal weight gain and especially the diet during pregnancy influence fetal growth and development, these being mirrored in birth size [3]. According to the Barker hypothesis, limited maternal nutrition may permanently alter the physiology and metabolism of the fetus, impacting on the child's later risk of chronic diseases, in a situation when the child is exposed to calorie-rich diet after birth, the phenomenon known as the programming effect [1–3]. The clinical consequences include obesity, type 2 diabetes, insulin resistance, hypertension, and hyperlipidemia with complications such as coronary heart disease. The effect appears to depend on the stage of pregnancy, and particularly vulnerable is the female fetus [4]. A vicious circle of an intergenerational transmission of noncommunicable diseases (NCDs) may thus ensue [5].

Today, overweight or obesity is more common than undernutrition. The WHO defined obesity as an epidemic in 1997. A recent NCD risk factor collaboration report documents that there are currently more obese and overweight people than underweight, whereas four decades ago the number of the overweight was half of that of underweight, reflecting the rapidity of the transformation [6]. This trend holds true in the population at reproductive age: it is estimated that two in three pregnant women are overweight or obese [7]. More recently, clinical conditions, phenotypes alternating from allergic, autoimmune, and inflammatory diseases to neurodevelopmental disorders, have been added to the list of NCDs with a potential starting point during the prenatal period. We need more profound understanding of the complex mechanisms underlying these heterogeneous manifestations that are currently seen as a major threat to human well-being.

One explanation for the rapid progress is the relatively recent environmental change and the emerged affluent lifestyle, to which our genome has not had time to attune. Undeniably, multifaceted preventive approaches are urgently called for to halt the cascade of diseases such as allergic and inflammatory diseases and obesity with its comorbidities.

Recent experimental and clinical data provide one new target for interventions aiming to close this vicious circle of current disease burden: the microbiome. Early interaction with the intestinal microbiome and its massive collective genome and a myriad of bioactive metabolites is a crucial component of healthy immune and metabolic programming. Recent scientific data suggest that alterations in the immune and metabolic phenotype can result from aberrant gut microbiome [8–10]. Furthermore, indirect evidence links a shift in the initial gut microbiome composition to a heightened risk of NCD: clinical practices interfering with the healthy host–microbe interaction, such as early antibiotic contact and cesarean section delivery, have been connected with immunoinflammatory diseases and obesity in epidemiological studies [11–14].

The dilemma of the neonate is to gradually generate an effective immunological responsiveness to pathogens while concomitantly developing a symbiotic coexistence with their individual environment- and age-adjusted microbiome and nutrients. This distinction is achieved by constant stimulation of the intestinal immune system by the indigenous microbiome, endorsing and maintaining the antiinflammatory milieu within the gut barrier functions [15–17]. However, the compositional development of the gut microbiome is highly sensitive to the environmental impacts including diet. Thus the cause and effect of the progression remain obscure: a shift in the gut microbiome may be a cause of NCD, or a consequence of the disease process initiated in utero, or the mechanism through which disadvantageous environmental influences provoke disease. An attractive vision arising from recent clinical intervention studies is repairing the gut microbiome composition in an attempt to reduce the risk of human disease.

Repairing the maternal gut microbiome: Successful microbiological programming in the child?

Mode of delivery and early environment

Although the *postpartum* colonization process in the infant gut has been intensively studied, the early events guiding the microbiome development have only recently become revealed. The most massive microbial exposure and a decisive inoculum for the establishment of the gut microbiome take place during vaginal delivery. It is estimated that 72% of early colonizers of vaginally delivered newborns correspond to those observed in their mothers' fecal samples [17,18]. The transmission of microbiome in neonates delivered by cesarean section appears to deviate from this pattern, as only a minority of species match with the maternal species present in the gut and birth canal [19]. In contrast, their microbiome originates from the oral or skin microbiome of the mother or the microbiome present in the hospital environment. A recent study applying whole genome metagenomic analysis confirms the earlier results [14]. Interestingly, the distinction in the gut microbiome composition between children delivered vaginally compared to those delivered by cesarean section has been reported to be a global phenomenon and extends beyond the neonatal period, underlining the instrumental impact of early microbial contact and the first colonizers [20,21]. Such a shift in the gut microbiome may be causally related to a sustained inflammatory tone in the feto–maternal immune interface and the neonatal gut barrier and thereby to an increased risk of chronic local and systemic inflammatory responses. This simplistic statement assumes a direct microbial connection between the mother and the fetus/neonate: the mother may provide the initial inoculum in microbial colonization process during vaginal delivery. However, the exposure to the complex microbial communities of the extrauterine world may be initiated already in utero, and furthermore, the maternal influence on the microbial programming may also be indirect.

In particular, probiotic research tends to support the indirect maternal influence. Specific probiotics have been shown to reverse increased intestinal permeability and to enhance specific IgA responses; clinical and experimental studies indicate that specific probiotics alleviate the inflammatory response both locally and systemically (reviewed in Ref. [12]), with an impact on the individual energy homeostasis with prevention of excessive body weight gain as end result (reviewed in Ref. [5]). These observations taken together indicate that the probiotic potential during pregnancy may not be limited to direct microbiome modulation and transfer. The effect may also be indirect via impacting on the maternal immunoinflammatory response and weight gain.

Pregnancy and maternal microbiome

There is a consensus in the scientific community that the initial colonization process directs the later microbiome succession and health of the infant [17,19]. However, the grounds for gut microbiome assembly may be generated already before birth and during the perinatal period. Indeed, a critical evaluation of the sources of the initial colonizers as the inoculum of the stepwise gut colonization is required: in addition, the health of the pregnant women is crucial for the microbiome profile that is transferred to the neonate. To take an example, the first intestinal discharge (meconium) was collected from newborns stratified by maternal diabetes status. The taxonomy analyses suggested that the overall bacterial content in meconium significantly differed by maternal diabetes status, and not by the mode of delivery [22].

There are several potential explanations for the apparently controversial findings. First, recent data tend to argue against the dogma concerning sterile fetal development, that is, the delivery representing the primary contact with the microbial world (discussed in previous chapters). Bacterial DNA has been detected in fetal environment, and this has been shown to represent a dynamic, viable microbiome [23]. The placenta has been shown to be colonized with bacteria [24]. Consequently, also the colonization of the fetus may be initiated already in utero, and this initial microbial contact may then be reinforced during delivery and breastfeeding [25,26]. Second, the gut microbiome composition changes during pregnancy, with a direct connection to the metabolic adjustments to the pregnant state [27]: pregnancy is associated with an increased production of pro-inflammatory factors due to increased adipose tissue, especially in the third trimester. The metabolic changes also affect the microbiome composition and activity, or vice versa. The net result at the end of normal pregnancy is a gut microbiome profile of elevated Proteobacteria and Actinobacteria and reduced bacterial richness [27]. The increase in pro-inflammatory profile of the female gut microbiome is exaggerated in obese pregnancy [28,29]. Overweight or obese women, again, more likely are subjects to cesarean section delivery due to heightened risk of pregnancy complications. Taking these observations together, the deviant microbiome observed in a cesarean section-delivered neonate may not directly originate from the cesarean section procedure per se, but from the events during pregnancy that have led to cesarean section. Aberrancies in the mother's microbiome composition and activity may be transferred to the fetus/neonate/infant by different routes: during pregnancy, at delivery via microbes in the mother's birth canal and gut, and by close contact between the mother and the newborn after delivery.

Breastfeeding

Maternal weight and weight gain as well as the mode of delivery not only impact on the gut microbiome composition of the pregnant and lactating women but also guide the breast milk composition, rich in active immunomodulatory compounds, including a unique microbiome, their growth factors, and metabolites. It is therefore noteworthy that the mode of delivery has an influence on breast milk microbiome composition, as distinct microbiome profiles have been documented between mothers delivering vaginally compared to those undergoing cesarean section delivery, but also between the types of cesarean section, elective versus nonelective cesarean section [30–32]. Emergency or nonelective cesarean section was shown not to diverge the early colonizers from the profile detected in vaginal delivery. These findings emphasize the impact of the physiological labor process, stress, and/or hormonal signals on the microbiome.

An infant's probability of being colonized by bifidobacteria is low if the mother has a higher body mass index and weight gain during pregnancy, and the neonate is delivered by cesarean section. Moreover, higher microbial diversity and abundance of *Bifidobacterium* and *Lactobacillus* characterize the breast milk microbiome of mothers after vaginal deliveries as compared to those delivering by (elective) cesarean section, but not consistently in different populations [30–34]. The breast milk microbiome profile is distinctive. To take an example, infants solely breastfed by their allergic and skin prick test-positive mothers had lower numbers of bifidobacteria compared to those receiving breast milk from nonallergic mothers [35].

The key health benefits of breastfeeding may result from promotion of environment-adjusted gut colonization. The individual composition of the mother's milk reflects its matrix of lipids and protein components and thereby the immediate environment and diet. All these components interact; the microbial adhesion to mucosal surfaces, important prerequisite for microbes to exert their immunomodulatory function, is modulated by polyunsaturated fatty acids [36]. Conversely, the gut microbiome is able to influence the nutritional status of the host (reviewed in Ref. [17]). Therefore the microbiome modulation needs to be considered in the close connection with the entire nutrition matrix; dissecting their impacts may prove impossible.

Pregnancy, mode of birth, and breast milk: A triad impacting on the gut colonization process

The initial inoculum before, at, and following birth generates the microbial bond between the mother and the child. Recent scientific advances indicate that microbial colonization of the gut may start already during the fetal period; these colonizing microbes present in amniotic fluid can also be recovered in meconium. Thereafter, the exposure to specific species in neonates is facilitated by the mode of delivery, early environment, and breastfeeding. This stepwise colonization process directs the later microbiome succession and health of the infant, and none of the stages can be considered exclusively fundamental in the process. Notwithstanding the recent demonstrations on the impact of transient gestational colonization in germ-free mice on the offspring phenotype (reviewed in Ref. [16]), this continuum of the microbiome succession is fundamental in human situation (Fig. 12.1).

It has become evident that *Bifidobacterium* colonization frequencies and counts among mother–infant pairs show a relationship [35]. *Bifidobacterium* species, again, appear major determinants in the context of risk of NCDs in child; lower numbers of bifidobacteria typify the early microbiome of an infant later developing atopic disease or excessive weight gain [8,9]. Conversely, clinical evidence has accumulated to indicate that a microbiome profile favoring bifidobacteria at an early age provides protection also against obesity development [8,9,37].

In an attempt to target dysbiosis, the purpose is not to alter the gut microecology per se, but to adjust pro-inflammatory signals to the immature gut barrier by the microbiome before altered structure and function in the target organ become consolidated. Thus the microbial programming resembles the idea of programming by early nutrition: markers of early nutrition are linked to the risk of chronic diseases. In fact, the gut microbiome, early environment, and nutrition are mutually dependent elements, which underlie the risk of disease in childhood and beyond. Maternal undernutrition modifies the epigenetic programming and has a negative

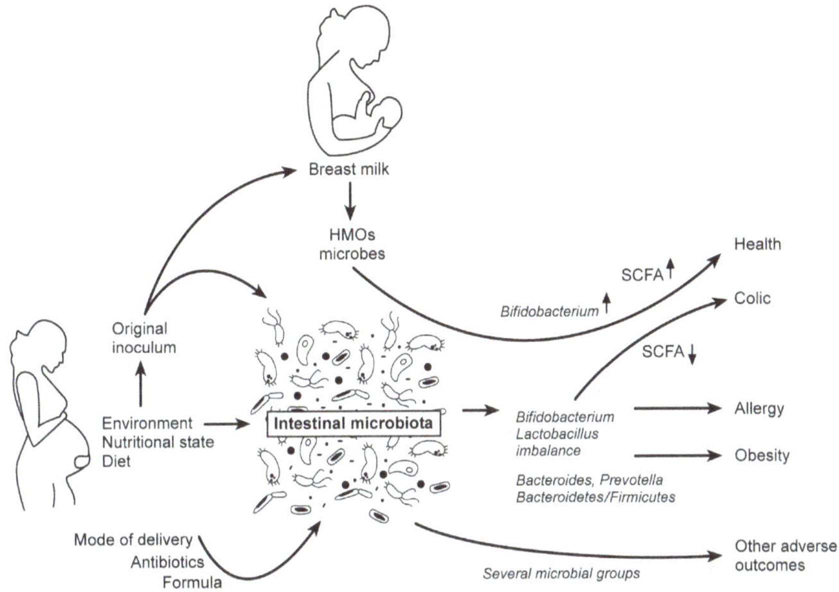

FIGURE 12.1 The establishment of the gut microbiome is a stepwise process during pregnancy, at delivery and close contact between the mother and the newborn after delivery- environmental and maternal influences on the pioneer colonizers in the microbial programming of health and disease. *SCFA*, short-chain fatty acid. Source: *Modified from Gómez-Gallego C, Garcia-Mantrana I, Martinez-Costa C, Salminen S, Isolauri E. Collado MC. The microbiome and malnutrition: changes influenced by nutritional status in early life. Ann Rev Nutr 2019;39:267–90.*

impact on fetal gene expression, thereby resulting in lasting imprint on the health of the descendants [38]. Analogous observations are emerging for the obese constitution [39]. Taken together, pregnancy is according to the current theory and evidence an optimal target for interventions aiming to reduce the risk of disease in future generations. By promoting the nutrition and health of the pregnant and lactating women and fostering the initial gut microbiome composition, the health of the next generation may be successfully improved (reviewed in Ref. [5]).

Evidence from randomized clinical trials: Underscoring the importance of the prenatal period

The gold standards in efforts to obtain recommendations to reduce the risk of disease are well-designed randomized placebo-controlled human intervention studies. The first reliable evidence suggesting that probiotic intervention may reduce the risk of chronic disease was provided by a randomized controlled trial assessing the efficacy of probiotics in reducing the occurrence of atopic eczema in high-risk children [40]. Altogether 159 pregnant women were randomized to receive either the probiotic *Lactobacillus rhamnosus* GG or placebo 2–4 weeks before expected delivery and subsequently during breastfeeding until the child was 6 months of age. The intervention was administered directly to the infant if breastfeeding was discontinued before the age of 6 months. The probiotic intervention resulted in a significant reduction in the occurrence of atopic eczema by the age of 2 years [40], and the effect has subsequently been reported to extend to the age of 4 and 7 years [41,42]. Interestingly, when only the subgroup of mother–infant pairs in whom the probiotic intervention was administered exclusively to the mother was analyzed, and maternal probiotic administration was observed to be associated with significant reduction in the risk of atopic eczema in the infant at the age of 2 years [43].

These results have subsequently been replicated in two independent studies. *Bifidobacterium longum* BL999 in combination with either *L. rhamnosus* LPR or *Lactobacillus paracasei* ST11 administered to the mother during the last 2 months of pregnancy, and the first 2 months of breastfeeding were shown to significantly reduce the risk of atopic eczema as compared to placebo in infants of mothers with atopic disease [44]. On the other hand, a probiotic milk containing *L. rhamnosus* GG, *Lactobacillus acidophilus* La-5, and *Bifidobacterium animalis* subsp. *lactis* Bb-12 consumed from 36 weeks of gestation to 3 months postnatally during breastfeeding has been reported to reduce the occurrence of atopic eczema compared to placebo until the age of 2 years in an unselected population [45].

Clinical trials assessing the efficacy of probiotic intervention in the prevention of atopic disease have yielded partially conflicting results. This may be due to several reasons including differences in the study populations and probiotic strains as well as the precise administration protocols used. According to a systematic review and meta-analysis of randomized controlled trials, probiotic intervention is only effective if administered both pre- and postnatally [46]. Interestingly, maternal probiotic intervention with *L. rhamnosus* HN001 from the first trimester of pregnancy until 6 months postnatally was ineffective in reducing the prevalence of atopic eczema by the age of 12 months in a randomized controlled trial of 423 infants with high hereditary risk [47] despite a previous trial by the same authors indicating that the same strain is associated with a significant reduction in atopic eczema if given both to the mother from week 35 of gestation until 6 months if breastfeeding and to their infants from birth to the age of 2 years. Taken together, these data underscore the notion that early microbial exposures during pregnancy, birth, breastfeeding, and early infancy form a continuum, and probiotic interventions aiming to improve long-term child health must be devised to appreciate this transitional period [48].

The mechanisms by which probiotic intervention during pregnancy and breastfeeding modify atopic disease risk remain poorly understood. As discussed earlier, maternal intervention with the combination of *L. rhamnosus* GG, *L. acidophilus* La-5, and *B. animalis* subsp. *lactis* Bb-12 was clinically efficacious [49]. When maternal and infant fecal samples were analyzed, all the probiotic bacteria were enriched in the maternal gut microbiome, but only *L. rhamnosus* was detected to colonize the infants. Moreover, no differences in overall gut microbiome composition were detected between the infants in the intervention and placebo groups [49]. These data may be interpreted to suggest that specific probiotics might not primarily exert their effect through modulating the composition of the gut microbiome. In support of this notion, several potential direct mechanisms of action have been reported. Maternal administration of *B. animalis* subsp. *lactis* Bb-12 alone or in combination with *L. rhamnosus* GG 2 weeks before delivery has been reported to modulate innate immune receptor expression in the placenta and the fetal gut [25]. Some evidence for systemic fetal immune modulation resulting from maternal probiotic supplementation is also available, as intervention with *L. rhamnosus* HN001 or *B. lactis* HN019 beginning 2–5 weeks before delivery reportedly leads to increased cord blood interferon-γ concentrations [50]. Postnatally, maternal probiotic supplementation has been reported to modulate the concentrations of immunomodulatory molecules and particularly that of transforming growth factor-β in breast milk, albeit the results in the clinical studies using different probiotic strains vary [43,50,51].

The efficacy of probiotic interventions in reducing pregnancy complications has also been investigated. Preterm birth is a serious global health concern, which affects approximately 10% of all pregnancies [52] and constitutes the most important cause of child mortality globally [53]. Recently, a systematic review and meta-analysis of randomized controlled trials found no evidence to suggest that probiotic supplementation during pregnancy reduces the risk of preterm delivery [54]. Although meta-analyses are often considered the most reliable source of evidence, it is important to note that 20 out of the 21 studies included in this meta-analysis were not designed to investigate the impact of probiotic supplementation on the risk of preterm delivery as a primary outcome. The only study originally devised for this purpose was not sufficiently powered and only included women with bacterial vaginosis [55]. In contrast to the meta-analysis of randomized trials, data from a prospective observational cohort study of more than 70,000 pregnant women in Norway suggest that consumption of a probiotic milk product in early pregnancy was associated with reduced risk of preterm delivery [56]. The question of probiotic efficacy in reducing the risk of preterm delivery or its complications on the child remains open.

The intergenerational transfer of NCDs is perhaps best exemplified by the risk of obesity and metabolic disease in the offspring imposed by the same disorders in the mother. Given the well-established role of the gut microbiome in the pathogenesis of overweight, obesity, and their comorbidities, probiotic intervention during pregnancy may be hypothesized not only to improve pregnancy outcomes but also to improve child health. In a randomized controlled trial of 254 pregnant women, probiotic intervention with *L. rhamnosus* GG and *B. animalis* subsp. *lactis* Bb-12 together with dietary counseling significantly reduced the incidence of gestational diabetes mellitus (GDM) as compared to the control group or dietary counseling alone [57]. Subsequent clinical trials assessing the efficacy of probiotic supplementation in the prevention of GDM have yielded conflicting results. Callaway et al. found the same two probiotics to be ineffective in the prevention of GDM in a study of 411 overweight or obese pregnant women [58]. In a similar fashion, no reduction in the occurrence of GDM was detected in overweight or obese women after administration with *Lactobacillus salivarius* UCC118 [59] or *L. rhamnosus* HN001and *B. animalis* subsp. *lactis* [60]. On the other hand, Wickens et al. reported reduced occurrence of GDM in women receiving *L. rhamnosus* HN001 in a randomized controlled trial of 373 pregnant women unselected with regard to metabolic state [61]. The reasons for the discrepant data are currently not known. Interestingly, probiotic supplementation in pregnant women with GDM is associated with improved glycemic control and blood lipid profiles as well as reduced inflammation and oxidative stress according to a recent systematic review and meta-analysis of 11 clinical trials with a total of 719 subjects [62]. However, the authors raise concerns about the heterogeneity of the included studies, the surrogate nature of some of the outcomes, and the reliability and generalizability of the findings.

Whether the potential beneficial metabolic effects of probiotics on the pregnant mother are reflected in the later health of the offspring or not remain largely an open question. In the first intervention trial designed to assess the efficacy of perinatal probiotic intervention in reducing the risk of atopic eczema discussed in detail earlier [40], supplementation with *L. rhamnosus* GG 2–4 weeks before delivery and until the age of 6 months appeared to moderate the initial phase of excessive weight gain from fetal life to 24–48 months of life particularly in children who later became overweight. In addition, a tendency for a reduction in the birth weight-adjusted mean body mass index at the age of 4 years was observed [37]. Future research is required to corroborate these initial findings and to dissect the contributions of prenatal, maternal, and direct probiotic interventions.

Interventions during pregnancy: Efficacy and safety

Probiotic research has influenced strongly our conception of microbes. The traditional view is defensive; any microbe is considered a potential pathogen causing disease. Modern molecular techniques identify microbes that are distributed in site-specific communities [63], and the current approach is not unspecific eradication but to balance the microbiome, and specifically the host–microbe interaction. In reality, humans and microbes together create a supraorganism in which interactions consist of bidirectional exchange of endocrine, immune, and neural signals with targets in metabolic, immune, humoral, and neural pathways [10,64]. The balanced microbe–host interaction appears to be crucial in reinforcing the gut barrier, thereby actually preventing pathogenic bacteria from breaching the gut mucosal barrier [65].

Aberrant compositional development of the gut microbiome not only increases the risk of infection but is considered a necessary initial step in the development of immune-mediated health problems such as allergic, autoimmune, and inflammatory diseases, as well as obesity [10,12,65]. The risk of untoward consequences of dysbiosis culminates during pregnancy and infancy, at the time of the configuration of the organism's immune, metabolic, and microbiological constitution, according to the programming theory [12].

The idea of gut microbiome modulation is the attempt of repairing the composition back to age- and environment- adjusted balance, but also averting deviant immune responsiveness during the sensitive period. There are two angles that need to be considered: for a successful programming optimal timing is decisive, and during the receptive period the fetus/neonate/infant is immature, and also, due to the interaction of the gut microbiome, early environment, and nutrition, one mode of prevention would not suffice to counter the plethora of NCDs. On this basis, the question of safety is fundamental.

The definition of probiotics [66] discussed earlier inherently suggests that strain characterization, health effects, and safety need to be demonstrated. Specific probiotics have been shown to reverse increased intestinal permeability and to enhance specific IgA responses, frequently defective at an early age [67]. Promotion of gut barrier functions, previous normalization of the properties of aberrant indigenous microbiome, involves degradation or structural modification of enteral antigens, regulation of the secretion of inflammatory mediators, and direction of the development of the immune system during the critical period of life when the risk of disease is elevated, by alleviation of the inflammatory tone both locally and systemically. Moreover, recent evidence from experimental and clinical studies indicates that the gut microbiome is also involved in the control of energy metabolism affecting energy acquisition and storage and contributing to insulin resistance and the low-grade inflammatory state [12,68,69]. Thus manipulation of the initial gut microbiome by specific probiotics offers a promising strategy to reduce the risk of infectious and NCDs.

In general, specific probiotics have an excellent safety record, when used according to the definition and concepts that each probiotic strain is inherently different from others and that similar health effects cannot therefore be expected even from closely related strains [66]. The safety documentation includes epidemiological studies, postmarketing reports, randomized clinical trials in pregnancy, and early infancy with long-term follow-up and demonstration of clinical practice in preterm children [5,54,57,70–74].

A large observational study involving 33,399 primiparous women suggests an independent protective association of probiotic milk consumption and preeclampsia, especially severe preeclampsia [70]. The authors examined, by self-reported food frequency questionnaire, the consumption of milk-based probiotic products in pregnancy and linked the data to the Norwegian Medical Birth Registry for preeclampsia and its subtypes. With probiotic milk intakes divided into categories, lower risks for severe preeclampsia were observed for weekly and daily intakes.

Focusing of the question of safety, a randomized double-blind intervention study assessed the safety and efficacy of perinatal probiotic (*L. rhamnosus* GG ATCC 53103)—supplemented dietary counseling by evaluating pregnancy outcome and fetal and infant growth during the 24 months' follow-up [57]. The safety of this approach was attested by normal duration of pregnancies with no adverse events in mothers or children; the 5-minute Apgar scores were inline in all study groups. No significant differences in prenatal or postnatal growth rates among the study groups were detected. There were no perinatal deaths or serious adverse incidences such as sepsis in mothers or newborns.

Other ways to modulate the gut microbiome composition for clinical benefit include prebiotics and postbiotics (as described elsewhere in this book). Prebiotic is *a substrate that is selectively utilized by host microorganisms conferring a health benefit* [75]. Postbiotics aim to mimic the beneficial therapeutic effects of probiotics while avoiding the risk of administering live microorganisms to vulnerable infants with immature intestinal barriers or impaired immune defenses. However, gut barrier dysfunction has been reported in experimental studies for some prebiotics [76], and a probiotic strain effective in control of atopic eczema and local and systemic inflammatory responses was associated with adverse gastrointestinal symptoms and diarrhea when used in heat-killed inactive form [77]. These data taken together we need to acknowledge that each pro-, pre-, and postbiotic is a unique component itself; their properties cannot be extrapolated or generalized to other substances.

Conclusions

During the first months of life, adaptation to the postnatal environment is guided in the gastrointestinal tract by massive antigen contacts in terms of microbial colonization and oral food intake. An important aspect to human development involves the establishment of the gut microbiome, which receives its original inoculum already before birth. The gut microbiome holds a key position with regard to the increasing burden of NCDs in the industrialized countries, manifestations ranging from allergic and inflammatory conditions as well as obesity to neuropsychiatric disorders. However, the microbiome composition as such is not the ultimate target of any intervention aiming to reduce the risk of these conditions, but instead the host–microbe interaction. Humans and the microbiome create a bidirectional exchange of endocrine, immune, and neural signals with targets in metabolic, immune, humoral, and neural pathways. Thus the gut microbiome is an important mediator between the host internal and external environments, and repairing the gut microbiome composition to an environment and age-adjusted balance may support the maturation of the key regulatory systems of the body in the direction of healthy growth and development.

Realization of this has led to the introduction of novel modes of intervention based on the consumption of beneficial live microorganisms as probiotics, their growth factors, as well as microbiome metabolites. Specific probiotics provide a safe microbial contact to avert deviant properties in the microbiome, gut barrier functions, and secretion of pro-inflammatory mediators. These aspects culminate in the pre- and perinatal period, when the key regulatory systems are sensitive to the programming phenomenon. Until recently, probiotic research in NCDs has remained mostly cross-sectional and descriptive; primary is the need to direct the focus to establishing cause and effect relationships [78]. The gold standard in efforts to provide clinical recommendations are well-designed randomized placebo-controlled human intervention studies. Presumably the best level of evidence can be obtained by subjecting their results to meta-analyses. Moreover, the probiotic definition [66] inherently suggests that strain characterization, health effects, and safety need to be demonstrated. There are well-controlled long-term intervention studies on record, but for a limited number of strains and populations. The transitional period from pregnancy, birth, and early infancy offers a promising time-window for probiotic interventions aiming to improve the long-term health of the child. The current body of evidence from clinical trials demonstrates a wide range of potential targets for intervention during this continuum. Future studies should ideally investigate specific interventions based on preclinical data and should aim at specific targets and outcome measures in well-defined populations.

References

[1] Barker DJ, Winter PD, Osmond C, Margetts B, Simmonds SJ. Weight in infancy and death from ischaemic heart disease. Lancet 1989;2(8663):577–80.

[2] Barker DJ. The developmental origins of adult disease. J Am Coll Nutr 2004;23 (Suppl. 6):588S–95S.

[3] Godfrey K, Robinson S, Barker DJ, Osmond C, Cox V. Maternal nutrition in early and late pregnancy in relation to placental and fetal growth. BMJ 1996;312 (7028):410–14.

[4] Ravelli AC, van Der Meulen JH, Osmond C, Barker DJ, Bleker OP. Obesity at the age of 50 y in men and women exposed to famine prenatally. Am J Clin Nutr 1999;70 (5):811–16.

[5] Isolauri E, Rautava S, Collado MC, Salminen S. Early microbe contact in defining child metabolic health and obesity risk. In: Green LR, Hester RL, editors. Parental obesity: intergenerational programming and consequences. Physiology in health and disease. New York: The American Physiological Society, Springer; 2016.

[6] NCD Risk Factor Collaboration. Trends in adult body-mass index in 200 countries from 1975 to 2014: a pooled analysis of 1698 population-based measurement studies with 19.2 million participants. Lancet 2016;387(10026):1377–96.

[7] Wankhade UD, Thakali KM, Shankar K. Persistent influence of maternal obesity on offspring health: mechanisms from animal models and clinical studies. Mol Cell Endocrinol 2016;435:7–19. Available from: https://doi.org/10.1016/j.mce.2016.07.001.

[8] Kalliomäki M, Kirjavainen P, Eerola E, Kero P, Salminen S, Isolauri E. Distinct patterns of neonatal gut microflora in infants in whom atopy was and was not developing. J Allergy Clin Immunol 2001;107(1):129–34.

[9] Kalliomäki M, Collado MC, Salminen S, Isolauri E. Early differences in fecal microbiota composition in children may predict overweight. Am J Clin Nutr 2008;87 (3):534–8.

[10] Kau AL, Ahern PP, Griffin NW, Goodman AL, Gordon JI. Human nutrition, the gut microbiome and the immune system. Nature 2011;474(7351):327–36.

[11] Kero J, Gissler M, Grönlund MM, Kero P, Koskinen P, Hemminki E, et al. Mode of delivery and asthma – is there a connection? Pediatr Res 2002;52(1):6–11.

[12] Rautava S, Luoto R, Salminen S, Isolauri E. Microbial contact during pregnancy, intestinal colonization and human disease. Nature Rev Gastroenterol Hepatol 2012;9 (19):565–76.

[13] Miettinen R, Hermansson H, Merkikukka M, Gissler M, Isolauri E. Mode of delivery—impact on risk of noncommunicable diseases. J Allergy Clin Immunol 2015;136:1398–9.

[14] Shao Y, Forster SC, Tsaliki E, Vervier K, Strang A, Simpson N, et al. Stunted microniota and opportunistic pathogen colonization in caesarean-section birth. Nature 2019;574(7776):117–21.

[15] Maynard CL, Elson CO, Hatton RD, Weaver CT. Reciprocal interactions of the intestinal microbiota and immune system. Nature 2012;489(7415):231–41. Available from: https://doi.org/10.1038/nature11551.

[16] Macpherson AJ, de Agüero MG, Ganal-Vonarburg C. How nutrition and the maternal microbiota shape the neonatal immune system. Nat Rev Immunol 2017;17(8):508–17.

[17] Gómez-Gallego C, Garcia-Mantrana I, Martinez-Costa C, Salminen S, Isolauri E, Collado MC. The microbiome and malnutrition: changes influenced by nutritional status in early life. Ann Rev Nutr 2019;39:267–90.

[18] Bäckhed F, Roswall J, Peng Y, Feng Q, Jia H, Kovatcheva-Datchary P, et al. Dynamics and stabilization of the human gut microbiome during the first year of life. Cell Host Microbe 2015;17(5):690–703.

[19] Koenig JE, Spor A, Scalfone N, Fricker AD, Stombaugh J, Knight R, et al. Succession of microbial consortia in the developing infant gut microbiome. Proc Natl Acad Sci USA 2011;108(Suppl. 1):4578–85.
[20] Salminen S, Gibson GR, McCartney AL, Isolauri E. Influence of mode of delivery on gut microbiota composition in seven year old children. Gut 2004;53(9):1388–9.
[21] Kumbhare SV, Kumar H, Chowdhury SP, Dhotre DP, Endo A, Mättö J, et al. A cross sectional comparative study of gut bacterial community of Indian and Finnish children. Sci Rep 2017;7(1):10555.
[22] Hu J, Nomura Y, Bashir A, Fernandez-Hernandez H, Itzkowitz S, Pei Z, et al. Diversified microbiota of meconium is affected by maternal diabetes status. PLOS One 2013;8:e78257. Available from: https://doi.org/10.1371/journal.pone.0078257.
[23] Younge N, McCann JR, Ballard J, Plunkett C, Akhtar S, Araújo-Perez F, et al. Fetal exposre to maternal microbiota in humans and mice. JCI Insight 2019;4:pii127806. Available from: https://doi.org/10.1172/jci.insight.127806.
[24] Aagaard K, Ma J, Antony KM, Ganu R, Petrosino J, Versalovic J. The placenta harbors a unique microbiome. Sci Transl Med 2014;6(237):237ra65.
[25] Rautava S, Collado MC, Salminen S, Isolauri E. Probiotics modulate host-microbe interaction in the placenta and fetal gut: a randomized, double-blind, placebo-controlled trial. Neonatology 2012;102(3):178–84.
[26] Collado MC, Rautava S, Aakko J, Isolauri E, Salminen S. Human gut colonization may be initiated in utero by distinct microbial communities in the placenta and amniotic fluid. Sci Rep 2016;6:23129. Available from: https://doi.org/10.1038/srep23129.
[27] Koren O, Goodrich JK, Cullender TC, Spor A, Laitinen K, Kling Bäckhed H, et al. Host remodeling of the gut microbiome and metabolic changes during pregnancy. Cell 2012;150(3):470–80.
[28] Collado MC, Isolauri E, Laitinen K, Salminen S. Distinct composition of gut microbiota during pregnancy in overweight and normal weight women. Am J Clin Nutr 2008;88(4):894–9.
[29] Collado MC, Isolauri E, Laitinen K, Salminen S. Impact of mother's weight on infant's microbiota acquisition, composition and activity during early infancy: a prospective follow up study initiated in early pregnancy. Am J Clin Nutr 2010;92(5):1023–30.
[30] Cabrera-Rubio R, Collado MC, Laitinen K, Salminen S, Isolauri E, Mira A. The human milk microbiome changes over lactation and is shaped by maternal weight and mode of delivery. Am J Clin Nutr 2012;96(3):544–51.
[31] Collado MC, Laitinen K, Salminen S, Isolauri E. Maternal weight and weight gain during pregnancy modify the immunomodulatory potential of breast milk. Pediatr Res 2012;72(1):77–85.
[32] Cabrera-Rubio R, Mira-Pascual L, Mira A, Collado MC. Impact of mode of delivery on the milk microbiota composition of healthy women. J Dev Orig Health Dis 2016;7(1):54–60.
[33] Urbaniak C, Angelini M, Gloor GB, Reid G. Human milk microbiota profiles in relation to birthing method, gestation and infant gender. Microbiome 2016;4:1. Available from: https://doi.org/10.1186/s40168-015-0145-y.
[34] Sakwinska O, Moine D, Delley M, Combremont S, Rezzonico E, Descombes P, et al. Microbiota in breast milk of chinese lactating mothers. PLoS One 2016;11:e0160856.
[35] Grönlund M-M, Gueimonde M, Laitinen K, Kociubinski G, Grönroos T, Salminen S, et al. Maternal breast-milk and intestinal bifidobacteria guide the compositional development of the Bifidobacterium microbiota in infants at risk of allergic disease. Clin Exp Allergy 2007;37(12):1764–72.
[36] Kankaanpää P, Salminen S, Isolauri E, Lee Y. The influence of polyunsaturated fatty acids on probiotic growth and adhesion. FEMS Microbiol Lett 2001;194(2):149–53.

[37] Luoto R, Kalliomäki M, Laitinen K, Isolauri E. The impact of perinatal probiotic intervention on the development of overweight and obesity: follow-up study from birth to 10 years. Int J Obes 2010;34(19):1531–7.
[38] Li CC, Maloney CA, Cropley JE, Suter CM. Epigenetic programming by maternal nutrition: shaping future generations. Epigenomics 2010;2(4):539–49.
[39] Vähämiko S, Laiho A, Lund R, Isolauri E, Salminen S, Laitinen K. The impact of probiotic supplementation during pregnancy on DNA methylation of obesity-related genes in mothers and their children. Eur J Clin Nutr 2019;58(1):367–77.
[40] Kalliomäki M, Salminen S, Arvilommi H, Kero P, Koskinen P, Isolauri E. Probiotics in primary prevention of atopic disease: a randomised placebo-controlled trial. Lancet 2001;357(9262):1076–9.
[41] Kalliomäki M, Salminen S, Poussa T, Arvilommi H, Isolauri E. Probiotics and prevention of atopic disease: 4-year follow-up of a randomised placebo-controlled trial. Lancet 2003;361(9372):1869–71.
[42] Kalliomäki M, Salminen S, Poussa T, Isolauri E. Probiotics during the first 7 years of life: a cumulative risk reduction of eczema in a randomized, placebo-controlled trial. J Allergy Clin Immunol 2007;119(4):1019–21.
[43] Rautava S, Kalliomäki M, Isolauri E. Probiotics during pregnancy and breast-feeding might confer immunomodulatory protection against atopic disease in the infant. J Allergy Clin Immunol 2002;109(1):119–21.
[44] Rautava S, Kainonen E, Salminen S, Isolauri E. Maternal probiotic supplementation during pregnancy and breast-feeding reduces the risk of eczema in the infant. J Allergy Clin Immunol 2012;130(6):1355–60.
[45] Dotterud CK, Storrø O, Johnsen R, Oien T. Probiotics in pregnant women to prevent allergic disease: a randomized, double-blind trial. Br J Dermatol 2010;163(3):616–23.
[46] Panduru M, Panduru NM, Salavastru CM, Tiplica GS. Probiotics and primary prevention of atopic dermatitis: a meta-analysis of randomized controlled studies. J Eur Acad Dermatol Venereol 2015;29(2):232–42.
[47] Wickens K, Barthow C, Mitchell EA, Stanley TV, Purdie G, Rowden J, et al. Maternal supplementation alone *with Lactobacillus rhamnosus* HN001 during pregnancy and breast-feeding does not reduce infant eczema. Pediatr Allergy Immunol 2018;29(3):296–302.
[48] Wickens K, Black PN, Stanley TV, Mitchell E, Fitzharris P, Tannock GW, et al. A differential effect of 2 probiotics in the prevention of eczema and atopy: a double-blind, randomized, placebo-controlled trial. J Allergy Clin Immunol 2008;122(4):788–94.
[49] Dotterud CK, Avershina E, Sekelja M, Simpson MR, Rudi K, Storrø O, et al. Does maternal perinatal probiotic supplementation alter the intestinal microbiota of mother and child? J Pediatr Gastroenterol Nutr 2015;61(2):200–7.
[50] Prescott SL, Wickens K, Westcott L, Jung W, Currie H, Black PN, et al. Supplementation with *Lactobacillus rhamnosus* or *Bifidobacterium lactis* probiotics in pregnancy increases cord blood interferon-gamma and breast milk transforming growth factor-beta and immunoglobin A detection. Clin Exp Allergy 2008;38(10):1606–14.
[51] Böttcher MF, Abrahamsson TR, Fredriksson M, Jakobsson T, Björkstén B. Low breast milk TGF-beta2 is induced by Lactobacillus reuteri supplementation and associates with reduced risk of sensitization during infancy. Pediatr Allergy Immunol 2008;19(6):497–504.
[52] Blencowe H, Cousens S, Oestergaard MZ, Chou D, Moller AB, Narwal R, et al. National, regional, and worldwide estimates of preterm birth rates in the year 2010 with time trends since 1990 for selected countries: a systematic analysis and implications. Lancet 2012;379(9832):2162–72.
[53] Liu L, Oza S, Hogan D, Chu Y, Perin J, Zhu J, et al. Global, regional, and national causes of under-5 mortality in 2000-15: an updated systematic analysis with implications for the Sustainable Development Goals. Lancet 2016;388(10063):3027–35.

[54] Jarde A, Lewis-Mikhael AM, Moayyedi P, Stearns JC, Collins SM, Beyene J, et al. Pregnancy outcomes in women taking probiotics or prebiotics: a systematic review and meta-analysis. BMC Pregnancy Childbirth 2018;18(1):14.
[55] Krauss-Silva L, Moreira ME, Alves MB, Braga A, Camacho KG, Batista MR, et al. A randomised controlled trial of probiotics for the prevention of spontaneous preterm delivery associated with bacterial vaginosis: preliminary results. Trials 2011;12:239.
[56] Nordqvist M, Jacobsson B, Brantsæter AL, Myhre R, Nilsson S, Sengpiel V. Timing of probiotic milk consumption during pregnancy and effects on the incidence of preeclampsia and preterm delivery: a prospective observational cohort study in Norway. BMJ Open 2018;8(1):e018021.
[57] Luoto R, Laitinen K, Nermes M, Isolauri E. Impact of maternal probiotic-supplemented dietary counselling on pregnancy outcome and prenatal and postnatal growth: a double-blind, placebo-controlled study. Br J Nutr 2010;103(12):1792–9.
[58] Callaway LK, McIntyre HD, Barrett HL, Foxcroft K, Tremellen A, Lingwood BE, et al. Probiotics for the prevention of gestational diabetes mellitus in overweight and obese women: findings from the SPRING double-blind randomized controlled trial. Diabetes Care 2019;42(3):364–71.
[59] Lindsay KL, Kennelly M, Culliton M, Smith T, Maguire OC, Shanahan F, et al. Probiotics in obese pregnancy do not reduce maternal fasting glucose: a double-blind, placebo-controlled, randomized trial (Probiotics in Pregnancy Study). Am J Clin Nutr 2014;99(6):1432–9.
[60] Pellonperä O, Mokkala K, Houttu N, Vahlberg T, Koivuniemi E, Tertti K, et al. Efficacy of fish oil and/or probiotic intervention on the incidence of gestational diabetes mellitus in an at-risk group of overweight and obese women: a randomized, placebo-controlled, double-blind clinical trial. Diabetes Care 2019;42(6):1009–17.
[61] Wickens KL, Barthow CA, Murphy R, Abels PR, Maude RM, Stone PR, et al. Early pregnancy probiotic supplementation with *Lactobacillus rhamnosus* HN001 may reduce the prevalence of gestational diabetes mellitus: a randomised controlled trial. Br J Nutr 2017;117(6):804–13.
[62] Zhang J, Ma S, Wu S, Guo C, Long S, Tan H. Effects of probiotic supplement in pregnant women with gestational diabetes mellitus: a systematic review and meta-analysis of randomized controlled trials. J Diabetes Res 2019;2019:5364730.
[63] Turroni F, Ventura M, Buttó LF, Duranti S, O'Toole PW, Motherway MO, et al. Molecular dialogue between the human gut microbiota and the host: a *Lactobacillus* and *Bifidobacterium* perspective. Cell Mol Life Sci 2014;71(2):183–203.
[64] Moloney RD, Desbonnet L, Clarke G, Dinan TG, Cryan JF. The microbiome: stress, health and disease. Mamm Genome 2014;25(1-2):49–74.
[65] Isolauri E, Kalliomäki M, Laitinen K, Salminen S. Modulation of the maturing gut barrier and microbiota: a novel target in allergic disease. Curr Pharm Des 2008;14:1368–75.
[66] Hill C, Guarner F, Reid G, Gibson GR, Merenstein DJ, Pot B, et al. Expert consensus document: The International Scientific Association for Probiotics and Prebiotics consensus statement on the scope and appropriate use of the term probiotic. Nat Rev Gastroenterol Hepatol 2014;11:506–14. Available from: https://doi.org/10.1038/nrgastro.2014.66.
[67] Isolauri E, Kirjavainen PV, Salminen S. Probiotics: a role in the treatment of intestinal infection and inflammation? Gut 2002;50(Suppl. III):iii54–9.
[68] Bäckhed F, Ding H, Wang T, Hooper LV, Koh GY, Nagy A, et al. The gut microbiota as an environmental factor that regulates fat storage. Proc Natl Acad Sci USA 2004;101(44):15718–23.
[69] Cani PD, Amar J, Iglesias MA, Poggi M, Knauf C, Bastelica D, et al. Metabolic endotoxemia initiates obesity and insulin resistance. Diabetes 2007;56(7):1761–72.

[70] Brantsaeter AL, Myhre R, Haugen M, Myking S, Sengpiel V, Magnus P, et al. Intake of probiotic food and risk of preeclampsia in primiparous women: the Norwegian Mother and Child Cohort Study. Am J Epidemiol 2011;174(7):807–15.
[71] Saxelin M, Chuang NH, Chassy B, Rautelin H, Mäkelä PH, Salminen S, et al. Lactobacilli and bacteremia in southern Finland 1989–1992. Clin Infect Dis 1996;22(3):564–6.
[72] Luoto R, Isolauri E, Lehtonen L. Safety of *Lactobacillus* GG probiotic in infants with very low birth weight: twelve years of experience. Clin Infect Dis 2010;50(9):1327–8.
[73] Collado MC, Rautava S, Isolauri E, Salminen S. Gut microbiota: a source of novel tools to reduce the risk of human disease? Pediatr Res 2015;77(1-2):182–8.
[74] Lundelin K, Poussa T, Salminen S, Isolauri E. Long-term safety and efficacy of perinatal probiotic intervention: evidence from a follow-up study of four randomized, double-blind, placebo-controlled trials. Pediatr Allergy Immunol 2017;28(2):170–5.
[75] Gibson GR, Hutkins R, Sanders ME, Prescott SL, Reimer RA, Salminen SJ, et al. Expert consensus document: The International Scientific Association for Probiotics and Prebiotics (ISAPP) consensus statement on the definition and scope of prebiotics. Nat Rev Gastroenterol Hepatol 2017;14(8):491–502.
[76] Ten Bruggencate SJ, Bovee-Oudenhoven IM, Lettink-Wissink ML, Katan MB, Van Der Meer R. Dietary fructo-oligosaccharides and inulin decrease resistance of rats to salmonella: protective role of calcium. Gut 2004;53(4):530–5.
[77] Kirjavainen PV, Salminen SJ, Isolauri E. Probiotic bacteria in the management of atopic disease: underscoring the importance of viability. J Pediatr Gastroenterol Nutr 2003;36(2):223–7.
[78] Zhao L. The gut microbiota and obesity: from correlation to causality. Nat Rev Microbiol 2013;11(9):639–47.

CHAPTER

13

Clinical data from postnatal interventions

Hania Szajewska

Department of Paediatrics, Medical University of Warsaw, Warsaw, Poland

Introduction

Early gut microbial colonization plays an important role in health and disease. Factors such as the delivery mode (vaginal vs cesarean section), type of infant feeding (breastfeeding or formula feeding), medications used during early life (antibiotics, proton pump inhibitors), and the environment are thought to affect the microbial colonization pattern in early infancy [1]. Alterations in the gut microbiome (dysbiosis) during this period, at least in part, contribute to the development and progression of diseases such as allergy, obesity, functional gastrointestinal disorders (infant colic, irritable bowel syndrome), necrotizing enterocolitis (NEC), type 1 diabetes, and autism spectrum disorders [2,3]. With the growing recognition of the role of gut microbiome in early life and its impact on health and subsequent diseases, discussed in detail elsewhere in this book, a number of interventions targeting the gut microbiome are increasingly being used. Among others, these include probiotics, prebiotics, and a combination of both (synbiotics). These interventions are currently gaining worldwide popularity and are increasingly being used in the pediatric population, including very young infants, despite some reservations regarding their efficacy and safety in this vulnerable population. Indeed, there are a number of issues and concerns related to the administration of these agents, particularly probiotics, early in life. The administration begins in early infancy, sometimes at birth, at a time when the gut microbiome is not fully established, and factors that influence the microbiome may potentially permanently affect the development

The Human Microbiome in Early Life
DOI: https://doi.org/10.1016/B978-0-12-818097-6.00013-4

of the ecosystem. The daily administration may be over a long time (several weeks or months). Lastly, evidence from at least one study with a long follow-up period documented that beneficial as well as unfavorable effects are seen years after the administration of probiotics during pregnancy and during the first months of life [4,5]. Such an observation raises concern that other long-term effects, such as immunosuppression, may also occur in later life.

In this chapter, data concerning the short- and long-term health effects of probiotics, prebiotics, and synbiotics that are administered during early postnatal life, that is, during the first weeks/months of life, both to preterm and term infants, are summarized. Data were identified by searches of the Cochrane Database of Systematic Reviews, the Cochrane Controlled Trials Register, and MEDLINE databases (all up to July 2019) as well as references from relevant articles. The search was limited to randomized controlled trials (RCTs) or their systematic reviews/meta-analyses, which were published in English. Data reporting the effects of gut microbiome modifiers administered prenatally or beyond early infancy, whether for prophylactic or therapeutic reasons, have been discussed in detail elsewhere and are not reviewed here.

Probiotics

A 2014 definition developed by the International Scientific Association for Probiotics and Prebiotics (ISAPP) defines probiotics as "live microorganisms that, when administered in adequate amounts, confer a health benefit on the host" [6]. The most commonly used probiotics are bacteria from the genus *Lactobacillus* or *Bifidobacterium* and the yeast, *Saccharomyces boulardii*. However, novel probiotics are an area of current investigation [7].

Probiotics in infant formulae

Probiotics have been added to many infant formulae with the aim of shifting the microbiome to match that of a breastfed infant with potential benefits attributed to breastfeeding. Infants fed with such formulae are exposed to a daily intake of probiotic strains. This is in contrast to older children and adults in whom consumption of a probiotic product constitutes only a portion of their diets. Hence, both safety and efficacy are crucial. A 2017 systematic review concluded that the administration of so far evaluated probiotic-supplemented formulae to healthy infants does not raise safety concerns with regard to growth and adverse effects [8]. Whereas some beneficial clinical effects of probiotics are possible, there is no existing robust evidence to recommend their routine use. The latter conclusion may reflect the small amount of data on a specific probiotic strain(s) and outcomes, rather than a genuine lack of an effect. These conclusions were in line with a 2011 position paper of the Committee on Nutrition of the European Society for Paediatric Gastroenterology, Hepatology and Nutrition (ESPGHAN) which, at least at that time, did not support the routine use of probiotic-supplemented formulae in infants [9]. Similarly, according to a 2014 position of the European Food Safety Authority (EFSA), there is no necessity to add probiotics to infant and follow-on formulae [10].

Probiotics for preventing allergy

It has been hypothesized that, among other causes, aberrant gut microbiome, due to factors such as mode of delivery (vaginal vs cesarean), use of antibiotics during the early neonatal period, and mode of feeding (breast vs formula feedings), contributes to the development of allergic diseases [1,11]. Prevention of allergic disorders through microbiome-focused interventions such as the provision of probiotics (and/or prebiotics) was evaluated by the World Allergy Organization (WAO) [12]. The WAO experts concluded that probiotic supplementation cannot be recommended for reducing the risk of allergy in children. However, the WAO considered that there is likely a net benefit in using probiotics for preventing eczema. Specifically, the WAO suggested: "a) using probiotics in pregnant women at high risk for having an allergic child; b) using probiotics in women who breastfeed infants at high risk of developing allergy; and c) using probiotics in infants at high risk of developing allergy." High risk was defined as the presence of a biologic parent or sibling with asthma, allergic rhinitis, eczema, or food allergy. All recommendations were conditional and supported by a very low quality of evidence. The recommendations were based on the findings of a 2015 meta-analysis, which pooled data on all probiotics [13]. Similarly, a 2018 meta-analysis, which also pooled data of all probiotics, showed that oral supplementation with probiotics during late pregnancy and lactation may reduce the risk of eczema (absolute risk reduction is 44 cases per 1000; 95% CI 20–64) [14]. However, a 2018 strain-specific meta-analysis of RCTs showed that prenatal or postnatal administration of *Lactobacillus rhamnosus* GG (LGG), the only probiotic studied in more than one RCT, was ineffective in reducing eczema [15]. This systematic review did not support the general recommendation to use probiotics for preventing eczema. LGG also had no effect on the risk of wheezing/asthma, with the exception of one trial that reported an increased risk of asthma at 7 years in the LGG group compared with the placebo group.

There was no effect of LGG administration on allergic rhinitis. The effect of LGG on food allergy was not reported in any of the trials.

Overall, while some probiotics may be effective in reducing some allergic symptoms, a number of questions remain unanswered. Which probiotic(s) should be used to reduce the risk of eczema? When should one start the administration of probiotics with proven efficacy? When should one stop? What is the appropriate dose of an effective probiotic?

Probiotics for treating infant colic

Infant colic (typically defined as crying for >3 hours daily for >3 days per week) is usually self-limited, with no long-term adverse effects; however, it may be very distressing to infants and their parents [16]. The therapeutic options for this common problem remain limited [17]. Evidence suggests that the gut microbiome in subjects with infantile colic differs from the gut microbiome in an unaffected population. In colicky infants, dysbiosis may affect gut motor function and gas production, leading to abdominal pain/colic [18].

A 2018 individual participant data meta-analysis, which included data from 4 RCTs involving 345 infants with colic, documented that the administration of *Lactobacillus reuteri* DSM 17938 at a dose of 1×10^8 colony-forming units (CFU) is likely to reduce crying and/or fussing time in breastfed infants with infantile colic, but its role in formula-fed infants is less clear [19]. Whether the type of feeding (i.e. breastfeeding versus formula-feeding) matters is uncertain, as the incidence of infant colic is similar among breastfed and formula-fed infants [20]. Data on other probiotics, either positive or negative, are too limited to allow one to draw reliable conclusions [21].

Probiotics for preventing infant colic

A 2019 Cochrane review identified six RCTs ($n = 1886$), which compared probiotics with placebo for preventing infant colic [22]. The pooled results of three RCTs ($n = 1148$) in which LGG and two multistrain products (one included four strains of lactobacilli, three strains of bifidobacteria, and *Streptococcus thermophilus* DSM 24731; and another included LGG, LC705, *Bifidobacterium breve* Bb99, and *Propionibacterium freudenreichii* subsp. *shermanii*) were assessed found a similar occurrence of new cases of colic in the probiotics and placebo groups (RR 0.46, 95% CI 0.18–1.19). The pooled results of three other RCTs ($n = 707$) found a reduced duration in crying time at study end in the probiotics group compared with the placebo group (MD −32.6 min/day, 95% CI −55.6 to −9.5). However, one of the included studies evaluated a prebiotic formula with added probiotic strains; thus, this was a synbiotic intervention. At the strain level, the effect was particularly evident for *L. reuteri* DSM 17938 (MD −44.3 min/day, 95% CI −66.6 to −21.9).

Probiotics for preventing necrotizing enterocolitis

Should probiotics be routinely used in preterm infants to reduce the risk of NEC? This is one of the most common questions asked with regard to probiotics. The factors involved in the pathogenesis of NEC, one of the most severe and life-threatening gastrointestinal diseases occurring in preterm infants, particularly with birth weight <1000 g, include formula feeding rather than breastfeeding, intestinal hypoxia−ischemia, and colonization of the gut with pathogenic microbiome [23]. A number of meta-analyses consistently showed that the administration of probiotics was associated with reduced risk for NEC or death [24−26]. However, all probiotics have been evaluated together. In contrast, a 2018 network meta-analysis aimed to identify strains with greatest efficacy [27]. Fifty-one RCTs involving more than 11,000 preterm infants were included. Based on the findings of the latter meta-analysis, the ESPGHAN recently formulated recommendations on the use of probiotics for preterm infants. In brief, if all safety conditions are met, the ESPGHAN recommends the use of *L. rhamnosus* GG ATCC 53103 at a dose ranging from 1×10^9 CFU to 6×10^9 CFU, as it might reduce NEC stage 2 or 3, or the combination of *Bifidobacterium infantis* Bb-02, *Bifidobacterium lactis* Bb-12, and *S. thermophilus* TH-4 at a dose of $3.0–3.5 \times 10^8$ CFU (of each strain), as it might reduce NEC stage 2 or 3 (both are conditional recommendations, with low certainty of evidence). No recommendation was made in either direction regarding the use of *L. reuteri* DSM 17938 (conditional recommendation; very low certainty of evidence). Similarly, no recommendation can be made in either direction regarding the use of the combination of *Bifidobacterium bifidum* NCDO 1453 (currently reclassified as *Bifidobacterium longum*) with *Lactobacillus acidophilus* NCDO 1748 (ATCC 4356, LA37, or NCIMB 30316) in preterm infants to reduce the risk of mortality, NEC stage 2 or 3, or sepsis (very low to moderate certainty of evidence). The ESPGHAN recommends against using *B. breve* BBG-001 to reduce the risk of mortality, NEC stage 2 or 3, or sepsis (conditional recommendation; low to moderate certainty of evidence) and against the use of *S. boulardii* for safety reasons [28].

Prebiotics

A 2015 ISAPP consensus group defined a prebiotic as "a non-digestible compound that, through its metabolization by microorganisms in the gut, modulates composition and/or activity of the gut microbiome, thus conferring a beneficial physiological effect on the host" [29]. The experts highlighted the importance of the selectivity of their metabolism, that is, as different prebiotics pass through the gastrointestinal tract to the colon, they are selectively fermented by specific health-promoting bacteria. An example of naturally occurring prebiotics is human milk oligosaccharides (HMOs). Other examples include inulin, oligofructose, fructo-oligosaccharides (FOS), galactofructose, and galacto-oligosaccharides (GOS). Studies have consistently shown that prebiotics increase the fecal counts of bacteria that are thought to be beneficial, such as bifidobacteria or certain butyrate producers [29].

Prebiotics in infant formulae

Like probiotics, prebiotics have been added to many infant formulae, with the same aim of shifting the microbiome to match that of a breastfed infant with potential benefits attributed to breastfeeding. A 2018 systematic review updated evidence on the effects of the administration of prebiotic-supplemented infant formulae compared with unsupplemented infant formulae [8]. Five databases were searched up to March 2017 for RCTs, and 41 publications were identified. The most studied prebiotic was a 9:1 mixture of short-chain (sc) GOS and long-chain (lc) FOS. This systematic review concluded that the administration of currently evaluated prebiotic-supplemented formulae to healthy infants does not raise safety concerns with regard to growth and adverse effects. Some favorable clinical effects are possible, primarily stool softening, which may be beneficial in some infants. However, no robust evidence supports the routine use of prebiotic-supplemented formulae. These conclusions were in line with a 2011 position paper of the ESPGHAN Committee on Nutrition that the administration of currently evaluated prebiotic-supplemented formulae to healthy infants does not raise safety concerns with regard to growth and adverse effects [9]. From the regulatory point of view, the European Commission allows the use of scGOS and lcFOS with a specific ratio of 9:1 and a maximal concentration of 0.8 g/dL. Furthermore, the safety of scGOS/lcFOS is endorsed by the EFSA, and it is an approved ingredient, identified as generally recognized as safe (GRAS; US FDA GRAS Notice [GRN] No. 477).

Prebiotics for treating infant colic

Oligosaccharide prebiotics (scGOS/lcFOS) may be the effective treatments for crying in formula-fed infants with colic [30–32]. However, these studies used various additional interventions, such as fermented milk formula [31], partially hydrolyzed whey protein and β-palmitate [30], or long-chain polyunsaturated fatty acids [32], making it difficult to formulate firm conclusions on the effects of scGOS/lcFOS specifically.

Prebiotics for prevention of allergy

A 2017 systematic review identified 22 RCTs that assessed the effects of prebiotic supplementation in infants on the prevention of allergies [33]. Supplementation with prebiotics compared with no supplementation resulted in a reduced risk of developing asthma or recurrent wheezing (RR 0.37, 95% CI 0.17–0.8), a reduced risk of food allergy (RR 0.28, 95% CI 0.08–1.00), and a similar risk of eczema (RR 0.68, 95% CI 0.4–1.15). As with probiotics, the effects of different prebiotics are not equivalent. It remains unclear which prebiotic(s) to use; however, in 15 studies, a mixture of FOS/GOS was used. The WAO suggested using prebiotic supplementation in not-exclusively breastfed infants, both at high and at low risk for developing allergy, and not using prebiotic supplementation in exclusively breastfed infants. Both recommendations are conditional and are based on very low certainty of the evidence [34].

2′FL and/or LNnT added to infant formulae

A number of benefits of breastfeeding, particularly protection against childhood infections, may be, at least partially, related to the presence of HMOs. In addition to well-documented prebiotic effects, the postulated effects of HMOs include antiadhesive/antimicrobial effects, modulation of intestinal epithelial cells, and immune modulation [35]. Currently, progress in biotechnology nowadays allows the production of at least some HMOs. Two HMOs, 2′-fucosyllactose (2′-FL) and lacto-*N*-neotetraose (LNnT), have recently been added to infant formulae, either alone or in combination [36–38]. Limited available evidence documents that the administration of currently evaluated formulae supplemented with 2′FL with/without LNnT supports normal growth, is safe, and is well tolerated. Some favorable clinical effects are possible, as one of these studies also found that the administration of 2′FL + LNnT-supplemented formula reduced parent-reported infant morbidity and medication use. However, further high-quality RCTs are needed and are underway (clinicaltrials.gov, NCT03722550). Future research should help to define the optimal composition and dosage for HMO interventions and their effectiveness in different settings/populations.

Synbiotics

The term "synbiotic" refers to a product that includes both a probiotic and a prebiotic [39]. However, there is a lack of an exact description as to what criteria a synbiotic should fulfill, which prompted the ISAPP to convene a consensus panel on synbiotics (https://isappscience.org). The ISAPP consensus statement is likely to be published in 2020. Compared to probiotics or prebiotics, data on the efficacy of synbiotics are scarce. Potential beneficial effects of synbiotics include bifidogenic effects, immunomodulatory effects, enhanced short-chain fatty acid production, improved stool characteristics, and improved viability of probiotics.

Synbiotics in infant formulae

In 2011, the ESPGHAN Committee on Nutrition concluded that only a limited number of synbiotic preparations administered in infant formulae have been studied in the context of a formal RCT. As with pro- and prebiotic-supplemented formulae, the Committee considered that, although the available data suggest that synbiotic-supplemented formulae are safe, the limited data call for caution in overinterpretation of these results. The committee did not recommend the routine use of formula supplemented with synbiotics in infants [9]. While new evidence is available for some synbiotic-supplemented formulae [40], it is clear that the efficacy and safety of each formula should be considered separately.

Synbiotics and infections

In low- and middle-income countries, the burden of sepsis and death in infants remains high [41], but may be reduced. A 2017 randomized, double-blind, placebo-controlled trial involving 4556 healthy newborns (at least 2000 g at birth, at least 35 weeks of gestation), carried out in rural India, found that compared to placebo, the administration of a synbiotic (*Lactobacillus plantarum* ATCC-202195 plus FOS), starting from day 2 to 4 of life, for 7 days, reduced the risk of sepsis and death (the combined primary outcome, RR 0.6, 95% CI 0.48–0.74) [42]. There was also a significant reduction in the lower respiratory tract infections.

Synbiotics and infantile colic

One RCT found that compared with placebo, the administration of a synbiotic (*Lactobacillus casei, L. rhamnosus, S. thermophilus, B. breve, L. acidophilus, B. infantis, Lactobacillus bulgaricus*, and FOS) for 30 days in 50 breastfed infants with infantile colic increased the successful treatment at the end of the intervention (RR 1.96, 95% CI 1.18–3.24) [43].

Synbiotics for preventing allergy

A 2016 meta-analysis (search date: October 2015) identified two RCTs ($n = 1320$; aged up to 6 months in one study and term neonates aged <3 days in the other) that evaluated the effects of synbiotics in the prevention of atopic dermatitis. The pooled results of these trials showed that, compared with placebo, the administration of synbiotics had no effect on the risk of atopic dermatitis (RR 0.44, 95% CI 0.11–1.83) [44]. However, pooling data on various synbiotics can be questioned. Hence, it is of note that individually, specific synbiotics in these two prevention trials significantly reduced eczema. One of these trials compared standard cow milk-based formula with α-lactalbumin-enriched formula supplemented with *L. rhamnosus* LCS-742, *B. longum* subsp. *infantis* M63 + 96% GOS, and 4% scFOS (OR = 0.11; 95% CI 0.01–0.94); thus, there were two interventions administered at the same time [45]; the second study assessed prenatal and postnatal administration of *L. rhamnosus* GG (ATCC 53103) and LC705 (DSM 7061), *B. breve* Bb99 (DSM 13692), *P. freudenreichii* subsp. *shermanii* JS (DSM 7076) + GOS (OR = 0.74; 95% CI 0.55–0.98) [46]. The latter study also found that the synbiotic intervention reduced IgE-associated allergy until 5 years of age in cesarean-delivered children but not in the total cohort [47].

Synbiotics and cesarean section

Worldwide, there are large differences in the rates of cesarean section (c-section). However, as documented in a 2016 The Lancet report, the rates of c-section deliveries are high in many regions [48]. Compared to the 10%–15% of deliveries as recommended by the WHO, c-sections account for 27.3% of deliveries in Eastern Europe and Central Asia; 26.9% in Western Europe; 32% in North America; and 44% in Latin America and Caribbean, and rates continue to rise in many settings. It has been hypothesized that the different gut microbiome of infants born via c-section may contribute to the risk of diseases later in life. Indeed, a 2018 systematic review found that children delivered by c-section had increased risk of asthma up to age of 12 years (OR 1.21, 95% CI 1.11–1.32) and obesity up to the age of 5 years (OR 1.59, 95% CI 1.33–1.9) [49]. At least two studies have found that normalization of the gut microbiome composition is possible by the administration of synbiotics. First, it was documented in a subset of infants from the previously described study [46,47]. More recently, a 2017 RCT found that infant formula supplemented with scGOS/lcFOS and *B. breve* M-16V administered from birth until 16 weeks of age may modulate gut microbiome by increasing levels of *Bifidobacterium*, and thus making it more similar to that in vaginally born infants [50]. However, it remains to be determined whether these microbiome changes impact clinical outcomes later in life.

Safety and quality of interventions targeting microbiome

Overall, interventions targeting microbiome such as probiotics and/or prebiotics are considered safe for use in otherwise healthy populations. Still, if they are to be given early in life, for a prolonged time, particularly live probiotic microorganisms, safety issues need special attention. A 2018 systematic review found that more than a third of the 384 trials evaluated did not provide information on harms, and only 2% adequately reported adverse events [51]. Thus uncertainty remains due to a lack and/or inadequate reporting of harms. There is also concern with regard to the quality of products targeting microbiome, probiotics in particular. This issue was recently reviewed by the ESPGHAN Working Group on Probiotics [52]. The Working Group suggested a more stringent quality control process to ensure "that the probiotic content as mentioned on the label meets the actual content throughout the shelf life of the product, while no contamination is present."

Conclusions

Interventions modifying the gut microbiome during the postnatal period have the potential to prevent and treat diseases in the pediatric population. However, guidance is needed regarding which interventions to be used for which clinical conditions, as well as the timing, dosage, and mode of administration. Despite the evidence to support the use of specific interventions in some clinical situations, further studies confirming the effect and defining the type, dose, and timing are still often required. As not all interventions within groups of probiotics, prebiotics, and synbiotics are equal, the efficacy as well as the safety should be established individually for each specific intervention. The use of interventions targeting the gut microbiome with no documented health benefits should be discouraged.

References

[1] Tanaka M, Nakayama J. Development of the gut microbiota in infancy and its impact on health in later life. Allergol Int 2017;66(4):515–22.

[2] Lynch SV, Pedersen O. The human intestinal microbiome in health and disease. N Engl J Med 2016;375(24):2369–79.

[3] Marchesi JR, et al. The gut microbiota and host health: a new clinical frontier. Gut 2016;65(2):330–9.

[4] Kalliomaki M, et al. Probiotics in primary prevention of atopic disease: a randomised placebo-controlled trial. Lancet 2001;357(9262):1076–9.

[5] Kalliomaki M, et al. Probiotics and prevention of atopic disease: 4-year follow-up of a randomised placebo-controlled trial. Lancet 2003;361(9372):1869–71.

[6] Hill C, et al. Expert consensus document. The International Scientific Association for Probiotics and Prebiotics consensus statement on the scope and appropriate use of the term probiotic. Nat Rev Gastroenterol Hepatol 2014;11(8):506–14.
[7] Kumar H, et al. Novel probiotics and prebiotics: road to the market. Curr Opin Biotechnol 2015;32:99–103.
[8] Skorka A, et al. Infant formulae supplemented with prebiotics: are they better than unsupplemented formulae? An updated systematic review. Br J Nutr 2018;119 (7):810–25.
[9] Braegger C, et al. Supplementation of infant formula with probiotics and/or prebiotics: a systematic review and comment by the ESPGHAN committee on nutrition. J Pediatr Gastroenterol Nutr 2011;52(2):238–50.
[10] Szajewska H, Szajewski T. Saturated fat controversy: importance of systematic reviews and meta-analyses. Crit Rev Food Sci Nutr 2016;56(12):1947–51.
[11] Simonyte Sjodin K, et al. Emerging evidence of the role of gut microbiota in the development of allergic diseases. Curr Opin Allergy Clin Immunol 2016;16(4):390–5.
[12] Fiocchi A, et al. World Allergy Organization-McMaster University Guidelines for Allergic Disease Prevention (GLAD-P): probiotics. World Allergy Organ J 2015;8(1):4.
[13] Cuello-Garcia CA, et al. Probiotics for the prevention of allergy: a systematic review and meta-analysis of randomized controlled trials. J Allergy Clin Immunol 2015;136 (4):952–61.
[14] Garcia-Larsen V, et al. Diet during pregnancy and infancy and risk of allergic or autoimmune disease: a systematic review and meta-analysis. PLoS Med 2018;15(2): e1002507.
[15] Szajewska H, Horvath A. Lactobacillus rhamnosus GG in the primary prevention of eczema in children: a systematic review and meta-analysis. Nutrients 2018;10(9):1319.
[16] Benninga MA, et al. Childhood functional gastrointestinal disorders: neonate/toddler. Gastroenterology 2016, S0016-5085(16)00182-7.
[17] Gordon M, et al. Dietary modifications for infantile colic. Cochrane Database Syst Rev 2018;10:CD011029.
[18] Sung V, Cabana MD. Probiotics for colic—is the gut responsible for infant crying after all? J Pediatr 2017;191:6–8.
[19] Sung V, et al. Lactobacillus reuteri DSM 17938 for managing infant colic: protocol for an individual participant data meta-analysis. BMJ Open 2014;4(12):e006475.
[20] Johnson JD, Cocker K, Chang E. Infantile colic: recognition and treatment. Am Fam Physician 2015;92(7):577–82.
[21] Dryl R, Szajewska H. Probiotics for management of infantile colic: a systematic review of randomized controlled trials. Arch Med Sci 2018;14(5):1137–43.
[22] Ong TG, et al. Probiotics to prevent infantile colic. Cochrane Database Syst Rev 2019;3:Cd012473.
[23] Neu J, Walker WA. Necrotizing enterocolitis. N Engl J Med 2011;364(3):255–64.
[24] AlFaleh K, Anabrees J. Probiotics for prevention of necrotizing enterocolitis in preterm infants. Cochrane Database Syst Rev 2014;(4):CD005496.
[25] Deshpande G, et al. Benefits of probiotics in preterm neonates in low-income and medium-income countries: a systematic review of randomised controlled trials. BMJ Open 2017;7(12):e017638.
[26] Dermyshi E, et al. The "golden age" of probiotics: a systematic review and meta-analysis of randomized and observational studies in preterm infants. Neonatology 2017;112(1):9–23.
[27] van den Akker CHP, et al. Probiotics for preterm infants: a strain-specific systematic review and network meta-analysis. J Pediatr Gastroenterol Nutr 2018;67(1):103–22.
[28] van den Akker CHP, et al. Probiotics and preterm infants: a position paper by the ESPGHAN Committee on Nutrition and the ESPGHAN Working Group for

Probiotics and Prebiotics. J Pediatr Gastr Nutr 2020. Available from: https://doi.org/10.1097/MPG.0000000000002655.
[29] Gibson GR, et al. Expert consensus document: the International Scientific Association for Probiotics and Prebiotics (ISAPP) consensus statement on the definition and scope of prebiotics. Nat Rev Gastroenterol Hepatol 2017;14(8):491–502.
[30] Savino F, et al. Reduction of crying episodes owing to infantile colic: a randomized controlled study on the efficacy of a new infant formula. Eur J Clin Nutr 2006;60(11):1304–10.
[31] Vandenplas Y, et al. Randomised controlled trial demonstrates that fermented infant formula with short-chain galacto-oligosaccharides and long-chain fructo-oligosaccharides reduces the incidence of infantile colic. Acta Paediatr 2017;106(7):1150–8.
[32] Vivatvakin B, et al. Effect of a whey-predominant starter formula containing LCPUFAs and oligosaccharides (FOS/GOS) on gastrointestinal comfort in infants. Asia Pac J Clin Nutr 2010;19(4):473–80.
[33] Cuello-Garcia C, et al. Prebiotics for the prevention of allergies: a systematic review and meta-analysis of randomized controlled trials. Clin Exp Allergy 2017;47(11):1468–77.
[34] Cuello-Garcia CA, et al. World Allergy Organization-McMaster University Guidelines for Allergic Disease Prevention (GLAD-P): prebiotics. World Allergy Organ J 2016;9:10.
[35] Bode L. Human milk oligosaccharides: next-generation functions and questions. Nestle Nutr Inst Workshop Ser 2019;90:191–201.
[36] Puccio G, et al. Effects of infant formula with human milk oligosaccharides on growth and morbidity: a randomized multicenter trial. J Pediatr Gastroenterol Nutr 2017;64(4):624–31.
[37] Marriage BJ, et al. Infants fed a lower calorie formula with 2′FL show growth and 2′FL uptake like breast-fed infants. J Pediatr Gastroenterol Nutr 2015;61(6):649–58.
[38] Goehring KC, et al. Similar to those who are breastfed, infants fed a formula containing 2′-fucosyllactose have lower inflammatory cytokines in a randomized controlled trial. J Nutr 2016;146(12):2559–66.
[39] Pandey KR, Naik SR, Vakil BV. Probiotics, prebiotics and synbiotics - a review. J Food Sci Technol 2015;52(12):7577–87.
[40] Vandenplas Y, et al. Safety of a new synbiotic starter formula. Pediatr Gastroenterol Hepatol Nutr 2017;20(3):167–77.
[41] Ehret DY, Patterson JK, Bose CL. Improving neonatal care: a global perspective. Clin Perinatol 2017;44(3):567–82.
[42] Panigrahi P, et al. A randomized synbiotic trial to prevent sepsis among infants in rural India. Nature 2017;548(7668):407–12.
[43] Kianifar H, et al. Synbiotic in the management of infantile colic: a randomised controlled trial. J Paediatr Child Health 2014;50(10):801–5.
[44] Chang YS, et al. Synbiotics for prevention and treatment of atopic dermatitis: a meta-analysis of randomized clinical trials. JAMA Pediatr 2016;170(3):236–42.
[45] Roze JC, et al. An alpha-lactalbumin-enriched and symbiotic-supplemented v. a standard infant formula: a multicentre, double-blind, randomised trial. Br J Nutr 2012;107(11):1616–22.
[46] Kukkonen K, et al. Probiotics and prebiotic galacto-oligosaccharides in the prevention of allergic diseases: a randomized, double-blind, placebo-controlled trial. J Allergy Clin Immunol 2007;119(1):192–8.
[47] Kuitunen M, et al. Probiotics prevent IgE-associated allergy until age 5 years in cesarean-delivered children but not in the total cohort. J Allergy Clin Immunol 2009;123(2):335–41.

[48] Boerma T, et al. Global epidemiology of use of and disparities in caesarean sections. Lancet 2018;392(10155):1341–8.
[49] Keag OE, Norman JE, Stock SJ. Long-term risks and benefits associated with cesarean delivery for mother, baby, and subsequent pregnancies: systematic review and meta-analysis. PLoS Med 2018;15(1):e1002494.
[50] Chua MC, et al. Effect of synbiotic on the gut microbiota of cesarean delivered infants: a randomized, double-blind, multicenter study. J Pediatr Gastroenterol Nutr 2017;65(1):102–6.
[51] Bafeta A, et al. Harms reporting in randomized controlled trials of interventions aimed at modifying microbiota: a systematic review. Ann Intern Med 2018;169 (4):240–7.
[52] Kolacek S, et al. Commercial probiotic products: a call for improved quality control. A position paper by the ESPGHAN Working Group for Probiotics and Prebiotics. J Pediatr Gastroenterol Nutr 2017;65(1):117–24.

Index

Note: Page numbers followed by "*f*" and "*t*" refer to figures and tables, respectively.

C

D

R

W

X

Kolibri 360 GmbH
Ferdinand-Jühlke-Straße 7
99095 Erfurt, Deutschland
produktsicherheit@kolibri360.de